CHEMICAL
PUBLICATIONS

CHEMICAL PUBLICATIONS

THEIR NATURE AND USE

FIFTH EDITION

M. G. Mellon, Ph.D., Sc.D.
Professor Emeritus of Analytical Chemistry
Purdue University

McGRAW-HILL BOOK COMPANY
New York St. Louis San Francisco Auckland Bogotá
Hamburg Johannesburg London Madrid Mexico Montreal New Delhi
Panama Paris São Paulo Singapore Sydney Tokyo Toronto

This book was set in Times Roman by Black Dot, Inc. (ECU).
The editors were Jay Ricci and James S. Amar;
the production supervisor was Phil Galea.
New drawings were done by Burmar.
The cover was designed by Janice Noto.
R. R. Donnelley & Sons Company was printer and binder.

CHEMICAL PUBLICATIONS

Their Nature and Use

1234567890DODO898765432

ISBN 0-07-041514-5

Library of Congress Cataloging in Publication Data

Mellon, M. G. (Melvin Guy), date
   Chemical publications, their nature and use.

   Bibliography: p.
   Includes indexes.
   1. Chemical literature. 2. Chemistry—Study and teaching. 3. Bibliography—Best books—Chemistry.
I. Title.
QD8.5.M44 1982   540'.72   81-20947
ISBN 0-07-041514-5   AACR2

THE CHEMICAL LITERATURE

[With apologies to Kipling]

The Literature,
The chemical literature—
When in doubt look it up in the literature.
Every question that man can raise, every phrase of every phase
of that question is on record
In the literature;
Thrashed out threadbare pro and con
In the literature.
Did the universe at large once carry a positive charge?
Why aren't the holes in macaroni square?
From Avogadro's number to the analysis of cucumber,
if you're interested you'll find it, for it's there
In the literature.
In *Journal* this or *Zeitschrift* that, *Comptes rendus* or *Zentralblatt,*
It's somewhere
in the literature.

—P. G. Horton

CONTENTS

PREFACE

If a book does not apologize for itself, it is in vain for the author to attempt it by a preface. I shall, therefore, only declare the nature and the intent of this publication.

Thomas Scott

A need to update its material is the usual reason for presenting a new edition of a textbook in science. In the present case this need involved making numerous changes, a few of them major and many of them minor in nature. Obsolete publications have been deleted. Some now considered relatively unimportant have been replaced with newer ones, if available. A few new publications have been added. Along with all these changes the author has tried to summarize the nature and relevance of important developments related to the storage and retrieval of information.

The general nature of the different kinds of sources has remained relatively unchanged. Primary periodicals, technical reports, and patents retain their essential characteristics. There are more of all of them. Many specialized periodicals have appeared. The size of many others has increased. Need for conserving shelf space and browser's time has inspired experiments in changing present methods of publication, especially research journals.

In spite of the vast expansion of the primary literature, particularly the research publications, the function of the secondary sources is still the collection, classification, organization, and dissemination of the new information as efficiently and rapidly as possible to render the data available for use.

There are more secondary periodicals—for indexing, abstracting, and reviewing. Many new ones are specialized. In contrast, *Chemical Abstracts* has broadened its coverage and services until it is now *the* general abstracting journal in the world. A separate chapter (7) has been allotted to this famous periodical.

The endless production of meter readings of great variety has resulted in the collection of specific kinds of data in various forms. Spectroscopic data are examples. Dictionaries and encyclopedias cannot keep pace with the deluge of new compounds, products, and processes, and the means developed for producing them.

Growth in the multivolume treatises, such as those of Beilstein and Gmelin, present both spatial and financial problems. Before chemists face submersion in the rising tide of information, is it not time for us to restrain our scientific nationalisms and pool our national resources and efforts to bring, and keep, such invaluable sources up to date?

The specific illustrative examples mentioned for several kinds of publications are intended to be reasonably late editions representative of their types. Some familiar works will be missed.

No doubt the most significant developments of the past decade have been automation of methods of storing and retrieving data. What deserves a separate book is limited here to a new chapter (15).

Development of familiarity with, and facility in using, the resources of the different kinds of chemical literature come chiefly from using the publications. To get students involved seems to necessitate assigning them individual problems in the library. The kinds of problems included in previous editions have been changed. The new problems are intended merely to be suggestive of workable assignments. Individual instructors should formulate their own specific problems to take into account the resources of their libraries and the general nature of the interests, and status of training, of their students.

Once again, valuable suggestions and information have come from many individuals, particularly those in various governmental agencies and those who reviewed the entire manuscript. The author is very indebted to the members of the staff of Chemical Abstracts Service who checked the manuscript for the chapter on *Chemical Abstracts*. Thanks are expressed to all who cooperated so generously.

Based on their wide experience in the Information Systems Department of E. I. du Pont de Nemours and Company, E. P. Bartkus and D. M. Krentz have contributed the most significant new material in the book. Their ready cooperation is deeply appreciated.

Finally, the author is very grateful to John Pinzelik, professor of library science and librarian of the M. G. Mellon Library of Chemistry. Since 1962, he has taught the course on chemical literature at Purdue. For more than 15 years he has directed attention to new publications and to the various changes in the different kinds of publications. He has checked endless bibliographical details during revision of the manuscript for this edition.

As with previous editions of the book, any comments on errors of commission or omission will be welcome.

M. G. Mellon

PART ONE

PUBLICATIONS:
KINDS AND NATURE

INTRODUCTION AND GENERAL OUTLINE

The past can teach us a great deal about the present by giving us the perspective of detachment.

J. F. White

Someone has stated that information of a scientific nature can be obtained, in many cases at least, by one or more of the following procedures: first, by inquiring of the individual who knows; second,[1] by performing the experimental investigations necessary to ascertain the desired facts; and, third, by consulting the scientific literature, where a record may be found of the published reports of others' work upon the subject in question.

Although it may be taking the path of least resistance to resort to the first possibility, provided an individual possessing the information is available, and although it is frequently very desirable to obtain experimental facts firsthand, usually recourse to either one of these alternatives is unnecessary or impracticable. In such instances, the chemist turns to the chemical library. The solubility data for sodium chloride in water, for example, can probably be given by various individuals, or it may be determined with fair accuracy by rather simple means; but for ordinary purposes anyone requiring such information would consult solubility tables. It is a matter of utilizing recorded chemistry.

Before beginning a journey to some distant point, one usually gives at least passing consideration to the reason for going and to the means to be employed in

---

[1]Compare B. Yates, p. 236, in R. T. Bottle (ed.), "Use of Chemical Literature," Butterworths, London, 1979.

reaching the desired destination. Similarly, before taking up the question of how to use the resources of a library, it seems desirable to give some attention to the kinds of inquiries which one takes to such a place and to the nature of the sources to be examined when one arrives. Since our concern is with chemistry and, therefore, chemical or technical libraries, we should have in mind the kinds of questions which a chemist takes to the library. A searcher, having become familiar with the different sources of information relating to the several types of questions, is then in a position to make effective use of the material.

At this point, therefore, there is presented a classification of the types of many questions for which the material in our various chemical publications may be expected to help in providing answers. The outline proposed[2] is based upon questions presented to the technological division of a large public library, where, for some 10 years, a record was kept of the more important inquiries submitted.

The inquirers ranged in their chemical interests all the way from commercial research and consulting chemists to children seeking directions for some chemical trick or civic groups requiring popular presentations of chemical subjects for club meetings. Considering their sources, it is not surprising that the questions varied widely in character, just as the questioners varied in their chemical interests. One individual wanted something very specific, such as the spectral transmission curve for a 10 percent aqueous solution of cupric nitrate, while another wanted to know "all about cement." Some wanted only a popular article or book, while others were satisfied with nothing less than the latest technical data.

An examination of the hundreds of questions indicates that most of them fell readily into rather well-defined divisions. The scheme formulated for this purpose is given below. In it no particular significance is attached to the order in which the various divisions have been placed. There is included for each division a statement of the general nature of the inquiries belonging to it, together with several current examples of typical questions.

Types of Questions

A *Specific.* Those in which the information desired relates to a single kind of chemical activity. The following areas are examples:
   1 *Bibliography.* Partial or complete lists of references, with or without annotations:
   The physiological action of plutonium 239
   The production of shale oil
   Popular articles on atomic fusion
   2 *History and biography.* Events in the life of an individual or in the development of an industry; the influences operating, and contributions

---

[2]M. G. Mellon, *Special Libraries,* **17,** 275 (1926); see also "Kinds of Literature Questions," R. Fugman, W. Braun, and W. Vaupei, *Angew. Chem., ***73,** 745 (1961).

made, during certain periods; the beginning and development of a theory or an industry:

Evolution of chemical symbols

Life of T. W. Richards

Development of nitroparaffins

3 *Existence, occurrence, and source.* The location of raw material; its form; compounds which are known:

Occurrence of chromium in Africa

Commercial sources of beryllium

Bromine substitution products of bibenzene

4 *Composition.* Natural materials and artificial products; specifications and standards; formulas and workshop recipes:

Formulas for mothproofing textiles

Composition of precious stones

Analyses of Indiana coals

5 *Methods of production, preparation, and manipulation.* Laboratory and commercial processes; details of procedure; materials required; apparatus employed:

Manufacture of polypropylene

Preparation of antibiotics

Fluorination of organic compounds

6 *Properties.* Physical and chemical (including physiological action); general and specific reactions:

Physiological effect of dihydrogen sulfide

Heat of dissolution of calcium chloride

Action of limestone in a blast furnace

7 *Uses.* Laboratory and industrial; general and special applications:

Uses of 2,2'-bipyridine

Industrial applications of selenium

Employment of dimethylsulfoxide in synthetic chemistry

8 *Identification, testing, and analysis.* Methods available; interpretation of results:

Determination of polychlorinated biphenyls (PCBs) in milk

Testing of glass fibers

Analysis of automotive engine emissions

9 *Patents and trademarks.* Date of expiration; details of specifications; objects previously protected:

Patent claims for making tetracycline

Specifications for ripening oranges

Date of expiration of patent on nitroethane

10 *Statistical data.* Production; consumption; cost; supply; price; market:

Production and supply of helium

Statistics on the uranium industry

Foreign activities in the drug industry

B *General.* Those in which the information desired relates to more than one of the above-mentioned classes. In this case we encounter two variations in the questions.

    1 Those in which there is clear indication of the particular classes which are involved:

    Preparation, properties, and uses of TNT

    Occurrence and composition of natural emeralds

    2 Those in which no such limitations are expressed or implied:

    Embrittlement of metals

    Desulfurization of coal

THE ORIGIN AND DEVELOPMENT OF CHEMICAL LITERATURE

Having in mind the kinds of questions a chemist takes to the library, we are next interested in the sources available which are likely to contain information of value for answering these questions. In considering this part of our problem, attention may be directed briefly to the origin and development of those publications which have come commonly to be designated as the "chemical literature" and to certain points regarding the state of knowledge and the conditions prevailing at the time the publications appeared. Such an examination furnishes some perspective for judging the comparative value of the records for present-day work.

We find ourselves today in the midst of a great volume of published material of chemical importance. What was its origin? Why do we have it? Who was responsible for it? How did it reach its present form? These and similar questions probably do not often disturb the average chemist. They do have some interest, however, in spite of the fact that often there are no definite answers.

Numerous works have been published dealing with the historical aspect of various phases of chemistry. Some publications have been devoted to the activities and contributions of a certain individual; some deal only with the origin, development, and significance of an idea or theory; some deal with special industries, with divisions of chemistry, or with countries; and some aim to present a more or less comprehensive view of the whole field. But none has as its primary aim the presentation of an account of the development of chemical publications as such.[3] In most works of a historical nature, at least occasional references are found to a considerable number of publications. Usually, however, the aim is to direct attention to some individual's idea or contribution rather than to the publication itself.

Even a superficial examination of the material available indicates a long

---

[3]See, however, W. Ostwald, "Handbuch der allgemeinen Chemie," vol. I. "Die chemische Literatur und die Organisation der Wissenschaft," Akademische Verlagsgesellschaft m.b.H., Leipzig, 1919.

course in humanity's development to the point of appreciating the significance of the modern scientific method and of applying it to a study of ourselves and of our environment. Even now, in the words of R. A. Millikan:

> Man himself is just . . . emerging from the jungle. It was only a few hundred years ago that he began to try to use the experimental and the objective method, to try to set aside all his prejudices and his preconceptions,[4] to suspend his judgment until he had all the facts before him, to spare no pains to see first all sides of the situation and then to let his reason and his intelligence, instead of his passion and his prejudice, control his decisions. That is called the scientific method.

As a result of the combined contributions of many individuals, there has been acquired a powerful tool in the scientific method, and there has been accumulated an enormous collection of facts, recorded in many places and touching many phases of human activities. The greatest number of significant contributions have been made in the fields of knowledge encompassed by the natural sciences, in which chemistry occupies a commanding position. The written records of these developments include a multitude of facts, experimentally determined, together with discussions and interpretations involving these facts.

At present our interest centers on *publications containing accounts of facts and theories,* rather than on the facts themselves; on the methods by which they were discovered or obtained; or on the ultimate effect of the theories developed to account for the facts.

When did people begin to apply to ideas the test of experimental verification? The answer must remain uncertain. Perhaps it is even not significant. We are convinced, however, that the adoption of the experimental method marked a very important advance in the attainments of the human race; and we are partially aware of the marvelous material changes that have accompanied and grown out of its application in industrial and scientific research work.

What disposal the first enthusiastic workers made of the results obtained by this method we can only guess. It seems entirely probable that no organized work, as we now undertake it, was done and that no written records were made, perhaps for centuries.

Historical records are meager for this development,[5] but we do know that Francis Bacon, about 1600, formulated a program for establishing research as a means of regenerating learning in the service of humanity. This included the foundation of a college of research, whose function was to foster the New Philosophy (the experimental and scientific method), and the provision for the publication of such discoveries as might be revealed.

Although Bacon did not see his project actively undertaken, his work did produce noteworthy results. The trend of the times became directed more and more toward intelligent experimentation. His "New Atlantis" was an imaginary

---

[4]To reject the idolatry of the traditional (Hugh Black, Scottish theologian).
[5]See, however, D. J. de S. Price, "Science Since Babylon," chap. 3, Yale University Press, New Haven, CT, 1961.

picture of his college of research in practical operation. This book, according to H. G. Wells, may be considered as one of the ten which have been most instrumental in shaping and directing human activities. Wells writes that it

> formulated the conception of a House of Science, incessantly inquiring and criticising and publishing, that should continually extend the boundaries of human knowledge; it replaced unorganized by organized scientific research, and did so much to insure the unending continuity of scientific inquiries; and it contains the essential ideas of the modern scientific process—the organized collection, publication and criticism of fact. . . . Of the supreme importance of the book itself as the seed of the Royal Society and most European academies, there can be little dispute. Like the rod of Moses, it strikes the rock of human capacity, and thereafter the waters of knowledge flow freely and steadily.

According to M. H. Liddell,[6] about 15 years after Bacon's death a group of

> divers worthy persons inquisitive into Natural Philosophy and other parts of humane learning, and particularly what has been called the "New Philosophy" . . . began by agreement to meet weekly in London to treat and discourse of such affairs.

Out of the efforts of this group, meeting for the purpose of discussing and sharing their intellectual interests, there was founded the Royal Society.[7] Commenting on the significance of this event, L. Strachey stated:

> If one were asked to choose a date for the beginning of the modern world, probably July 15, 1662, would be the best to fix upon. For on that day the Royal Society was founded, and the place of science in civilization became a definite and recognized thing. The sun had risen above the horizon; and yet, before that, there had been streaks of light in the sky. The great age of Newton was preceded by a curious twilight period—a period of gestation and preparation, confused, and only dimly conscious of the end toward which it was moving.

The Royal Society stands first in at least two respects: it was the first society of its kind to survive, and it was the first to publish the proceedings of its meetings.[8] This publication was first printed May 6, 1665, as the *Philosophical Transactions* of the Royal Society, and it has had a continuous existence since that date. Even today the first volume is interesting.

Of particular significance to chemists is the fact that Robert Boyle was one of the most active pioneer members. To quote again from Liddell,[9] Boyle

---

[6]M. H. Liddell, *Science,* **60,** 25 (1924).

[7]See Sir Harold Hartley (ed.), "The Royal Society: Its Origins and Founders," The Royal Society, London, 1960; also E. N. da C. Andrade, "A Brief History of the Royal Society," The Royal Society, London, 1960; and J. I. Cope and H. W. Jones (eds.), "Sprat's History of the Royal Society," Washington University Studies, St. Louis, MO, 1958.

[8]The Accademia del Cimento, founded at Florence, in 1657, . . . was the first scientific society of any importance. . . . Although it lived but ten years, it enriched the world by leaving a volume of important records of experiments, chiefly in pneumatics." *Saggi di Naturali Experienze Fatti.*

The *Journal des savans* began January 5, 1665 (M. de Sallo as editor), with reports of new scientific equipment and ideas. Later, papers on new work were included. After a break in the 1790s, the title was changed to *Journal des savants.* Since 1816, it has not been a scientific journal.

[9]Loc. cit.

refers to himself as a member of the Invisible College, composed originally of scientific men bound together as an esoteric sodality without name or meeting place and having for its sole end the alleviation of the physical and spiritual ills of humanity.

The time when Boyle and his associates formed the Royal Society may be taken as the date of the dawn of chemical literature as we now know it. A simple beginning it was, this first systematic recording of scientific papers and discussions, and the preservation of these contributions for posterity; but now the practice is so universal, and the contributions are so numerous, that one is amazed to find over 14,000 periodicals publishing information more or less closely related to the work of the individual engaged in chemical and related pursuits. E. J. Crane, in 1922, stated[10]:

> Since 1665, when this first scientific journal put in its appearance there has been an accelerated increase in the number of journals of interest to the chemist, and the bulk of the accumulated material has become almost staggering in its proportions. . . . The vastness of the chemical journal literature, its rapid and continuous advance upon the frontiers of our knowledge, and its essentially unorganized state, are . . . arguments why the chemist should see to it that he learns how best to make use of the means which have been provided to make this sea of information navigable.

In 1650, there were no scientific journals. By 1950 there were about 100,000.[11] The data of Fig. 1 show that, for the past 200 years, the growth rate is an exponential function of time. Each 15 years, approximately, the number of periodicals has doubled. This rapid increase has characterized practically all science and technology. The exponential rate of increase may now be decreasing.[12] Contrary to many statements, there has been no *explosion* in the amount of the literature.

It has already been noted that several methods of communicating ideas were probably used during the centuries preceding the appearance of the *Philosophical Transactions* of the Royal Society. Following the inception of this initial serial publication, however, the practice of making a permanent record of individuals' contributions started to spread. Gradually, various European academies came into existence, each founding, sooner or later, its own publication.[13] The general condition of affairs during this period necessarily made progress slow. There was still much confusion in the ideas regarding chemical matters. Over a century had to elapse between the founding of the Royal Society and the work of Lavoisier.

---

[10]*Ind. Eng. Chem.*, **14**, 901 (1922).

[11]D. J. de S. Price, op. cit.; "Little Science Big Science," Columbia University Press, New York, 1963; and *Int. Sci. Techn.* (March, 1963), p. 37.

For recent trends in the growth of the chemical literature, see D. B. Baker, *Chem. Eng. News*, **54**(20), 23 (1976); **59**(22), 29 (1980).

[12]J. Walsh, *Science*, **199**, 1188 (1978).

[13]See M. Ornstein, "The Rôle of Scientific Societies in the Seventeenth Century," Ph.D. dissertation, Columbia University, New York, 1913, for developments in England, France, Germany, and Italy; see also R. J. Moore, "The Rôle of the Scientific Society in Chemistry," *Chemist*, **16**:327 (1939).

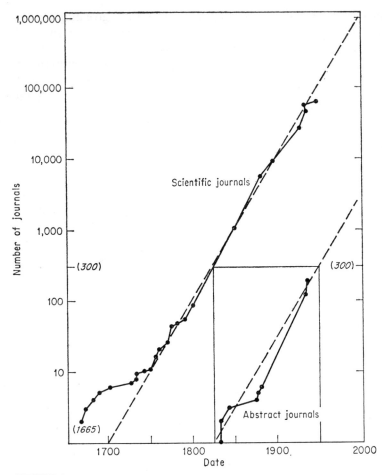

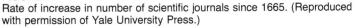

FIGURE 1
Rate of increase in number of scientific journals since 1665. (Reproduced
with permission of Yale University Press.)

Following his discoveries, there appeared, in 1778, the first chemical journal,
Crell's *Chemisches Journal*. Others followed some years later.

By 1820, the volume of the material appearing had reached a stage where it
seemed desirable to collect and summarize the contributions for each year.
Accordingly, in 1821, J. J. Berzelius began his famous *Jahresbericht über die
Fortschritte der physischen Wissenschaften*. This publication represents the first
chemical review serial.[14] Not long afterward, a journal seemed justified whose
purpose would be the publication of short summaries of each article as soon as
possible after the article appeared. *Pharmaceutisches Centralblatt* was begun in

[14]*Berlinisches Jahrbuch für die Pharmacie und die damit verbunden Wissenschaften* was started in
1795 and continued until 1840.

1830 to meet this need.[15] Developments of the present century give ample proof of the vision which these early chemists had regarding the place for such publications.

With a place to publish articles of their own and a source of summaries of others' work, together with reports of yearly developments, the chemists' needs in this direction seemed to be supplied for the time being. But with the appearance of more and more journals and the constant accumulation of facts in the rapidly widening range of chemical activities, there arose the need for other types of publications. Not more or different journals were needed this time, but rather publications in which the material published could be gathered together and arranged according to definite schemes. Reference works were the answer— digests and treatises, including indexes of known compounds, compilations of chemical properties, discussions of definite fields of the science, dictionaries, and other works.

EARLY BOOKS ON CHEMISTRY

One might well conclude from the preceding statements that there are no records of chemical matters available before the founding of the Royal Society. Such is not the case. Although this event did result in the establishment of the first scientific journal to survive, various textbooks or treatises on chemistry, pharmacy, and metallurgy were already well known.[16]

On considering that modern chemical ideas began with Lavoisier's work, it seems that before 1600 the state of chemical knowledge must have been such as to make these books of doubtful value to anyone except the chemical historian.[17] Many facts mentioned stand out as accurate observations of the early worker; but the records of processes are meager, and the recipes of the alchemist are often mysterious, inaccurate, and misleading, in the light of our present knowledge. Thus, T. Bergman wrote, "The history of chemistry is properly divided into the mythological, the obscure, and the certain."[18]

For the purpose of reference there is included a list of some of these early publications.[19] The arrangement is based upon the divisions of chemical activity used by J. C. Brown.[20] The student should bear in mind that these works were

---

[15]*Journal de pharmacie et de chemie* and *Annales de chimie* had been doing some of this kind of work.

[16]See S. A. Ives and A. J. Idhe, *J. Chem. Educ.*, **29,** 244 (1952), for a description of the books of alchemy and the history of chemistry contained in the Duveen Library (D. I. Duveen, "Bibliotheca alchemica et chemica," E. Weil, London, 1949).

[17]See, however, J. Read, *J. Chem. Educ.*, **37,** 111 (1960), who mentions the Ebers Papyrus, some 35 centuries old, which is a kind of primitive pharmacopoeia containing over 800 prescriptions and remedies. Also see E. A. Caley, *J. Chem. Educ.*, 3:1149 (1926), and 4:979 (1927), for English translations of two Greek papyri, which the translator describes as being by far the most ancient documents we possess dealing with chemical arts and operations as such. They appear to have been written about the end of the third century.

[18]"Disertatio gradualis de primordiis chemiae," Upsaliae, Sweden, 1779.

[19]See also J. W. Mellor, "Treatise on Inorganic and Theoretical Chemistry," vol. 1, pp. 19–73, Longmans, Green & Co., Inc., New York, 1922.

[20]J. C. Brown, "A History of Chemistry from the Earliest Times," McGraw-Hill Book Company, New York, 1920.

textbooks or short treatises on chemical knowledge and practice of their day and, in general, not reports of experimental work and discoveries resulting therefrom.

It should be noted, too, that there were no books, as we know them today, prior to the invention of printing with movable type. Earlier "books" were hand-produced manuscripts, many in the form of tracts, essays, and pamphlets. Stone, clay tablets, papyrus, vellum, parchment, and, later, paper were used at different times. Early contributions in science, including chemistry, are summarized by J. L. Thornton and R. I. J. Tully in "Scientific Books, Libraries, and Collectors."[21]

Prehistoric Period (up to 1500 B.C.)

As an art, chemistry dates far back of the Christian era. There are no records of actual practice as an industry, but mention is made of various chemical facts by a number of writers, as in the Bible. The writings may be largely mythology, superstition, philosophy, or religion. Among the writers are the Chaldean, Egyptian, Byzantine, Persian, Indian, Roman, and Greek. J. R. Partington cites an amazing number of such items in his "History of Chemistry."[22]

Alchemical Period (1500 B.C. to A.D.1650)

In many ways the activity of the period was directed toward the search for gold and for the philosopher's stone. Included was the possibility of transmuting base to noble metals. Many practical facts accumulated as a result of amalgamations, calcinations, combustions, cupellations, dissolutions, distillations, extractions, evaporations, fusions, percolations, refinings, roastings, smeltings, and probably other processes.[23]

Out of all these operations came considerable practical knowledge. Production of glass and pottery became established. Soap was made. Dyes and dyeing came along later, as did the recognition of certain acids and salts. The pursuit of medicines yielded various products, some probably of value and some not.

Recognition of a few metals came early, particularly those found in nature or easily recovered, such as gold (first to be recognized), silver, copper, mercury, and later iron. There was a Bronze Age, succeeded by an Iron Age. Mercury and antimony appeared among the medicines. Arsenic, sulfur, and a few other elements became known.

The alchemists must early have noted the souring of milk, the fermentation of

[21]The Library Association, London, 1971.

[22]Macmillan and Company, Ltd., London. Volume 1 (part I, 1970; part II, not published yet) covers the earliest information available up to about 1500 A.D. Volumes 2 (1961) and 3 (1962) cover approximately the next 300 years, that is, into the beginning of the modern quantitative period. Then vol. 4 (1964) continues to the date of publication.

[23]See Martin Levey, "Chemistry and Chemical Technology in Ancient Mesopotamia," Elsevier Publishing Corporation, New York, 1959.

various products to yield contaminated ethanol, and a number of other like transformations.

The Egyptians acquired much knowledge from the Chaldeans. The Arabians appropriated what was known to the Egyptians and the Greeks. According to chemical historians, various Latin translations of writings by Arabians and others of this period are known, such as "Aurifera Artis Quam Chemiam Vocant" (1572) and "Theatrum Chemicum" (1659–1661). These profess to be teachings of alchemists.

Among the Greek writers (speculative philosophers) should be mentioned Heraclitus (540–475 B.C.), who, like Bergson, declared that "becoming" or eternal change is the sole actuality; Aristotle (384–322 B.C.), who formulated the laws of deductive reasoning, that is, from the general to the specific; and Pliny (A.D. 23–79), who wrote the "Historia Naturalis" of 160 books, the last five of which are an account of the known chemical knowledge. Much later Roger Bacon (1214–1294) advocated the inductive system of philosophy, that is, reasoning from the specific to the general, and noted the importance of experiment as a tool of research. His "Opus Majus" was published in 1733.

Partington mentions a long list of alchemical writers. The names of a few of them, with reported life dates, follow: Galen (130–201), Geber (Hayyan) (720–815), Al-Kindi (800–870), Al-Razi (Rhazes) (866–925), Albertus Magnus (1193–1280), and Aquinas (1225–1274). Historians differ on some spellings and dates. Thus, Lull may be Lully or Lullus. Likewise, Geber is much more commonly found than Hayyan. Titles of books may differ in spelling or language, as in different editions and translations. In some cases authors used shortened titles.

Rudolph Hirsch listed 98 alchemical writers, including titles of their works, up to 1536.[24] J. W. Stillman has presented "The Story of Alchemy and Early Chemistry."[25]

During the prehistoric, alchemical, iatrochemical, and phlogiston periods, chemical writers devised and used many symbols for substances and apparatus. Table I shows 60 such symbols from the Fisher Scientific Collection of Alchemical and Historical Pictures.[26]

The 84 alchemical symbols listed by C. J. S. Thompson[27] are stated to be some of those used during the medieval period. Also he included a secret alphabet used during the fourteenth and fifteenth centuries.

Iatrochemical Period (1500 to 1700)

The preceding period included the Middle Ages, while this period includes the Renaissance, when the search for medicinal substances became an important

---

[24]*Chymia*, **3**, 115 (1950).
[25]Dover Publications, Inc., New York, 1960.
[26]Fisher Scientific Company, Pittsburgh, PA.
[27]"The Lure and Romance of Alchemy," pp. 131–132, G. G. Harrap and Company, London, 1932.

TABLE I

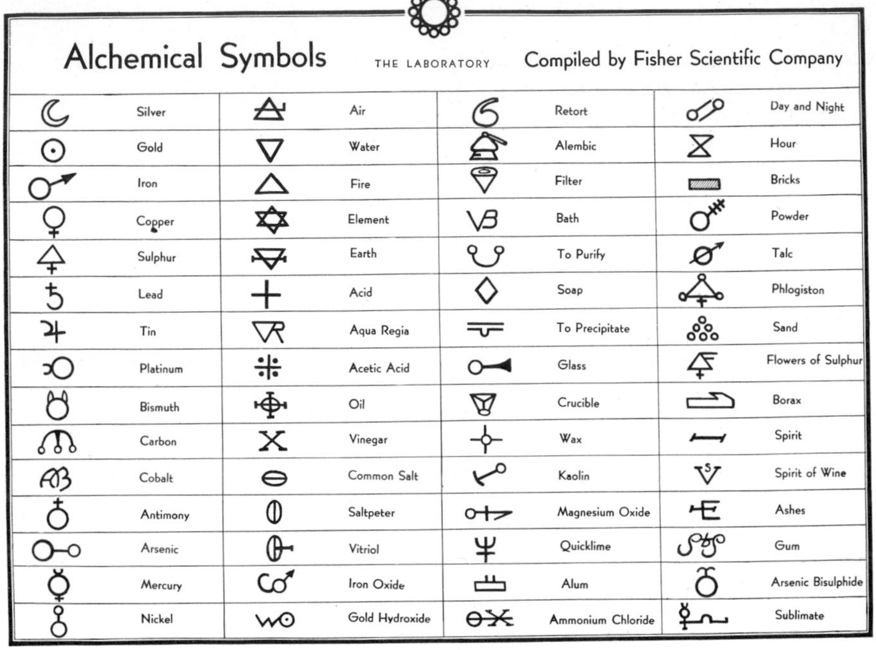
Alchemical Symbols
THE LABORATORY
Compiled by Fisher Scientific Company
Silver
Gold
Iron
Copper
Sulphur
Lead
Tin
Platinum
Bismuth
Carbon
Cobalt
Antimony
Arsenic
Mercury
Nickel
Air
Water
Fire
Element
Earth
Acid
Aqua Regia
Acetic Acid
Oil
Vinegar
Common Salt
Saltpeter
Vitriol
Iron Oxide
Gold Hydroxide
Retort
Alembic
Filter
Bath
To Purify
Soap
To Precipitate
Glass
Crucible
Wax
Kaolin
Magnesium Oxide
Quicklime
Alum
Ammonium Chloride
Day and Night
Hour
Bricks
Powder
Talc
Phlogiston
Sand
Flowers of Sulphur
Borax
Spirit
Spirit of Wine
Ashes
Gum
Arsenic Bisulphide
Sublimate

aim. Of course, the hope of transmuting base metals into gold persisted. The following list includes the names of some of the contemporary writers and the titles of certain writings ascribed to them.[28]

Agricola (Bauer), G. (1494–1555): "De Re Metallica Libri XII," 1556; "Furni Novi Philosophii," 1646–1649.

Anthony, F. (1550–1623): "Medicinae Chymicae et Veri Potabilis Auri Assertio," 1610.

Ashmole, E. (1617–1692): "Theatrum Chemicum Britannicum," 1651.

Avicenna (ibn-Sina), (980–1037): "Artis Chemicae Principes," 1572.

Balduin, C. A. (1632–1682): "Aurum Superius et Inferius Aurae . . . ," 1675.

Barba, A. A. (1575–1650): "El Arte de los Metales," 1640.

[28]The names, titles, and dates in these lists were taken largely from the following sources: "Catalogue Générale des Livres Imprimes," Bibliothèque Nationale, Paris; "Catalogue of Printed Books" and "General Catalogue of Printed Books," British Museum, London; D. I. Duveen, "Bibliotheca Alchimica et Chemica," E. Weil, London; J. R. Partington, "A History of Chemistry," 4 v., Macmillan and Company, Ltd., London; and J. C. Poggendorff, "Biographisch-Literarisches Handwörterbuch zur Geschichte der Exacten Wissenschaften," Barth, Leipzig.

Published more recently is the privately printed "Alchemy and the Occult." This meticulously produced catalogue lists part of the Paul and Mary Conover Mellon collection in the Beinicke Rare Book and Manuscript Library of Yale University.

Volumes 1 and 2 (1968) list 160 books. The arrangement is chronological, beginning with Saint Isodorus, "De Responsione Mundi et Astrorum Ordinatione," 1472. This volume includes beautiful reproductions of woodcuts.

Volumes 3 and 4 (1977) list 149 manuscripts, dating from 1225. In an article on "The Mellon Collection of Alchemy and the Occult," Ian MacPhail, *Ambix,* **14,** 198 (1967), states that this collection "is probably the finest single collection of alchemical manuscripts in existence."

Barchusen, J. C. (1666–1723): "Elementa Chemiae," 1718.

Beguin, J. (–1618): "Tyrocinium Chymicum," 1610; "Les Élémens de Chymie," 1615.

Biringuccio, V. (1480–1539): "De la Pirotechnia," 1540.

Croll, O. (1580–1609): "Basilica Chymica," 1608.

Du Chesne, J. (1544–1609): "Opera Medica," 1614.

Escker, L. (1530–1593): "Probierbuchlein," 1556; "Aula Subterrannea," 1574.

Glaser, C. (1615–1673): "Traité la Chymie," 1663.

Glauber, J. R. (1604–1670): "Furni Novi Philosophici," 1648; "Operis Mineralis," 1651; "Opera Chymica," 2 v., 1651.

Hales, S. (1677–1761): "Vegetable Staticks," 1727; "Statitical Essays," 1733.

Helmont, J. B. van (1577–1644): "Opuscula Medica Inaudita," 1644; Ortus Medicinae vel Opera et Opuscula Omnia," 1648.

Kircher, A. (1602–1680): "Oedipus Aegyptiacus," 3 v., 1652–1653; "Mundus Subterraneus," 2 v., 1665.

Kunckel, J. (1630–1703): "Laboratorium Chymicum," 1716; "Ars Vitraria Experimentalis," 1679.

Lemery, N. (1645–1715): "Cours de Chymie," 1675; "Pharmacopeé Universelle," 1697.

Libavius, A. (1540–1616): "Alchemia," 1597; "Alchymistische Practic," 1600; "Alchymia," 1606.

Lull, R. (1232–1315): "De Secretis Naturae," 1518.

Marggraf, A. S. (1709–1782): "Chemische Schriften," 2 v., 1761.

Mayow, J. (1641–1679): "Tractus Quinque Medico-Physici," 1673.

Mynsicht, A. von (1603–1638): "Thesaurus et Armentarium Medico-Chymicus," 1631.

Neri, A. (–1614): "L'Arte Vetraria," 1612.

Nicols, T. (1600–1660): "A Lapidary," 1652.

Palissy, B. (1510–1589): "De L'Arte de la Terre," 1580.

Paracelsus, P. A. (1493–1541): "De Natura Rerum," 1525; "Archidoxorum," 1525; "Opera Medico-Chimica," 1603–1605.

Plattes, G. (1575–1650): "Discovery of Subterranean Treasure," 1639.

Ripley, G. (1415–1490): "The Compound of Alchemy," 1591.

Rolfinck, W. (1599–1673): "Chimia in Artis Formam Redacta," 1662.

Rülein, U. (1465–1523): "Bergbüchlein," 1500.

Sala, A. (1575–1637): "Opera Medico-Chymica," 1647.

Sennert, D. (1572–1637): "Epitome Naturalis Scientiae," 1618.

Sylvius, F. (1614–1672): "Opera Medica," 1671.[29]

Valintinus, B.: "Les Douze Clefs de Philosophe," 1624.

Webster, J. (1610–1682): "Metallographia," 1671.

Phlogiston Period (1650 to 1775)

According to chemical historians, the general aim of this period was the development of a rational theory of chemistry. However, in his essay, "The Incompleat Chymist," J. B. Eklund[30] emphasizes the evolvement and develop-

[29]E. A. Underwood, *Endeavor,* **31,** 73 (1972), states that Sylvius had the first chemical laboratory in any university. It was at the University of Leyden and was first used Aug. 8, 1669.

[30]Smithsonian Studies in History and Technology, No. 33, Smithsonian Press, Washington, DC, 1975.

ment of many details of practice involved in making and studying a considerable number of substances in the seventeenth and eighteenth centuries. Included in his essay is a dictionary of obsolete terms used in this period.

The following list includes the names of selected writers of this period and the titles of some of their publications:

Becher, J. J. (1635–1682): "Naturkündigung der Metallen," 1661; "Physicae Subterrannae," 1669.

Bergman, T. (1735–1784): "De Tubo Feruminatorio," 1779; "Opuscula Physica et Chemica," 3 v., 1779–1783; "De Minerarum Docimasia Humida," 1780.

Boerhaave, H. (1668–1738): "A New Method of Chemistry," 1727; "Elementa Chemiae," 2 v., 1732.

Boyle, R. (1626–1691): "The Sceptical Chymist," 1661; "Experimenta et Considerationes de Coloribus," 1680.

French, J. (1616–1657): "The Art of Distillation," 1653.

Geoffroy, E. F. (1672–1731): "Table des Differentes Rapports Observes en Chimie entre Differentes Substances," 1718.

Hoffmann, F. (1660–1742): "Opuscula Physico-Medicaantehac Deorsim Edita de Elementis, Viribus, Utilitate et Usu Medicatorium Fontium," 2 v., 1725; "Opera Omnia Physico-Medica," 6 v., 1740.

Hooke, R. (1635–1703): "Micrographia," 1665.

Juncker, J. (1679–1759): "Conspectus Chemiae Theoretico-Practicae," 2 v., 1730–1738.

Macquer, P. J. (1718–1784): "Elemens de Chymie Théoretique," 1749; "Elemens de Chymie Pratique," 1751; "Dictionnaire de Chymie," 2 v., 1766.

Manget, J. J. (1652–1742): "Bibliotheca Chemica Curiosa," 1702.

Marggraf, A. S. (1709–1782): "Chymischer Schriften," 2 v., 1761.

Neumann, C. (1683–1737): "Lectiones Chymicae," 1727.

Pott, J. H. (1692–1777): "Chymische Untersuchungen," 1746.

Roth-Scholtz, F. (1681–1736): "Deutsches Theatrum Chemicum," 3 v., 1728–1732.

Stahl, G. E. (1660–1734): "Chymia Rationalis et Experimentalis," 1720; "Fundamenta Chymiae et Experimentalis," 1723.

Tachinius, O. (1620–1690): "Epistola de Famoso Liquore Alkahest," 1655.

Quantitative Period (1775 to 1900+)

The general aim of the period was to extend and apply chemical knowledge on the basis of the rational theory available. The time coincides with the development of modern chemistry. As many are unaware of the books written during the early part of the period, the following list is included.

Accum, F. (1769–1838): "System of Theoretical and Practical Chemistry," 2 v., 1803; "A Practical Essay on the Analysis of Minerals," 1816.

Berthollet, C. L. (1748–1822): "Essai de Statique Chymique," 1803.

Berzelius, J. J. (1779–1848): "Föreläsurgar i Djurkemien," 1806; "Lärbok i Kemien," 1808.

Black, J. (1728–1799): "Lectures on the Elements of Chemistry," 1803.

Chaptal, J. A. C. (1756–1832): "Tableau Analytique du Cours de Chimie,"[31] 1783; "Elemens de Chimie," 3 v., 1790; "Chimie Applique aux Arts," 4 v., 1807.

---

[31]French spelling varies—*chimie* or *chymie*.

Dalton, J. (1766–1844): "Elements of Chemistry," 1799; "A New System of Chemical Philosophy," vol. 1, 1808–1810; vol. 2, 1827.

Davy, H. (1778–1829): "Elements of Chemical Philosophy," 1812; "Elements of Agricultural Chemistry," 1813.

Dumas, J. B. A. (1800–1884): "Traité de Chimie Applique aux Arts," 1828–1846.

Faraday, M. (1791–1867): "Chemical Manipulation," 1827; "Experimental Researches in Chemistry and Physics," 1859.

Fourcroy, A. F. de (1755–1809): "Philosophie Chimique," 1792; "Système des Connaissances Chimiques," 11 v., 1801–1802.

Fremy, E.: "Traité de Chimie Generale," 6 v., 1865–1866.

Gmelin, L. (1788–1853): "Handbuch der Theoretischen Chemie," 11 v., 1817–1819.

Götting, J. F. (1755–1809): "Vollstandiges Chemisches Probekabinett," 1790.

Jacquin, J. F. E. von (1766–1839): "Lehrbuch der Allgemeinen und Medicinischen Chemie," 2 v., 1783.

Klaproth, M. H. (1743–1817): "Beitrage zur Chemischen Kenntnis der Mineralkörper," 1795; "Chemisches Wörterbuch," 5 v., 1807–1810.

Lampadius, W. A. (1772–1842): "Handbuch der Chemischen Analyse der Mineralkörper," 1801.

Lavoisier, A. L. (1743–1794): "Opuscules Physiques et Chymiques," 1774; "Traité Élémentaire de Chymie," 1789; "Memoires de Chimie," 1805.

Mitscherlich, E. (1794–1863): "Lehrbuch der Chemie," 1829.

Morveau, G. de (1737–1816): "Elemens de Chimie," 1771; "Élémens de Chymie Théorique et Pratique," 3 v., 1777–1778.

Pfaff, C. H. (1773–1852): "Handbuch der Analytischen Chemie," 1821.

Richter, J. B. (1762–1807): "Anfangsgründe der Stöchymetrie," 3 v., 1792; "Ueber die neuren Gegenstände der Chemie," 11 pts., 1791–1802; "Chemisches Handwörterbuch," 6 v., 1802–1805.

Rose, H. (1795–1864): "Ausfuhrliches Handbuch der Analytischen Chemie," 1829.

Thénard, L. J. (1777–1857): "Traité de Chimie Élémentaire, Théorique, et Pratique," 4 v., 1813–1816.

Thomson, T. (1773–1852): "A System of Chemistry," 4 v., 1802; "Elements of Chemistry," 1810; and others.

Wenzel, C. F. (1740–1793): "Einleitung zur Höhren Chemie, 1774; "Lehre von der Verwandschaft der Körper," 1777.

Following these writers came Berthelot, Bunsen, Gerhardt, Laurent, Liebig, Mendeleev, Meyer, Pasteur, Stas, Wöhler, and a host of others up to the present time.

In his "Books on Chemistry Printed in the United States, 1755–1900," W. Miles reported his study of their origin. It was published in *Lib. Chron.* **18,** 51 (1952).

OUTLINE OF CHEMICAL PUBLICATIONS

With this brief indication of the origin and development of early chemical publications, next in order is an outline covering the material available at the present time. Subsequent chapters deal with the different kinds of publications and their use.

In a systematic study of the literature of chemistry and chemical technology, some classification is needed for the numerous sources composing the published information relating to this area. The material may be classified either according to the nature of the subject matter, as is the system in *Chemical Abstracts*, or according to the nature of the publication in which the information appears. This second basis of classification seems the logical one to adopt for those who want practical assistance in using a chemical library, since periodicals, bulletins, patents, and books are the units of a library.

The arrangement of this book is one of publications rather than of information. However, the two bases of classification are combined in a way, as there is included a statement of the kind of information which each type of publication contains. Although an attempt has been made to define the different classes of publications to make them reasonably inclusive and exclusive, not all the different sections are differentiated from each other by sharp lines of demarcation. There is some overlapping and occasional gradual merging of one into the other.

For some time the author divided publications into (1) those containing chiefly original contributions and (2) those containing chiefly compilations and discussions of facts already recorded.[32] A third group has been suggested,[33] that is, those containing lists of people engaged in the field of chemistry and guides, catalogues, and directories to their publications, laboratories, plants, products, and services. These three groups may be more definitely designated as indicated.

Primary Sources

These publications contain chiefly new material or new presentations and discussions of known material. In general, they contain the latest published information, which is essentially unorganized. Examples are most periodicals, governmental bulletins, patents, dissertations, and manufacturers' technical pamphlets.

Secondary Sources

These publications contain chiefly known material that is arranged and organized according to some plan. Thus, the physiocochemical constants in the tables of Landolt-Börnstein are systematic compilations of data previously published but so widely scattered as to be of little practical use unless collected and classified. Examples of this kind of publication are a few periodicals; bibliographies; reference works, such as tabular compilations, dictionaries, encyclopedias, formularies, and treatises; and monographs and textbooks.

---

[32]M. G. Mellon, *Chem. Met. Eng.,* **33:**97 (1926).
[33]J. F. Smith and J. D. Scott, "Literature of Chemical Technology," in R. E. Kirk and D. F. Othmer (eds.), "Encyclopedia of Chemical Technology," vol. 8, pp. 418–467, Interscience Publishers, Inc., New York, 1952.

Tertiary Sources

These publications contain compilations of information about chemists and chemical technologists and what they do. Such sources are organized to aid the searcher in using both primary and secondary sources and in following the production of and trade in chemicals and chemical equipment. Examples of this kind of publication are library guides, trade catalogues, directories, and biographies.

PRIMARY SOURCES— PERIODICALS

By a fiction as remarkable as any to be found in law, what has once been published (no matter what the language) is usually spoken of as known, and it is often forgotten that the rediscovery in the library may be a more difficult and uncertain process than the first discovery in the laboratory.

John William Strutt, Lord Rayleigh

The publications which contain new material and which are issued, in general, at regular intervals of time, are designated as periodicals. They contain much of the latest published information. It is largely from them that the facts are gathered for the preparation of the various publications discussed later as secondary sources.

These periodicals in many cases appear under such titles as "bulletin," "journal," "proceedings," and "transactions" (or equivalent terms in foreign languages), particularly if they are the publications of scientific or technical societies.

The amount of this journal literature is enormous. The German periodical *Liebig's Annalen der Chemie* reached 766 volumes in 1972, when volume numbers were discontinued. After one journal, *Philosophical Transactions of the Royal Society* (London), was started in 1665, more have followed until now some 14,000 are monitored by abstractors for articles of chemical interest. Most of the titles of the present list are additions of the twentieth century. It is well to recall, also, that a complete chemical library would contain many more than those currently being published. Hundreds survived for a time and then were discontinued. In 1893 Bolton[1] listed over 400 periodicals of chemical interest. In 1927 Crane and Patterson[2] listed 1,889 periodicals, of which 1,263 were then appearing.

Producers of Periodicals

Practically all periodicals are produced by a scientific society, a private individual or company, or a commercial publishing concern.[3] The first of these producers includes a large number, while individuals' publications are rare. The general object of the first two has been to furnish a place for authors to publish their papers and to disseminate other information of interest to the subscribers or members of the societies.

The material in commercial publications is generally very practical in nature, as indicated later. The aim is to keep industrial chemists and chemical engineers informed of developments of interest to them.

With over 100,000 members, the American Chemical Society (ACS) is the largest specialized producer of periodicals. The current list, in order of appearance, includes the titles listed below. (The letters in parentheses refer to frequency of appearance: w, weekly; bw, biweekly; m, monthly; bm, bimonthly; and q, quarterly.)

1876 *Journal of the American Chemical Society* (bw) (*American Chemical Journal*, 1879–1913, incorporated with it in 1913)
1896 *Journal of Physical Chemistry* (bw)
1907 *Chemical Abstracts* (bw)
1924 *Chemical Reviews* (bm)
1936 *Journal of Organic Chemistry* (bw)
1940 *Chemical & Engineering News* (w), supersedes *Industrial and Engineering Chemistry, News Edition*, 1923–1939
1947 *Analytical Chemistry* (m), supersedes *Industrial and Engineering Chemistry, Analytical Edition*, 1929–1946
1953 *Journal of Agricultural and Food Chemistry* (bm)
1956 *Journal of Chemical and Engineering Data* (q)
1959 *Journal of Medicinal Chemistry* (m)
1960 *Chemical Titles* (bw)
1962 *Biochemistry* (bw)
1962 *Inorganic Chemistry* (m)
1962 *Industrial and Engineering Chemistry, Fundamentals; Industrial and Engineering Chemistry, Process Design and Development;* and *Industrial and Engineering Chemistry, Product Research and Development* (q)

[1]H. C. Bolton, "Select Bibliography of Chemistry," sec. 1, Smithsonian Misc. Coll. No. 850, Smithsonian Institution, Washington, DC, 1893.
[2]E. J. Crane and A. M. Patterson, "A Guide to the Literature of Chemistry," appendix 6, John Wiley & Sons, Inc., New York, 1927.
[3]*Journal of Research of the National Bureau of Standards* and the *Journal of Research of the U.S. Geological Survey* are published by the United States government; the *Canadian Journal of Chemistry* is issued by the National Research Council of Canada.

1964 *Chemistry* (m), supersedes a periodical privately published first in 1928, as *The Chemistry Leaflet;* changed to *SciQuest,* 1979

1968 *Accounts of Chemical Research* (m)

1968 *Macromolecules* (bm)

1971 *CHEMTECH* (m), supersedes *Industrial and Engineering Chemistry, Industrial Edition,* which had superseded the *Journal of Industrial and Engineering Chemistry,* 1909–1922.

1972 *Journal of Physical and Chemical Reference Data* (q)[4]

Several divisions of the American Chemical Society publish their own periodicals. Examples are the *Journal of Chemical Education* (Division of Chemical Education) and the *Chemical Information Bulletin* (Division of Chemical Information).

Some of these publications, such as the *Journal of the American Chemical Society,* have changed little in nature since their start. Others have evolved in various ways, a prime example being the *Journal of Industrial and Engineering Chemistry,* as shown in the above listing.

A foreign example of changing titles is a British journal devoted very largely to "pure" chemistry; which was started in 1859 as the *Quarterly Journal of the Chemical Society.* The word "quarterly" was dropped from the title in 1862. In 1966, after more than a century, the journal appeared in four parts: (A) *Inorganic, Physical, and Theoretical Chemistry;* (B) *Physical-Organic Chemistry;* (C) *Organic Chemistry;* and (D) *Chemical Communications.* Further changes followed in 1972, with the distribution in 1981 being as follows: *Dalton Transactions,* Inorganic Chemistry; *Perkin Transactions* 1, Organic and Bio-organic Chemistry; *Perkin Transactions* 2, Physical Organic Chemistry; *Faraday Transactions* 1, Physical Chemistry; and *Faraday Transactions* 2, Chemical Physics. These are all parts of the previous *Journal of the Chemical Society (London).*

Additional changes in titles may follow the merger, in 1980, of the Chemical Society (London) with the Royal Institute of Chemistry to form the Royal Society of Chemistry.

Cornubert[5] lists 13 changes of title preceding the adoption, in 1834, of the present title, *Journal für Praktische Chemie.*

The American Chemical Society operates a customized service for biochemistry, environmental chemistry, and medicinal chemistry. Twice a month subscribers are sent (1) the *ACS Single Article Announcement,* and (2), the key feature, a selection of complete articles taken from 18 journals of the Society on microfiche. The service also includes (3) a subscription to one of the three journals: *Biochemistry; Environmental Science and Technology;* or *Journal of Medicinal Chemistry.*

---

[4]Published in cooperation with the American Institute of Physics and the National Bureau of Standards.

[5]See page 33.

A number of periodicals were started in the nineteenth century by individuals whose names became so closely associated with the publication that one finds, especially in the older literature, many references including these names, such as Liebig's *Annalen der Chemie,* Fresenius's *Zeitschrift für analytische Chemie,* and Hoppe-Seyler's *Zeitschrift für physiologische Chemie.* The person's name is usually that of the first editor. In the *Annalen der Physik,* however, the name changed with each new editor, which accounts for the large number of references such as *Gilbert Ann., Pogg. Ann., Wied. Ann.,* and *Drude Ann.,* referring respectively to L. W. Gilbert, J. C. Poggendorff, G. H. Wiedemann, and P. K. L. Drude, the early editors.

The number of scientific societies of one variety or another which have been formed is surprisingly large. Even those which may properly be designated as chemical make up a considerable group. Bolton's list[6] includes 56 such organizations. Reid[7] included a later list of chemical societies and their publications. Still more comprehensive are the lists of Crane and Patterson[8] and Hull and Kohr.[9]

The first chemical society was founded in 1785 with the name Edinburgh University Chemical Society. In the Western Hemisphere the first chemical societies originated in Philadelphia.[10] They were the Chemical Society of Philadelphia, founded in 1792, and the Columbian Society of Philadelphia, founded in 1811.

Frequency of Appearance

Chemical periodicals are published at widely varying intervals of time, with individual examples ranging from publications appearing annually to those issued every week (occasionally more frequently). Proceedings of annual meetings of societies are examples of the former, while the latter include certain trade journals and chemical news weeklies. In general, the publications of national societies appear each month.

Promptness of publication is desired by editor, author, and reader. Usually several months elapse in this country from the time a manuscript is received until it appears in published form. Many journals note the date a communication is received. Publication at a considerably later date indicates revision of the original manuscript or failure to issue it promptly. In some periodicals provision is made for more prompt issuance of summaries.

---

[6]H. C. Bolton, "Chemical Societies of the Nineteenth Century," Smithsonian Misc. Coll. No. 1314, Smithsonian Institution, Washington, DC, 1902.

[7]E. E. Reid, "Introduction to Organic Research," p. 73, D. Van Nostrand Company, Inc., Princeton, NJ, 1924.

[8]Op. cit., appendix 5.

[9]C. Hull et al. (part 1) and J. R. Kohr (part 2), "Scientific and Technical Societies of the United States and Canada," National Academy of Sciences and National Research Council, Washington, DC, 1955.

[10]W. Miles, *Chymia,* **3,** 95 (1950).

Distribution by Language and Country

There is little published information on the number of periodicals published in different countries in the various fields of chemistry, including the distribution by languages.

The following data from Chemical Abstracts Service (see Chapter 7) on the primary sources being abstracted are of interest. The statistics are for 1980, and the figures are percentages of the total.

1 Language of journal articles: English, 64.7; Russian, 17.8; Japanese, 5.2; German, 4.0; French, 2.0; Polish, 1.1; Italian, 0.8; 43 other languages, 4.4.

2 National sources of the journal literature: United States, 26.2; USSR, 19.0; Japan, 10.4; Germany, 7.0; United Kingdom, 5.9; France, 4.2; India, 3.4; Canada, 2.6; Italy, 2.4; Poland, 2.1; Czechoslovakia, 1.4; China, 0.7; all others, 15.4.

The percentage of English articles is an indication of the increasing dominance of this language for chemical publications. No doubt there are various reasons for the dominance of English. It is a very hybridized language, many words having origins in German, French, Italian, Spanish, Dutch, Scandinavian, Latin, Celtic, Anglo-Saxon, and others.[11] One professor of English gives a popular talk, "English, a Foreign Language."

The Content of Periodicals

The material composing the various periodicals shows great variation not only in the kind of information but also in the quality and quantity of certain types. G. B. Shaw's comment that some literary contributions are extensive and some intensive applies also to productions dealing with chemistry. It might be added in this case that some are both intensive and extensive, while many could not be correctly designated as possessing either characteristic. The wide range covered varies from epoch-making papers to mere bits of trade and personal information.

Increasing chemical activity, accompanied by a rise in costs of publication, has made it necessary for editors to stress brevity of presentation until many papers are now little more than long abstracts.

The character of this material is so diverse and the manner of publication results in such an intermingling of one type of information with another that there is no very satisfactory scheme of classification. The following scheme is proposed as a possibility which may aid in keeping in mind the general nature of the content of periodicals:

1 *Reports of researches and discussions thereof.* These would include most of

---

[11]See G. Wranglen, *J. Electrochem. Soc.*, **123**, 36C (1976). P. K. Jones, *Inf. Proc. Manage.*, **14**, 363 (1978), advocates the use of Esperanto for improved information management because of its neutrality, rational structure, clarity, and expressive power.

the publications of scientific societies, covering both pure and applied chemistry. Some of the periodicals are limited to articles in a special field. Book reviews and notices of the activities of societies are often included. As examples of the nonindustrial type there may be mentioned the *Journal of Physical Chemistry* and the *Journal of the Chemical Society;* on the industrial side the *Journal of the Electrochemical Society* and *Chemical Engineering Progress* are typical.

Some periodicals have a section, "Letters to the Editor," to provide for preliminary announcements of research to establish priority. The implication is that details will appear later. The practice has spread to entire journals, examples being *Analytical Letters* (1967–   ), *Chemical Physics Letters* (1967–   ), *Physical Review Letters* (1958–   ), *Tetrahedron Letters* (1959–   ), and *Inorganica Chimica Acta Letters,* (1980–   ).

Because of the extent of the literature and the consequent difficulty individuals have in starting work in a new field, it is increasingly important to publish review articles which serve to orient the workers by giving them a bird's-eye picture of the present status of the field. Fortunately, such articles are becoming more common in a number of periodicals, and *Chemical Reviews* is devoted entirely to such material, as is *Reviews of Modern Physics* (see Chapter 6C).

2 *Chemical news.* Some periodicals are devoted largely to a consideration of current developments in either the general field or some special portion of it. The material consists, for the most part, of advertisements, market reports, trade announcements, short reviews of progress, special commercial developments, industrial notes, personal notes (including obituaries), and general news items; many of the items on industrial and scientific advances mention the means employed and the individuals concerned in accomplishing the advances. *Chemical Week, Chemiker-Zeitung, Chemical & Engineering News, Chemistry and Industry, Chemical Age, Chemistry in Britain, Chemisch Weekblad, Chemistry International,* and many trade journals illustrate this type.

In recording developments in the production of industrial chemicals, technical periodicals usually describe the plant and/or the process. As the descriptions are never released until operation has been carried on for some time, even the technical articles do not keep the reader entirely abreast of latest developments. It is necessary to follow advertisements for new products. Only the new products which are for sale are advertised, of course.

Special interest attaches to current prices for chemicals, oils, pharmaceuticals, pigments and dyes, raw materials, reagents, resins, and waxes. *Chemical & Engineering News* publishes a semiannual report on these commodities. *Chemical Marketing Reporter* is well known for such data.

Some journals carry alphabetical lists of products advertised. Reference to such a list indicates quickly whether a desired item is included. Lists of advertisers have been common, but one seldom knows that a given concern markets a given product. One practice is to include a key number with each advertisement. Then by encircling the appropriate key number on an included tear-out mailing card one may easily send for further information about desired items.

In some cases the "Proceedings" of a society may contain only detailed reports of business activities. Business reports of the American Chemical Society appear in *Chemical & Engineering News.*

3 *Reviews or abstracts of articles.* In a number of cases an important part of the periodicals is devoted to abstracts of papers appearing in other periodicals. A discussion of this part of chemical literature is reserved for Chapters 6B and 7.

4 *Combination of other types.* Some periodicals publish many articles on research together with a number of the other items such as editorials, reports of research, articles on commercial developments, addresses, book reviews, and advertisements.

Collected Papers of Scientific Meetings

Many scientific and technical societies sponsor sessions at which a series of more or less closely related papers are presented. Such sessions are usually designated as symposia, conferences, congresses, proceedings, or equivalent entities.

Papers so presented may appear together in a regular issue of the appropriate journal or, occasionally, as a special supplementary volume of the journal (*Festschrift* in German). More likely today, they are issued as separate volumes, unrelated to any journal. This may cause a classification problem for librarians.

To provide for such collections of papers, the American Chemical Society, for example, established in 1974 its "ACS Symposium Series."[12] To facilitate issuance of the papers in book form, they are reproduced as submitted by the authors in camera-ready form. They are not reviewed or edited, except by the presiding officer of the symposium, who becomes the editor of the book. The papers may include both reports of research and reviews. Examples of such publications, all appearing in 1977, follow:

Larsen, J. W. (ed.): "Organic Chemistry of Coal"
Pryor, W. A. (ed.): "Organic Free Radicals"
Strand, R. C. (ed.): "Modern Container Coatings"
Supran, M. K. (ed.): "Lipids as a Source of Flavor"

There are several indexes to such collections of papers. One is the *Directory of Published Proceedings,* a monthly bibliographical periodical issued by Inter-Dok Corporation, White Plains, NY. Acquisition service is offered for all publications cited. A second source, issued by the Institute for Scientific Information beginning in 1978, is entitled *Index to Scientific and Technical Proceedings.*

---

[12]Many "Advances" of the American Chemical Society consist of such a collection of papers (see Chapter 6C). An example is Positronium and Muonium Chemistry, *Adv. Chem. Ser.,* No. 175.

Preprints

Some groups prepare preprints of the papers scheduled for their programs. The chief argument for the practice is to give prospective attendants at the meeting an opportunity to read the papers and thus to prepare for informal discussion of the material. Many of these papers are published in the usual way. The ones of concern here are those not published. Probably they have not been reviewed, and they cannot have been revised to incorporate suggestions made in any discussion when presented. *Chemical Abstracts* has not abstracted preprints. An example of preprints is the collection of manuscripts for the 1978 meeting of the Division of Environmental Chemistry of the American Chemical Society.

Since 1937 the American Chemical Society has issued *Abstracts of Papers* preceding its meetings. Presumably the abstracts are those submitted by the authors.

Microform Editions

Pages of printed matter, drawings, and other items of publications may be reproduced photographically and reduced in size to microform. Employed primarily to conserve space, such miniature reproduction of publications is increasing in use. The material so copied consists of small photographic images, most often of pages of printed or typescript matter. There are two common kinds of microform.

Microfilm, in 16mm or 35mm size, contains one page per frame. In reels or cassettes, all primary periodical publications of the American Chemical Society are available back to the first volume (see "ACS Primary Publications in Microform," 1978).

Another example is the "American Institute of Physics Microform Catalogue," 1979. It lists as being available 27 journals of the American Institute of Physics, 17 journals of the Institute of Physics (United Kingdom), and 20 journals in translations.

Microfiche consists of sheets of microimages on photographic film, usually mounted on 3- by 5-inch or 4- by 6-inch cards, and available in positive or negative form. All current primary journals of the American Chemical Society are available in microfiche, beginning with January 1974.

Microopaque cards have images like microfiche, reproduced on photographic paper. Microprint, or miniprint, cards are similar.

More recent is the possible storage of data optically in the form of holograms. Some developments, with suggested advantages, are outlined in an article by T. H. Maugh II, entitled "Library in a Hologram."[13]

[13]*The Sciences*, **19**, No. 5, 20 (1979).

PERIODICALS CONTAINING MATTER OF CHEMICAL INTEREST

The preceding outline indicates the kind of chemical information to be found in periodicals. Not all this information appears in chemical journals, however. In fact, chemical articles are more or less widely scattered throughout scientific periodicals. Depending upon the proportion of their contents which is devoted to chemical matters, the periodicals of interest may be separated into three groups, which will now be considered.

Journals of General Science

As might be expected, the first scientific periodicals were not limited to any one division of science. No one of these divisions was sufficiently well defined in 1660 or could muster the necessary cooperation and financial support to warrant undertaking a separate publication. At any rate, in the journals of this early period appear contributions from the several sciences which were then sufficiently evolved to be called such.[14]

These publications are devoted, then, to the various fields of science, including chemistry, since the latter was one of the earliest to develop. Since 1800, the proportion of chemical articles appearing in these journals, compared with the total output of such articles, has become less and less. Although the number now appearing here is relatively small, their quality is often excellent. The names of some of the more important publications of this class are listed below, together with the name of the country in which they are published and their beginning date. The portion of the title in bold-faced type shows the standard abbreviation used in citing references to periodicals in *Chemical Abstracts*.[15]

List of Journals of General Science

Argentina

1876–   An*ales de la* **So***ciedad* **Cien***tifica* **Argen***tina*
1945–   **Cien***cia* **Invest***igacion*

Austria

1848–   **Oesterr***eichische* **Akad***emie der* **Wiss***enschaften,* **Mathematisch-***
Naturwissenschaftliche **Kl***asse,* **Sitzungsberichte**

---

[14]See D. A. Kronick, "A History of Scientific and Technical Periodicals, 1665–1790," Scarecrow Press, Inc., Metuchen, NJ, 1976. Also B. Houghton, "Scientific Periodicals," Linnet Books, Hamden, CT, 1975; A. J. Meadows (ed.), "The Scientific Journal," ASLIB, London, 1979.

[15]See Chemical Abstracts Service Source Index (CASSI), Chapter 7. Some discontinued journals are not included. For suggestions on the identification of less common abbreviations of titles of chemical journals, see T. G. Labov, *Adv. Chem. Ser.,* No. 30, p. 102 (1961).

1864–   Oesterreichische Akademie der Wissenschaften, Mathematisch-Naturwissenschaftliche Klasse, Anzeiger

Australia

1938–   Australian Journal of Science, The
1948–   Australian Journal of Scientific Research
1950–   Australian Journal of Applied Science

Belgium

1832–   Bulletin de la Classe des Sciences, Academie Royale de Belgique
1875–   Annales de la Societe Scientifique de Bruxelles

Brazil

1929–   Anais da Academia Brasileira de Ciencias

Canada

1882–   Transactions of the Royal Society of Canada
1929–1950   Canadian Journal of Research

Denmark

1814–   Kongelige Danske Videnskabernes Selskab, Oversigt over Selskabets Virksomhed
1917–   Kongelige Danske Videnskabernes Selskab, Matematisk-Fysiske Meddelelser

Finland

1909–   Annales Academiae Scientiarum Fennicae

France

1710–   Memoires de l'Academie des Sciences
1795–1835   Proces-Verbaux des Seances de l'Academie
1835–   Comptes Rendus Hebdomadaires des Seances de l'Academie des Sciences
1857–1926   Moniteur Scientifique du Docteur Quesneville
1946–   Journal des Recherches du Centre National de la Recherche Scientifique, Laboratoires de Bellevue (Paris)

Germany

1710–	*Sit*zungsber*ichte der* **Dtsch**sch*en* **Akad**em*ie der* **Wiss**ensch*aften zu* **Berlin, Kl**ass*e für* **Math**em*atik und* **Allgem**ein*e* **Naturwiss**ensch*aften*
1788–1795	**Bibl**i*othek der Neuesten* **Phy**si*sche-*Chem*ischen, Metallurgischen, Technologischen, und Pharmaceutischen* Lit*eratur*
1809–1813	**Bull**et*in des Neuesten und Wissenwuerdigsten aus den* **Natur**wiss*enschaften.* Superseded by:
1814–1818	**Museum** *des Neuesten und Wissenwuerdigsten aus dem Gebiete der* **Naturwiss**ensch*aft, der Kuenste, der Fabriken*
1820–1931	**Dinglers Polytech**nisch*es* **J**ourn*al*
1835–	Sit*zungsber*ichte der **Math**em*atisch-*Natur*wiss*ensch*aftlichen* **Kl**ass*e der* **Bayer**isch*en* **Akad**em*ie der* **Wiss**ensch*aften zu* **Muen-chen**
1897–	**Umsch**au *in* **Wiss**ensch*aft* u*nd* **Tech**ni*k*
1913–	**Naturwissenschaften**
1946–	*Z*eit*schrift für* **Naturforsch**u*ng*

Great Britain

1665–	**Philo**so*phical* **Trans**act*ions of the* **Roy**al **Soci**et*y of* **London**
1783–	**Trans**act*ions of the* **Roy**al **Soci**et*y of* **Edinburgh**
1798–	**Phil**oso*phical* **Mag**az*ine, The*
1820–1928	**Trans**act*ions of the* **Cambridge Phil**oso*phical* **Soci**et*y*
1831–1938	**Brit**ish **Assoc**iat*ion for the* **Advance**m*ent of* **Science, Repo**rt *of the Annual Meeting*
1832–	**Proc**eed*ings of the* **Roy**al **Soci**et*y of* **Edinburgh**
1843–	**Proc**eed*ings of the* **Cambridge Phil**oso*phical* **Soci**et*y*
1854–	**Proc**eed*ings of the* **Roy**al **Soci**et*y* (London), *Series* **A**
1869–	**Nature (London)**
1920–	**Discovery**
1925–	**Proc**eed*ings of the* **Leeds Phil**oso*phical and* **Liter**ary **Society,** **Sci**ent*ific* **Sect**io*n*
1939–	**Advan**c*ement of* **Science,** *The*
1942–	**Endeavour**
1956–	**New Sci**ent*ist*

India

1914–	**J**ourn*al of the* **Indian Inst**it*ute of* **Science**
1932–	**Curr**ent **Science**
1934–	**Proc**eed*ings of the* **Indian Acad**em*y of* **Sciences**
1936–	**Proc**eed*ings of the* **Nation**al **Acad**em*y of* **Sciences, India**
1950–	**J**ourn*al of* **Scientific Research** *of the* **Banaras Hindu University,** *The*

Ireland

1836–   Proceedings of the Royal Irish Academy
1877–   Scientific Proceedings of the Royal Dublin Society, The

Italy

1847–   Atti dell Accademia Nazionale dei Lincei, Memorie, Classe di Scienze Fisiche, Matematiche e Naturali
1865–   Atti della Accademia delle Scienze di Torino
1907–   Scientia (Milan)
1907–   Societa Italiana per il Progresso delle Scienze, Atti della Riunione
1930–   Ricerca Scientifica, La
1937–   Societa Italiana per il Progresso delle Scienze, Scienze e Tecnica

Japan

1887–   Journal of the Faculty of Science, Imperial University of Tokyo
1911–   Science Reports of the Imperial Tohoku University
1912–   Proceedings of the Japan Academy
1922–   Journal of the Scientific Research Institute, Tokyo
1930–   Journal of Science of the Hiroshima University
1930–   Science Reports of the Tokyo Kyoiku Daigaku
1949–   Science Reports of the Research Institutes, Tohoku University
1949–   Journal of the Osaka Institute of Science and Technology
1952–   Annual Report of Scientific Works, the Faculty of Science, Osaka University

Mexico

1887–   Memorias y Revista de la Academia Nacional de Ciencias
1940–   Ciencia (Mexico City)

The Netherlands

1866–   Archives Neerlandaises des Sciences Exactes et Naturelles
1898–   Koninklijke Nederlandse Akademie van Wetenschappen,

New Zealand

1868–   Transactions of the Royal Society of New Zealand

Norway

1761–   Kongelige Norske Videnskabers Selskabs, Skrifter

1926–   *K*onge*lige* **Norske Videnska***bers* **Selsk***abs,* **Forh***andlinger*

Poland

1889–1952   **Bull***etin* **Intern***ational de l'***Acad***emie* **Polon***aise des* **Sciences des**
                    Lettres. Superseded by:
1953–   **Bull***etin de l'***Acad***emie* **Polon***aise des* **Sci***ences*
1946–   **Ann***ales* **U***ni*ve*rsitatis* **Mariae Curie-Sklodowska, Sec***tio* **AA**
                    Physica et Chemia

Russia

1922–   **Dokl***ady* **Akad***emii* **Nauk SSSR**
1936–   **Iz***vestiya* **Akad***emii* **Nauk SSSR, Otd***elenie* **Khim***icheskikh*
                    Nauk

South Africa

1909–   *S*out*h* **Afr***ican* **J***ournal of* **Science**

Spain

1904–   **Rev***ista de la* **Real Acad***emia de* **Ciencias Exact***as,* **Fis***icas y*
                    Nat*urales de* **Madrid**
1947–   **Rev***ista de* **Cienc***ia* **Apl***icada*

Sweden

1739–   *K*ung*liga* **Svenska Vetenskapsakad***emiens,* **Handl***ingar*

Switzerland

1846–   **Arch***ives des* **Sciences**
1945–   **Experientia**

United States

1818–   **Am***erican* **J***ournal of* **Science**
1826–   **J***ournal of the* **Franklin Inst***itute*
1838–   **Proc***eedings of the* **Am***erican* **Phil***osophical* **Soci***ety*
1846–   **Proc***eedings of the* **Am***erican* **Acad***emy of* **Arts** *and* **Sciences**
1883–   **Science**
1913–   **Am***erican* **Scientist**
1915–   **Proc***eedings of the* **N***ational* **Acad***emy of* **Sciences** *of the* **United**
                    St*ates of America*

1945–   *Scientific American*
               Proceedings of the academies of the various states in the
               United States

Journals of Nonchemical Science

The growth and development of the several sciences were accompanied, sooner
or later, by the appearance of periodicals devoted to the interests of each of
these sciences. As we now well know, none of these fields stands apart,
unrelated to any of the others. Biology, for example, cannot detach itself from
physics, nor physics from chemistry. It is only to be expected that a journal
devoted primarily to one of these fields other than chemistry may contain an
occasional article of chemical interest. Investigations carried on by physicists,
with the results published in a physical periodical, are often of great importance
to physical chemists.

The number of these borderline publications is so large that no effort has
been made to compile a list. Individuals making searches of the literature must
use their own judgment in deciding where articles will be found that might
appear in various journals.

Journals of Chemistry

The remaining group of periodicals is devoted to chemistry and chemical
technology. Chemistry, both scientific and technologic, has expanded to cover
such a range of activities that we have now a basis for a further subdivision.
Periodicals which are devoted to several or all phases of the science may be
designated as "general journals"; the foremost of these include the main
publications of the various general chemical societies scattered throughout the
world. With the increasing development of specialized lines of chemical work,
there have appeared, especially since 1900, a large number of periodicals
devoted to special or limited fields. These may be designated as "specialized
journals."

The names of some of the more important journals of these two general
classes are listed below. In this compilation the aim was to select a representative
rather than a complete list.[16]

Of historial interest in this connection is R. Cornubert's "La Litterature
Chimique Mondiale," H. Dunod, Paris, 1943. Chapter I is a brief history of
chemical documentation. Chapter IV is a catalogue of the principal journals of
14 countries in Europe; of the United States, Canada, and Argentina in
America; of India and Japan in Asia; and the Phillipines in Oceania.

The data include changes of titles, abbreviations, and editors.

[16]For a list of discontinued journals, see E. E. Reid, "Introduction to Organic Research," chap.
5, D. Van Nostrand Company, New York, 1924; G. M. Dyson, *Adv. Chem. Ser.*, No. 30, p. 83
(1961), lists 93 obsolete chemical journals of the nineteenth century. Many early periodicals are
listed in the introduction to vol. 1 and 2 of J. R. Partington's "History of Chemistry."

General Chemical Journals   In the case of the general journals the earliest ones available have been included, together with the more important later ones, particularly the publications of all the larger, general chemical societies. The dates included are of interest in showing how one after another appeared and in indicating the mortality rate of such publications.

List of General Chemical Journals

1778–1781[17]	**Chem**isches **J**ournal *für die* **Freunde** *der Naturlehre*. Superseded by:
1781–1786	**Neues**ten **Ent**deckungen *in der* **Chem**ie
1783–1784	**Chem**isches **Arch**iv. Superseded by:
1784–1791	**Neues Chem**isches **Arch**iv. Superseded by:
1792–1798	**Neuestes Chem**isches **Archiv**
1784–1803	**Chem**ische **Ann**alen *für die Freunde der Naturlehre* (L. Crell)
1785–1795	**Beitraege** *zu den Chemischen* **Ann**alen *von Lorenz* **Crell**
1789–1815	**Ann**ales *de Chimie*. Changed to:
1816–1913	**Ann**ales *de Chim*ie *et de* **Phys**ique. Split into:
1914–	**Ann**ales *de Chim*ie **(Paris)** *and* **Ann**ales *de* **Phys**ique
1790–1802	**Ann**ali *de Chim*ica *e* **Storia Nat**urale
1793–	*Journal der* **Phar**macie *für* **Aerzte** *und* **Apoth**eker
1794–1815	*Journal des* **Mines.** Changed to:
1816–1851	**Ann**ales *des* **Mines**
1797–1813	*Journal of* **Nat**ural **Phil**osophy, **Chem**istry *and the* **Arts** (Nicholson). Merged into:
1798–	**Phil**osophical **Mag**azine, *The*
1798–1803	**Allgem**eines **J**ournal *der* **Chem**ie (Scherer). Changed to:
1803–1806	**Neues Allgem**eines **J**ournal *der* **Chem**ie (Gehlen). Changed to:
1807–1810	*Journal für die* **Chem**ie, **Physik** *und* **Mineral**ogie. Changed to:
1811–1833	*Journal für* **Chem**ie *und* **Physik** (Schweigger). Changed to:
1834–	*Journal für* **Prakt**ische **Chem**ie (Erdmann)
1800–1802	**Arch**iv *für* **Theoret**ische **Chem**ie
1803–1818	**Arch**iv *der* **Agrikulturchem**ie *für Denkende* **Landw**irthe
1809–1814	**Bull**etin *de* **Phar**macie *et des* **Sciences Accessoires.** Changed to:
1815–1841	*Journal de* **Phar**macie *et des* **Sciences Accessoires.** Changed to:
1842–1942	*Journal de* **Phar**macie *et de* **Chim**ie. Merged with: **Bull**etin *des* **Sci**ences **Pharmaco**logiques. To form:
1943–	**Ann**ales **Phar**maceutiques **Franc**aises
1813–1814	**Mem**oirs *of the* **Columbian Chem**ical **Soc**iety *of Philadelphia*
1817–	**Nord**ische **Blaetter** *für* **Chem**ie *(Scherer)*

---

[17]Overlapping dates indicate change of name during the year.

1822–	**Arch**iv der **Pharm**azie und Berichte der Deutschen Pharmazeutischen Gesellschaft
1823–1831	**Mag**azin für **Phar**macie. Merged with:
1822–1831	**Arch**iv der **Apoth**eker-Vereins im Noerdlichen Deutschland. To form:
1832–1839	**Ann**alen der **Phar**macie. Changed to:
1840–1873	**Ann**alen der **Chem**ie und **Phar**macie. Changed to:
1874–	**Ann**alen der **Chem**ie, **Liebigs**
1835–	**Arch**iv der **Pharm**azie
1824–1833	**Giorn**ale di **Farm**acia **Chim**ica e Scienze Accessorie. Changed to:
1834–1845	**Bibl**ioteca de **Farm**acia, Chimica, Fisica, Medicina. Changed to:
1845–1882	**Ann**ali di **Chim**ica **Appl**icata alla Medicina cioe alla Farmacia. Changed to:
1882–1884	**Ann**ali di **Chim**ica **Appl**icata alla Farmacia ed alla Medicina. Changed to:
1885–1897	**Ann**ali di **Chim**ica e di **Farm**acologia. Changed to:
1898–1900	**Ann**ali di **Farm**acoterapia e **Chim**ica **Biol**ogica
1825–1876	**Journal** de **Chim**ie **Med**icale, de **Pharm**acie, et de **Tox**icologie
1826–	**Bull**etin de la **Soc**iete **Ind**ustrielle de **Mulhouse**
1828–1833	**Journal** für **Tech**nische und **Oekon**omische **Chem**ie
1829–	**Amer**ican **J**ournal of **Pharm**acy
1831–1839	**Gazz**etta **Eclet**tica di **Chim**ica **Farm**aceutica
1840–1858	**Chemist, The**[18]
1841–	**Pharm**aceutical **J**ournal
1841–1843	**Proc**eedings of the **Chem**ical **Soc**iety of London. Merged with:
1841–1848	**Mem**oirs and **Proc**eedings of the **Chem**ical **Soc**iety of London. Changed to:
1849–1862	**Quart**erly **J**ournal of the **Chem**ical Society. Changed to:
1862–	**J**ournal of the **Chem**ical Society (Now in six parts)
1842–1859	**Chem**ical **Gaz**ette, The. Changed to:
1859–1932	**Chem**ical **News** and Journal of Industrial Science, The
1845–	**J**ournal de **Pharm**acie et de **Chim**ie
1846–1887	**Chem**isch-**tech**nischen **Mitt**heilungen der **Neuesten Zeit**
1855–1875	**Chem**ische **Ackersmann**
1858–1859	**Krit**ische **Z**eitschrift für **Chem**ie, **Phys**ik und **Math**ematik. Changed to:
1860–1864	**Z**eitschrift für **Chem**ie und **Phar**macie. Changed to:
1865–1871	**Z**eitschrift für **Chem**ie

---

[18]According to Florence E. Wall [*The Chemist*, **18**, 248 (1941)], there was a weekly periodical with this title as early as 1824, which she considered the earliest chemical periodical in English. However, James Kendall [*Nature*, **159**, 867 (1947)], considered the *Proceedings of the Chemical Society of the University of Edinburgh*, published in 1785, as the first chemical journal in English.

1858–1863	**Repertoire** *de* **Chimie Pure** *et* **Appliquee.** Merged into:
1858–1906	**Bulletin** *de la* **Societé Chimique** *de* **Paris.** Changed to:
1907–	**Bulletin** *de la* **Societé Chimique** *de* **France**
1862–1901	**Chemisch-technisches Repertorium** (Berlin)
1866–1880	**Boston Journal** *of* **Chemistry** (Nichol). Changed to:
1881–1883	**Boston Journal** *of* **Chemistry** *and* **Popular Science Review.** Changed to:
1883–1902	**Popular Science News** *and* **Boston Journal** *of* **Chemistry**
1868–1945	**Berichte** *der* **Deutschen Chemischen Gesellschaft.** Changed to:
1947–	**Chemische Berichte**
1869–1872	**Zhurnal Russkogo Khimicheskogo Obshchestva.** Changed to:
1873–1878	**Zhurnal Russkogo Khimicheskogo Obshchestva** *y* **Fizicheskogo Obshchestva.** Changed to:
1879–1930	**Zhurnal Russkogo Fiziko-Khimicheskogo Obshchestva.** Continued as:
1931–	**Zhurnal Obshchei Khimii** and **Zhurnal Fizicheskoi Khimii**
1870–1877	**American Chemist,** *The*
1871–1891	**Chemical Review.** (London)
1871–	**Gazzetta Chimica Italiana**
1877–	**Chemiker-Zeitung**
1879–1913	**American Chemical Journal.** Merged with:
1879–	**Journal** *of the* **American Chemical Society**
1880–	**Monatshefte** *für* **Chemie**
1880–1921	**Journal** *of the* **Chemical Society** *of* **Japan.** Changed to:
1921–1947	**Nippon Kagaku Kaishi.** Changed to:
1948–	**Nippon Kagaku Zaashi**
1882–	**Recueil** *des* **Travaux Chimiques** *des* **Pays-Bas**
1887–1903	**Bulletin** *de l'***Association Belge** *des* **Chimistes.** Changed to:
1904–1927	**Bulletin** *de la* **Societé Chimique** *de* **Belgique.** Changed to:
1927–1944	**Bulletin** *de la* **Societé Chimique** *de* **Belgique** *et* **Recueil** *des* *Travaux Chimiques Belges.* Changed to:
1945–	**Bulletin** *des* **Societes Chimiques Belges**
1889–	**Svensk Kemisk Tidskrift**
1903–1928	**Anales** *de la* **Sociedad Espanola** *de* **Fisica** *y* **Quimica.** Changed to:
1929–1940	**Anales** *de la* **Real Sociedad Espanola** *de* **Fisica** *y* **Quimica.** Changed to:
1941–1947	**Anales** *de* **Fisica** *y* **Quimica (Madrid).** Split into:
1948–	**Anales** *de la* **Real Sociedad Espanola** *de* **Fisica** *y* **Quimica. (Madrid). Serie A.** *Fisica* and **Serie B.** *Quimica*
1903–1948	**Arkiv** *für* **Kemi, Mineralogi och Geologi.** Split into:
1949–	**Arkiv** *für* **Kemi** and **Arkiv** *für* **Mineralogi och Geologi**
1903–	**Chemisch Weekblad**
1907–	**Chemicke Listy**
1917–1920	**Canadian Chemical Journal.** Changed to:

1921–1937	**Can**adian **Chem**istry and **Metall**urgy. Changed to:
1938–1951	**Can**adian **Chem**istry and **Process Ind**ustry. Changed to:
1951–	**Can**adian **Chemical Process**ing
1918–	**Helv**etica **Chim**ica **Acta**
1921–	**Roczniki Chem**ii
1922–1947	**Scientific Papers** of the **Inst**itute of **Phys**ical and **Chem**ical **Research (Tokyo).** Changed to:
1948–	**J**ournal of the **Scientific Research Inst**itute, **Tokyo**
1923–	**Rev**ista de la **Facultad** de **Cien**cias **Quim**icas **(Universidad Nacio**nal de **La Plata)**
1924–1927	**Quart**erly **J**ournal of the **Indian Chem**ical **Soc**iety. Changed to:
1928–	**J**ournal of the **Indian Chemical Society**
1926–	**Bull**etin of the **Chemical Soc**iety of **Japan**
1929–	**Collect**ion of **Czech**oslovak **Chem**ical **Commun**ications
1931–	**Zhurnal Obshchei Khim**ii
1934–1949	**Aust**ralian **Chem**ical **Inst**itute **Journal & Proceedings,** *The.* Changed to:
1949–1950	**Royal Australian Chemical Inst**itute **Journal & Proceedings,** *The.* Changed to:
1951–	**Proceedings** of the **Royal Aust**ralian **Chemical Institute**
1946–	**Wiadomosci Chemiczne**
1947–	**Acta Chem**ica **Scand**inavica
1948–	**Aust**ralian **J**ournal of **Chem**istry
1951–	**Acta Chim**ica **Academiae Scientiarum Hungaricae**
1951–	**Can**adian **Journal** of **Chem**istry[19]
1960–	**Pure** and **Applied Chemistry**[20]
1977–	**J**ournal of **Chemical Research**

Specialized Chemical Journals   In the first three editions of this book several titles of journals were included as examples for each of the sections in *Chemical Abstracts*.[21] As there are now 80 sections,[22] it is impractical to follow past practice. Instead, only six well-known, general subdivisions were selected, namely, analytical, applied, biological, inorganic, organic, and physical.

The titles selected are for journals often used in the given areas. Many others might well have been cited. In recent years many more specialized journals have appeared in these areas.

In many cases the present title of a journal is not the one originally used. Some have had several changes. The date included is that of the founding of the

[19]*Canadian Journal of Research* (sec. B) from 1923 to 1951.
[20]Official journal of the International Union of Pure and Applied Chemistry.
[21]See page 30 of the 3d ed. for the list compiled in 1957.
[22]See page 142 for a list of the 80 sections.

journal under its original title, in case there has been a change. The order in each subdivision is alphabetical by abbreviation.

Analytical Chemistry

Anal*ytica* Chim*ica* **Acta,** 1947
Anal*ytical* Chem*istry,* 1929
J*ournal of the* Asso*ciation of* Off*icial* Anal*ytical* Chemists, 1915
Mikrochim*ica* **Acta,** 1937
Zeitschrift für Anal*ytische* Chem*ie,* 1862
Zhurnal Anal*itischeskoi* Khim*ii,* 1946

Applied Chemistry

Chem*ical* Eng*ineering (*New Y*ork),* 1902
Chem*ical* Eng*ineering* Prog*ress,* 1908
Chem*ische* Ind*ustrie (***Duesseldorf***),* 1949
Chim*ica e l'***Industria (***Milan),* 1919
Chim*ie* & Ind*ustrie (***Paris***),* 1918
Ind*ustrial and* Eng*ineering* Chem*istry,* 1909; subdivided into three separate journals in 1962
J*ournal of* Chem*ical* Tech*nology and* Biotech*nology,* 1882
Nippon Kagaku Kaisi, 1880[23]
Trans*actions of the* Inst*itution of* Chem*ical* Engi*neers,* 1923
Zh*urnal* **Prikl***adnoi* **Khim***ii,* 1928

In Chapter 7, Table VI lists examples of industrial periodicals scanned by Chemical Abstracts Service for items for *Chemistry Industry Notes.*

Biochemistry (Biological Chemistry)

Biochem*ical* **J***ournal,* 1906
Biochem*ische* **Z***eitschrift,* 1906
Biochemistry, 1962
Biochim*ica et* **Biophys***ica* **Acta,** 1947
Biochimie, 1914
Biokhimiya, 1935
J*ournal of* Biochem*istry (*Tokyo), 1922
J*ournal of* Biol*ogical* Chem*istry,* 1905
Zeitschrift für Phys*iolo*gische Chem*ie,* 1877

Inorganic Chemistry

Inorg*anic* Chem*istry,* 1962
Polyhedron, 1955
J*ournal of* Organom*etallic* Chem*istry,* 1963

---

[23]Formerly Journal of the *Chemical Society of Japan.*

Zeitschrift für **Anorganische** *und* **Allgemeine Chem***ie,* 1892
Zhurnal **Neorganicheskoi Khim***ii,* 1956

Organic Chemistry

Angew*andte* **Chem***ie,* 1936
J*ournal of* **Organic Chem***istry,* 1936
J*ournal für* **Prakt***ische* **Chem***ie,* 1834
Liebigs Ann*alen der* **Chem***ie,* 1832
Macromolecules, 1968
Tetrahedron, 1957

Physical Chemistry

Journal of **Chem***ical* **Physics,** 1933
Journal of **Colloid Science** *and* **Interface Science,** 1946
Journal of **Physical Chem***istry,* 1896
Kolloid Z*eitschrift &* **Zeitschrift für Polym***ere,* 1906
Kolloid *and* **Polymer Science,** 1906
Theor*etica* **Chim***ica* **Acta,** 1963
Zeitschrift für **Physikalische Chem***ie (Leipzig),* 1887
Zhurnal **Fiz***icheskoi* **Khim***ii,* 1930

In addition to the specialized journals listed for inorganic, organic, and physical chemistry, many articles appear in the journals of the general chemical societies. For physical chemistry, journals of the physical societies should be included.

The "Science Citation Index 1981 Guide" (see Chapter 6A) contains a detailed subdivision of the scientific and technical journals covered in the *Science Citation Index.*

NOTES ON USING PERIODICALS

Photocopying

In various circumstances it may be desirable, or necessary, to obtain a photocopy of part or all of a journal article. This practice reduces the risk of loss or defacement of a borrowed copy of a journal. Copying machines have become common, and probably every significant chemical library has one or more.

To find the location of a desired journal, one may consult CASSI, as suggested in Chapter 7, which lists the chemical holdings of many libraries.

Full-size photocopies are known as photostats. They are easy to read, but bulky if there are many pages. Microfilm, usually 35 mm, is more commonly employed.

Instructions for ordering photocopies should be available from librarians. The joint copying service initiated in 1978 by Chemical Abstracts Service (CAS) and the National Technical Information Service (NTIS) has been discontinued.

Chemical Abstracts Service has issued the following announcement:

Now the CAS library will supply qualifying libraries and archives with single copies of papers and patents in accordance with the provisions of the Copyright Act of 1976. The service will be available . . . starting November 1, 1979.

The documents available include most papers and patents cited in *Chemical Abstracts, Chemical Titles, Chemical Industry Notes,* and *CA Selects* since 1972. Also Russian material published since 1972, most of the material published in the 1960s, and some 1950s material.

An optional flat-fee license to copy ACS primary journals is being offered to institutional subscribers on a trial basis.

Further information is available from Chemical Abstracts Service (see Chapter 7).

Copyrights

An important item to consider in making photocopies is what legally may be copied.[24] Most books and scientific periodicals are copyrighted. Thus, on the back of the title page of this book appears the following statement:

Copyright © 1982, 1965, 1958 by McGraw-Hill, Inc. All rights reserved. Copyright 1940, 1928 by McGraw-Hill, Inc. All rights reserved. Copyright renewed 1956, 1967 by M. G. Mellon. Printed in the United States of America. Except as permitted under the United States Copyright Act of 1976, no part of this publication may be reproduced or distributed in any form or by any means, or stored in a data base or retrieval system, without the prior written permission of the publisher.

Copyrights of journals, and usually of books, are held by the publishers rather than by the authors.

Because of increasing disregard of the old copyright law, a new United States law (Public Law 94-553) became effective January 1, 1978. Its general purpose is better to protect the rights of authors and publishers, and to provide for fair use of copyrighted works.

In an editorial[25] Herman Skolnik noted that two specific parts of the law, Sections 107 and 108, are relevant to document accessibility and to library services.

Section 107 concerns the purpose and the character of the uses. An infringement is considered to be "a use that supplants any part of the normal market for a copyrighted work." Fair use includes "reproduction for nonprofit purposes, such as criticism, news reporting, teaching, scholarship, and research."

Section 108 concerns limitations on reproducing copyrighted works. There

---

[24]See J. C. Stedman, *AAUP Bull.,* **64,** 142 (1978), for a discussion of photocopying and the law as it applies to academic reserves.
[25]*J. Chem. Inf. Comput. Sci.,* **17,** 2A (1978).

are several important conditions. It is not an infringement for a library to reproduce no more than one copy, or to distribute such copy, if (1) the copy is not for commercial advantage; (2) the library is open to the public or available to other persons doing research in a specialized field, and (3) the copy includes a notice of copyright.

In case of unsupervised use of reproducing equipment, a library is not liable for copyright infringement if the equipment displays a notice that making a copy may be subject to the copyright law.

Through the Association of American Publishers, the publishers of scientific and technical journals organized the Copyright Clearance Center to provide a centralized mechanism for users of reproducing equipment to pay copying fees to member publishers. For this service to publishers, the users (libraries) are charged a service fee in addition to what the publisher establishes as its copying charge.

Beginning January 1, 1978, publishers of journals in the United States include on the first page of each article a standardized code that identifies the article and publisher and states the fee charged for permission to copy. Thus, the first article in the January issue of *Analytical Chemistry* for 1978 has at the bottom of the page the code 0003-2700//8/0350-0002 $01.00/0. A copying organization, such as a library, retains an additional copy of the coded first page of a copied article, enters the organization's identification, and submits such first page to the Copyright Clearance Center with payment or for periodic billing.

Under the new United States law, the author becomes the owner of the copyright to his work immediately upon its completion. To follow established practice regarding copyrights, the American Chemical Society, for example, will not publish an article without transfer to it of the copyright. This means that the author(s) must submit, along with the manuscript, a signed copyright transfer statement for the manuscript.

Abbreviation

To save space, abbreviations of the word titles of journals are widely used. Over the years there have been variations in the abbreviations adopted, both among different countries and within a given country. For example, *J., Jr., Jnl.,* and *Jour.* have been used for *Journal*. The ultimate in condensation is likely to be found in treatises, such as those of Beilstein or Gmelin. In the former, for instance, *A* stands for *Justus Liebig's Annalen der Chemie,* and *B* for *Berichte der Deutschen Chemischen Gesellschaft. Ann.* and *Ber.* may appear as the respective abbreviations.

Identification of unfamiliar and less common forms of abbreviations for journal titles have been discussed by T. G. Labov[26] and G. M. Dyson.[27]

---

[26] *Adv. Chem. Ser.,* No. 4, p. 45 (1951), and No. 30, p. 102 (1961).
[27] Ibid., No. 30, p. 83 (1961).

In compiling reference lists involving abbreviations such as *J., Jr., Jnl.,* or *Jour.,* should one change an older abbreviation to the one now commonly used, or copy the one found in the original source cited?

Two generally approved practices are now in vogue. One applies to general usage in dealing with periodicals for publication. The other is a very compact form for computer use.

The International Organization for Standardization in 1973 adopted its International Standard ISO 833, "Documentation International List of Title Word Abbreviations." These standard abbreviations are embodied in the *Source Index* published by Chemical Abstracts Service (generally referred to as CASSI; see Chapter 7). The entries are listed alphabetically by *abbreviation.* The list includes primarily periodicals being published at the time of the issue of the list. However, there are many entries for changes of titles.

In the CASSI listing two usages are contrary to the usual practice in non-English languages: (1) all main words in the titles and subtitles are capitalized to simplify editing; and (2) all diacritical marks are omitted. In addition, the Germanic ä, ö, and ü, and the Norwegian ø, treated as diphthongs in alphabetizing, are transliterated to ae, oe, and ue, and oe, respectively. The diaresis is not used to indicate inflection.

Abbreviations of words not in titles often are used in the text of articles, as well as in other kinds of publications. *Chemical Abstracts,* with about 500 in current use, may well be the leader. In any case of uncertainty about the meaning, the list approved for a particular journal should be consulted.

The computer-readable abbreviations, known as CODEN, are unique, unambiguous, six-character codes assigned to titles of serial and nonserial publications in all subject areas.[28] They are designed for computer-readable files. The sixth character in each CODEN is a computer-calculated character, added to detect CODEN errors in computer-based systems. The CODEN for *Analytical Chemistry* is ANCHAM and for *Chemical Abstracts* CAPCB4.

Originated by the American Society for Testing and Materials, the International CODEN Service, known as INTERCODE, is now operated by Chemical Abstracts Service. CAS assigns and disseminates CODEN through the 1979 "International CODEN Directory" and its annual supplements (see Chapter 7).

This directory is an index to all CODEN assigned since 1954 (approximately 150,000). It consists of three parts: an alphabetical listing by CODEN; an alphabetical listing by titles of the publications to which CODEN have been assigned; and the Keyword-out-of-Context (KWOC) Index to these publication titles. It is published annually as microfiche and updated with a midyear supplement.

Important uses of the Directory are (1) to identify the complete title of a publication when only the CODEN is known, (2) to identify the CODEN for a publication title, (3) to determine whether a publication listed has ever been

---

[28]See A. L. Batik, "The CODEN System," *J. Chem. Doc.,* **13,** 111 (1973).

issued under a different title, and (4) to determine whether a listed publication appears in more than one language.

International Standard Numbers

Unique international systems have been devised for numbering books and serials: International Standard Book Numbers (ISBN) and International Standard Serial Numbers (ISSN). Like the CODEN abbreviations, ISBN and ISSN are designed for computer-readable purposes.

The ISBN in a book is assigned by its publishers. As an example, in the 1979 catalogue of John Wiley & Sons there is the number 0-471-02037-0. The first digit is the national, geographic, or language identifier; 0 is the English language. Next is the publisher prefix, John Wiley & Sons being 470 and 471. The third term is for the title, "Kirk-Othmer Encyclopedia of Chemical Technology" (3d ed., vol. 1). The final number is a computer check digit. Wiley advises using the ISBN when ordering a book.

The ISSN for a serial is assigned through the International Serials Data System. Examples are ISSN 0009-2258 for *Chemical Abstracts* and ISSN 0003-2700 for *Analytical Chemistry*.

Translations

Reference was made earlier to a report that, in 1980, 64.7 per cent of the journal literature being abstracted by *Chemical Abstracts* was published in English. This means that some 35 per cent appeared in about 50 other languages.

Because of the importance of many papers not in English, translations into English often are desired. Help may appear near at hand in various specific cases in institutions having staffs of diverse language backgrounds. Another possibility is to turn to literature consultants who can provide varied library services, such as making searches, building indexes, and making translations. Arrangements have to be made for the specific help needed. Apart from these possibilities, several general sources of translations are available.

First, there are the complete foreign journals issued in English.[29] These may be English editions, such as the German *Angewandte Chemie* or the most important Japanese periodicals. Some 50 Soviet journals are also being issued in English translation.

Second, in addition to these cover-to-cover translations, there are many individual translations being made available.

The most ambitious collective undertaking is the International Translation Centre, 101 Doelenstraat, Delft, The Netherlands. The primary function of this center is to maintain a central reference catalogue and to run an information bureau to handle requests for, and information on, translations.

[29]C. J. Himmelsbach and Grace E. Boyd, "A Guide to Scientific and Technical Journals in Translation," Special Libraries Association, New York, 1972.

This center issues a monthly publication, *World Index of Scientific Transla-tions and List of Translations Notified to the International Translations Centre.* The index lists the available translations from East European and Asiatic languages into Western languages of journal articles, patents, and standards relating to science and technology. There are quarterly and annual cumulations. The list announces the monthly acquisitions of translations by the center, with complete bibliographical citations.

In the United States there is the National Translations Center at the John Crerar Library, 35 West 33d Street, Chicago, IL 60616. Translations prepared by government agencies, industries, societies, academic institutions, and individu-als are contributed to the center. It indexes them and makes them available to others on request at a nominal photocopying and service fee. A semimonthly bulletin, *Translations-Register Index,* announces new acquisitions. There is a quarterly cumulative index.[30]

An example of a specialized source of information on translations is the *ASM Translations Index,* a quarterly publication of the American Society for Metals.

Citation of References

An important detail in journals is the method of handling references to other publications. Generally the citations are (1) located as footnotes at the bottom of the page to which they refer and in the order in which they are mentioned or (2) gathered together at the end of the article. In the latter case they may be arranged in the order cited or alphabetically by author. The latter scheme seems preferable when the number is large, in case a reader wants to check for the inclusion of a particular name.

In the treatises, e.g., those of Beilstein and of Gmelin, the reference is included within the text at the point of citing (see Chapter 10). In some areas the journals include within the text the name of the author cited, along with the date of publication.

The items comprising a reference to a journal may include the author's name, the abbreviation of the journal title, and the location in the journal of the item cited. The details in this citation should be just sufficient to enable the user to go with certainty to the publication sought.

Assuming bound volumes of a journal on the stack shelves of a library, one needs first the call number, usually Dewey or Library of Congress system (see Chapter 13), to locate the general area of the stacks. Next is the title of the journal. Then, for most journals, come, in order, the volume number, the page number, and the year of publication. An example of such a reference to *Analytical Chemistry* might be *Anal. Chem.,* **50,** 140 (1978). This practice assumes a periodical paged continuously throughout the year. If the issues are

---

[30]For sources in other countries, and for other information on translations, see R. T. Bottle, chap. 5, "Use of the Chemical Literature," Butterworths, London, 1979.

paged separately, the issue number should follow the volume number. Otherwise, the issue number is of interest only to one using unbound copies.

For some years Chemical Abstracts Service has been placing the year first, which would make the above reference *Anal. Chem.,* **1978,** 50, 140. Placing the year first, in boldface or italic, is appropriate for journals having no volume number, a typical reference being *J. Chem. Soc.,* **1956,** 85. Some editors would insist on the cited analytical reference being *Anal. Chem.,* **50,** (1978), 140.

In at least two cases a user does not want the year first. One example is a journal having several volumes per year, such as the *Journal of Biological Chemistry,* which issued eight volumes in 1977. A different case involves journals having series numbers, such as *Annales de Chimie.* In it, as soon as a series reaches Volume 30, a new series begins, with a new Volume 1. Thus, this famous journal is now in series 15. A recent reference might be *Ann. Chim.,* [15], **2,** 210 (1977). In such a case, one wants first the series number.

As viewed by a literature chemist, *Chemical & Engineering News* provides an annoying divergence from the general practice of citing volume, page, and year. For example, at the bottom of a page is the entry *C & EN,* June 5, 1978, with a page number. The volume number, **56,** and the issue number, 23, are not indicated. In using a bound volume one has to look for the month and the date of the month for the issue, and finally the page number.

The inclusion of advertising in some periodicals, particularly those slanted toward applied chemistry, is of some library concern. There are two practices: (1) the advertisements, entirely separate from the "editorial" matter, are discarded when binding the issues for the year; or (2) they are interspersed more or less throughout each issue, and remain so on binding.

The latter case is objectionable because of the extra stack space needed for essentially transient material, and because finding references to the "editorial" material is inconvenient.

Another practice, of bibliographical concern, involves placing special invited articles within the advertising section. If the advertisements are discarded on binding the "editorial" material, such an article disappears and cannot be cited.

If the advertisements are bound, there may be variation. Assuming monthly issues of a journal, binding of the advertisements along with the "editorial" matter of each issue, and separate paging of the advertising "A" sections, there are two possibilities: (1) If the citation gives merely the "A" page, such as 21A, without any issue number or month of publication, one has to look through the successive 21A pages until the article is found. (2) If the issue number or month of publication is part of the bibliographical data, one still has to locate the article, often in several bound parts of a journal.

Another practice is now being followed in *Analytical Chemistry.* The advertising sections are continuously paged through the 12 monthly issues. Then they can be separated and bound together to give a combined "A" volume (1478 pages in 1979).

Although this practice eases the problem of locating a specific reference, there remains the objection of having an article interspersed with advertisements. An important example, related to Chapter 15 of this book, is the 18-page article "Information Retrieval and Laboratory Data Management."[31] It is not only flanked by, but also split into four parts by more than seven pages of, advertisements.

Indexes

Most journals have author and subject indexes, arranged separately or in combined format. There is some variation in the location of these indexes. Usually they are at the end of a volume. If there is a single yearly volume, the index is annual. Such a volume may consist of several bound parts. Some journals have several volumes per year, with an index for each volume.

Many journals have collective or cumulative indexes covering various periods of time. Generally the period is 5 years or some multiple of 5.

If there is no index, usually at the beginning of each volume there is a table of contents which lists the titles of the articles.

An increasing practice in journals is to include suggested subject headings for indexing purposes. Examples are keywords, descriptors, and identifiers. Use of such entries by index builders should improve subject indexes. The annual subject index of *Analytical Chemistry* for 1981 is entitled Keyword Index.

The searcher must guard against relying too heavily on subject indexes. Too often they have been largely index titles or words, and they seldom contain entries for all the important points covered by an article. Chemists, as a class, have not shown great aptitude in selecting adequate titles for papers. Thus, a title such as "Chemical Affinity" reveals very little about the content of the article.

Evaluation of the Merits of Journals

Of the thousands available, which journals are best? To which ones should a library of limited means subscribe? Which ones in a given list should be discontinued if funds are reduced? As these and related questions do arise, individuals and groups face them.

Of much importance to consider is the nature of the journal and the situation for which its contents and quality are significant. Thus, the *Journal of Chemical Physics* may be fundamental to some theoretical physical chemists; but it can hardly be of value to an analyst with the problem of determining the contaminants in the water of the Ohio River at the intake of the purification plant at Cincinnati. Again, a journal's importance for the staff of an institute for cancer research would be of little or no interest to a group in ceramic engineering.

---

[31]R. E. Dessy and M. K. Starling, *Anal. Chem.,* **51,** 924A (1979).

Each library, whether academic, industrial, or institutional, has to decide which journals seem best for its situation and staff. There is no general formula applicable for all specific cases.

Two general ratings of many journals have been made, one by the Institute of Scientific Information (see Chapter 6B), and one by Chemical Abstracts Service (see Chapter 7).

An article by Eugene Garfield[32] describes a rating scheme of the Institute of Scientific Information. The primary listing is based on the number of citations found to individual journals covered by *Science Citation Index* (see Chapter 6B). Thus, a journal is considered important if it is cited often in other journals. This citing indicates that the cited journal contains articles of interest to the authors writing in the citing journal. Based on this study, covering the year 1974, the journal cited most often was the *Journal of the American Chemical Society*.

A section of *Science Citation Index* is the *Journal of Citation Reports*. It is an index of journal-journal links based on a grouping and summation of condensed citations using journals rather than authors as the primary sorting key. This second indicator of journal significance is a kind of index, known as impact. It is a measure of the relationship between citations and articles published. In the data reported above, the impact was calculated by dividing the number of 1974 citations of 1972 and 1973 articles by the number of articles published in 1972 and 1973. For the *Journal of the American Chemical Society* the impact was 4.38. In general, journals having impacts less than 2.0 are considered unimportant. Also, a relatively small percentage of total journals account for the great majority of citations.

Chemical Abstracts Service included in its 1975 edition of *Source Index* (CASSI) a list of the 1,000 journals considered most important among the more than 14,000 periodicals being monitored. In this case the rating was based on the frequency of citation in *Chemical Abstracts* for volumes 79 and 80. There are two lists, the first by title in alphabetical order, and the second by rank order. Here, again, the top-listed periodical was the *Journal of the American Chemical Society*.

The most important journals for semiconductors[33] and for electrochemistry[34] have been reported by D. T. Hawkins. Earlier[35] a study in 1961 listed the 100 most used journals in the Science and Technology Reference Department of the New York Public Library.

Techniques have been reported by G. Pinski[36] for constructing influence measures for individual journals and for the subfields of chemistry.

[32]*Nature,* **264,** 609 (1976).
[33]*J. Chem. Inf. Comput. Sci.,* **16,** 21 (1976).
[34]Ibid., **17,** 41 (1977).
[35]*Chem. Eng. News,* May 13, 1963, p. 68.
[36]*J. Chem. Inf. Comput. Sci.,* **17,** 67 (1977).

Dual Publication

Several problems with the production and storage of journals are increasingly troublesome. The budgets of publishers, libraries, and individuals are limited. Publishing expenses are rising for paper, printing, binding, and distribution. Circulation to libraries and individuals is declining.

For individuals, and especially for libraries, space to shelve bound journals is an increasing problem. The older journals are becoming larger and there seems no end to the introduction of new ones.

An additional problem concerns the individual reader. One paper refers to this problem as a blight having two aspects.[37] The first aspect is designated as dilution, or the presence in a given journal of too many papers with no interest to the subscriber. The second aspect is designated as dispersion, or the spreading of papers of interest through too many journals.

A suggested improvement (palliative), if not a complete remedy, is the adoption of a dual or two-edition system of publication. Details of operation might vary, but in general it would amount to a synopsis/microform production. The synopsis would consist of probably not over 1,000 words, including bibliographic details. The complete experimental and other details would be deposited as typescript in a designated location. The synopses would be printed and circulated in the usual way. Complete papers could then be reproduced at the depository in the form of microfilm, microfiche, or full size copies.

Trial experiments with such a system were reported by the American Chemical Society and the Royal Society of Chemistry.[38]

An international effort, launched in 1977, is represented by the *Journal of Chemical Research*. It is jointly sponsored by the Royal Society of Chemistry, the Gesellschaft Deutscher Chemiker, and the Société Chimique. Publication consists of two parts.

Part S consists of a synopsis of each research paper of one or two printed pages in length. Part M contains reproductions of the full texts of the original typescripts corresponding to the synopses appearing in Part S. Intended to fulfill its archival role, the journal appears in two formats: microfiche and miniprint. The latter is a conventional-sized journal with the typescript reproduced at a 3:1 linear reduction to give nine typescript pages on each journal page.

Editorial Policies

Users of different journals may note some variation in editorial practice.[39] Editors of most journals publish periodic recommendations for preparing

---

[37]L. C. Cross and I. A. Williams, *Chem. Brit.*, **11**, 224 (1975).

[38]For additional information see the following papers: J. A. Moore, *J. Chem. Doc.*, **12,** 75 (1972); E. L. Eliel, *J. Phys. Chem.*, **78**, 1339 (1974); W. D. Garvey and S. D. Gottfredson, *Inf. Process Manag.*, **12,** 165 (1976); M. J. Laflin, *J. Micrograph*, **10,** 281 (1977); S. W. Terrant and L. R. Garson, *J. Chem. Inf. Comput. Sci.*, **17,** 61 (1977).

[39]See Lois DeBakey, "The Scientific Journal: Editorial Policies and Practices," C. V. Mosby Company, St. Louis, MO, 1976.

manuscripts for consideration for publication in their journals. Accepted abbreviations and format may be included. In general, these suggestions are intended to facilitate editorial processing and to achieve acceptable uniformity in the printed products.

The latest recommendations for publications of the American Chemical Society are in its "Handbook for Authors," 1978. The following general sections are included: ACS Books and Journals; The Scientific Paper; The Manuscript; The Editorial Process; and eight appendixes. There is no mention of the ASME (American Society of Mechanical Engineers) recommended practice for preparing graphs of numerical data. Consequently, grid lines, for example, usually are not shown, except in engineering papers.

Nomenclature may be mentioned in such recommendations, but few editors are likely to take the initiative in developing and following consistent, logical usage.

List of General Periodicals

Mention has been made of the *Source Index* of Chemical Abstracts Service which lists the titles of all periodicals being monitored. An early list is H. C. Bolton's "A Catalogue of Scientific and Technical Periodicals," Smithsonian Institution, Washington, DC, 1897.

Many libraries have general compilations which contain national or international lists of periodicals. Supplements report the many changes taking place with time. The following titles illustrate lists having occasional value for checking items:

"British Union Catalogue of Periodicals"
"Guide to Scientific Periodicals"
"Ulrich's International Periodicals Directory"
"Union List of Serials in Libraries in the United States and Canada"
"World List of Scientific Periodicals"

Volume-Year Data

Frequently it is desirable to know the volume number of a journal for a given year, or the reverse of this. Tables containing such synchronistic data for a limited number of periodicals are included in the following books:

Atack, F. W.: "The Chemist's Yearbook," Sherratt, Altrincham, England, 1954.
Bolton, H. C.: "A Catalogue of Scientific and Technical Periodicals," Smithsonian Institution, Washington, 1897.
Comey, A. M.: "A Dictionary of Solubilities," The Macmillan Company, New York, 1921.
Dyson, G. M.: "A Short Guide to Chemical Literature," Longmans, Green & Co., Inc., New York, 1958.
Friend, J. A. N.: "A Textbook of Inorganic Chemistry," Chas. Griffin and Co., Ltd., London (only the later volumes).

Richter, F.: "Beilstein's Handbuch der organischen Chemie," vol. I, p. xxvi, Springer-Verlag OHG, Berlin, 1918.

————: "Erstes Ergänzungswerk," vol. I, p. x, Springer-Verlag OHG, Berlin, 1928.

————: "Zweites Ergänzungswerk," vol. I, p. xxvii, Springer-Verlag OHG, Berlin, 1941.

Roth, W. A., and K. Scheel: "Landolt-Börnstein's physikalisch-chemische Tabellen," p. 1634, Springer-Verlag OHG, Vienna, 1923.

Seidell, A.: "Solubilities of Inorganic and Organic Substances," D. Van Nostrand Company, Inc., Princeton, NJ, 1940–1941.

No general compilation of such data is known.

Errata

When an error that escapes an author is detected, correction is usually made in a subsequent issue of the periodical. Frequently this issue is the last for the year. To guard against missing such corrections, the searcher should examine the table of contents and the index of the journal for a year or two after publication of the original article. Preferably librarians should enter such corrections at the appropriate place as published.

Other Data

In many periodicals the professional connection or address of the author(s) is given either at the beginning or at the end of the article. Often the date of receipt and/or acceptance of the manuscript for publication is given to establish priority.

PRIMARY SOURCES— TECHNICAL REPORTS

There are many virtues in books, but the essential value is the adding of knowledge to our stock by the record of new facts.

R. W. Emerson

In recent decades there has been a great increase in publications which are nonperiodical in nature and which originate in institutions of some kind. These irregular publications are issued mainly by national, state, and municipal governmental agencies. By far the greatest production is national. There are a few privately endowed, nongovernmental agencies.

Following the present practice of Chemical Abstracts Service (Chapter 7), these publications are designated as technical reports. They should not be confused with manufacturers' technical publications (Chapter 5).

PUBLIC DOCUMENTS

Various Federal and state scientific organizations have been established for investigating those matters which concern the common welfare and for securing the desired publicity for the results obtained. The Government of the United States is the greatest of all publishers of scientific works.[1] The publications

---

[1] See B. W. Atkinson, "Two Centuries of Federal Information," Academic Press, New York, 1978; "Funding for Various Federal Agencies for Research and Development in 1978 and 1979," *Chem. Eng. News,* **56** (43), 18 (1978).

themselves include the results of the investigations of thousands of governmental workers conducting researches in many areas of scientific endeavor.

Large commercial concerns carry on independent research work, often on a large scale; but since such work is usually undertaken for gainful purposes, these concerns feel under no obligation to reveal the results of their investigations. The governmental bureaus, experiment stations, and laboratories aim to investigate and solve problems for the benefit of a whole industry, frequently under a cooperative arrangement with representatives of the industry. Thus, operators of limited means, individuals who cannot afford the experimentation necessary to demonstrate the advantages of adopting new procedures for their small-scale operations, have available for their use the results of various investigations. They have a place to take their difficulties where facilities and trained investigators may be able to offer valuable assistance.

Of great importance also are the collection and dissemination of statistics, such as those on production, sales, and uses of commodities.[2]

Nearly a century ago Disraeli spoke to the British House of Commons on the importance of public documents. "In my opinion," he said, discussing the money to be voted for official printing and publishing, "there is no vote to which the Committee has given its sanction which is more advantageous for the public service than the present one, which produces a body of information that guides the legislature and influences to a great degree the ultimate prosperity of the country."

Various countries are conducting this kind of governmental activity. It is believed, however, that the United States does far more in this direction than any other country; consequently, the present discussion is devoted largely to our own publications. It should not be forgotten, however, that very important contributions come from many foreign institutions.

The publications of these bureaus and laboratories are usually issued irregularly, except the annual reports, often in the form of circulars, pamphlets, or bulletins. D. E. Gray described these technical reports as "A form of publication which is characterized by its heterogeneity of style, professional stature, size, and form of reproduction, and by the absence of anything like volume and number relationships."[3]

There are two Government sources containing much information on the general organization and activities of many Federal agencies, along with some details on kinds of publications.

The first source, the *Government Manual*,[4] is an annual compilation. It lists all branches of the Government, and the various subdivisions. Many agencies,

[2]See annual *Statistical Abstract of the United States*, U.S. Department of Commerce, Washington, DC; J. L. Androit, "Guide to U.S. Government Statistics," Documents Index, McLean, VA, 1973; and annual *American Statistics Index*, Congressional Information Service, Washington, DC.

[3]*Bibliography of Technical Reports*, **16**, 141 (1951).

[4]1978/79 *United States Government Manual*, Government Printing Office, Washington, DC. Appendix *A* lists abolished and transferred agencies and functions. Appendix *B* lists commonly used abbreviations and acronyms, such as DNA, FDA, and COG.

offices, administrations, and other administrative units are described. Many are in one of the dozen departments, but a number are independent.

The information of chemical interest in the manual varies widely from agency to agency. It is reasonably informative in some cases, but in others one would get no idea that the agency maintains any laboratories.

The second source may be designated as the *Whitten Report*.[5] This 1,200-page study includes information for 779 Federal laboratories. Various items are listed, such as name, location, size, number of personnel, budget, and mission. The mission statement may be very specific ("molecular biology," for example) or as uninformative as many titles of Ph.D. theses.

Most of the statements included here for Federal agencies were contributed by them. In a few cases the information was obtained from one or more publications, such as the two already cited.

The material composing technical reports includes, for the most part, results of investigations. Occasionally mere compilations of matter published elsewhere, but not readily available, make up the publication, such as bulletins containing the various state mining laws. Frequently the results of investigations carried on in these laboratories are published as articles in government or other journals.[6]

PUBLICATIONS OF FEDERAL AGENCIES

The reports from Federal agencies are considered according to the division of the Government under which the bureaus come. Nearly every change in administration brings some departmental reorganization. Old agencies are discontinued, and new ones are organized. Others are changed from one department to another, and still others are given different names. It is difficult to know that a given bureau is in a given department, and of course, the bureau may no longer exist shortly after the statement is written.

At the present time (1978) the agencies selected for discussion seem of most interest to chemists.

Depository Libraries

Public Law 90-620 provides for a class of libraries in the United States in which certain Government publications are deposited for the use of the public. These are known as designated depository libraries. They are listed (revised to April, 1977) in a Joint Committee Print, First Session of the 95th Congress.

---

[5]Report of a subcommittee of the Committee on Appropriations, House of Representatives, 95th Cong., Jamie L. Whitten, chairman. Part 2 is the Investigative Report on the Utilization of Federal Laboratories, 1978, Government Printing Office, Washington, DC.

[6]"Government Periodicals," Price List 36, 174th ed., August 1979, lists all Government-sponsored periodicals and the subscription services. These publications are indexed under the various departments and agencies.

The total of such libraries is 1165. Seven states have fewer than 10, Alaska being lowest with 5, and California highest with 97.

A few of the designated libraries, known as Regional Depositories, are required to receive and retain one copy of all Government publications made available to depository libraries, either in print or microform. All others may select the classes of publications suited to their clientele. The distribution of Regional Depositories follows: 5 states have two each, 13 states have none, and the others have one each.

Locating Desired Publications

Without some general sense of direction, one's search for Federal publications on a given subject may be tedious, or even fruitless, because of the amount of material available and the problem of finding it.

First of all, there are certain general compilations which serve as guides and indexes to all such publications. They are probably less widely used than the more specific sources mentioned below. Of the comprehensive indexes the following, arranged chronologically, are the most important:[7]

1774–1881	"Descriptive Catalogue of the Government Publications of the United States" (Poore)
1789–1909	"Checklist of U.S. Public Documents," (U.S. Government Printing Office)
1881–1893	"Comprehensive Index to the Publications of the United States Government" (Ames)
1893–1940	"Document Catalogue" (a catalogue of the public documents of Congress and of all the departments of the government)
1895–1933	*Document Index* (an index to the reports and documents of Congress)
1895–	*Monthly Catalog of United States Government Publications*
1919–	*Monthly Checklist of State Publications*

The *Monthly Checklist* is a record of the documents and publications issued by the various states and received by the Library of Congress.

The *Monthly Catalog* lists the publications issued each month by the Federal government. It includes those for sale by the Superintendent of Documents, those for official use, and those sent to Depository Libraries. There are four

---

[7]For more detailed information on U.S. Government publications, see the following books:

A. M. Boyd, "United States Government Publications," H. W. Wilson Company, New York, 1952.

J. Morehead, "Introduction to United States Public Documents," Libraries Unlimited, Littleton, CO, 1975.

V. N. Palic, "Government Publications: A Guide to Bibliographic Tools," Library of Congress, Washington, DC, 1975.

L. P. Schmeckbier and R. B. Eastin, "Government Publications and Their Use," Brookings Institution, Washington, DC, 1969.

———, "Government Reference Books," Libraries Unlimited, Littleton, CO, biennial.

indexes, each arranged alphabetically: Authors, Titles, Subjects, and Series/ Reports. Cumulative indexes appear semiannually and annually.

The Government Printing Office (GPO) is the Federal printer and part of the Legislative Branch. The printed items are produced in response to print orders of Congress and Federal agencies. Any free price lists are available here.

The Superintendent of Documents administers the sales of the Government Printing Office, and sells only products of this office. It has no obligation to reprint or to provide documents out of print. New and stock titles are announced and sold through the *Monthly Catalog* and 19 regional bookstores.

Each bookstore has a complete microfiche catalogue of all the titles and subscriptions offered for sale. Most such stores are in the Federal buildings in the following cities: Atlanta, Birmingham, Boston, Chicago, Cleveland, Columbus (OH), Dallas, Denver, Detroit, Houston, Jacksonville, Kansas City, Los Angeles, Milwaukee, New York, Philadelphia, Pueblo, San Francisco, Seattle, and Washington (Government Printing Office, Pentagon, U.S. Information Agency, and the Departments of Commerce, State, and Health and Human Services).

Much useful information is contained in "Consumers Information," General Services Administration, Washington, DC, 1977.

How to Obtain Federal Documents

The following excerpts are from the instructions printed in the issue of the *Monthly Catalog* for January 1979:

Publications entered in the catalogue that are for sale by the Superintendent of Documents are indicated by a "for-sale" statement in the text entries. Those Congressional publications which are not for sale may be obtained from the responsible committee, or from the House or Senate Documents Room. Entries for publications sold by the National Technical Information Service include a statement to that effect. Unless otherwise indicated, publications are distributed by the issuing office which appears immediately before the publication date in each entry. Additional information necessary for procuring documents appears above or within text entries. Requests to issuing offices should contain titles and dates of publication.

Remittances for Government Printing Office sales documents should be made to the Superintendent of Documents, U.S. Government Printing Office, Washington, DC 20402, by check or money order. Foreign remittances should be made by International Postal Money Order, by draft on a United States or Canadian bank, or by UNESCO Coupons. All remittances should be made to the Superintendent of Documents. Arrangements may be made to use Visa and MasterCard accounts.

Persons who make frequent purchases from the Superintendent of Documents may find a deposit account convenient. Deposits of $50 or more are accepted against which orders may be placed without making individual remittances or first obtaining quotations. Order blanks are furnished for this purpose.

There is a special handling charge on all orders mailed to countries outside the United States and its possessions. This charge is one-fourth of the current selling price of the publications and subscription services ordered.

Because publications are stored by their stock numbers (S/N), orders can be processed more quickly and accurately if stock numbers and titles are used. If the stock number is not known, the order may be by title and the classification number which appears above each entry. Congressional documents and reports should be ordered by title together with the document or report number and the Congress number, e.g., 94-1: H, doc. 1.

The classification given above entries for departmental publication is that used in the Library Division of the Office of Superintendent of Documents. The shilling mark (/) is used before numbers or letters that ordinarily would be superior or inferior numbers or letters. It is not practical to print these numbers or letters as superiors or inferiors in the *Monthly Catalog*.

Some journals list selected Government publications. For example, the March/April, 1979, issue of *Applied Spectroscopy* has a four-page classified list on spectroscopy. The reports are nearly all available from the National Technical Information Service, most in printed copy (PC) or microfiche (MF). The bibliographical data include title, author(s), source, report number, and price.

The bibliographical identification of technical reports usually is made by a prefix consisting of a combination of letters and/or numbers and a serial number. The recurrent letters and numbers of the prefix are a series of codes. So many codes have been used that a dictionary was compiled to identify them with the agencies originating the reports and assigning the numbers.[8]

An introduction to the dictionary deals with various problems involving the evolution and use of such codes. There is a series of 48 reference notes. References 1 and 2 are Ambiguous Numerical Codes and Ambiguous Alphabetical Codes, respectively. Then follow 46 references for information on sources of codes, for example, the Tennessee Eastman Corporation and the University of California Radiation Laboratory.

Two recent examples of codes in *Chemical Abstracts* are Report 1976, IS-T-713, and Report 1977, ORNL/TM-5722.

Rather than await the appearance of the various indexes of reports and then rely upon finding information in them, often one may deal with individual bureaus. Many of them have their own lists of publications in print.

Next to having these separate lists at hand, it is very helpful to know the kind of work being done in the various bureaus. For example, one should refer to the U.S. Geological Survey for information on the surface waters of the Southwest, and to the National Bureau of Standards (NBS) for metric standards.

---

[8]Lois E. Godfrey and Helen F. Redman, "Dictionary of Report Series Codes," Special Libraries Association, New York, 1973.

THE AGENCIES

To provide a general idea of the nature of a number of agencies and to indicate the kinds of publications issued thereby, a brief statement follows concerning the chemical activities and publications most likely to be chemically important. More details may be found by consulting either the annual reports of the directors or chiefs of the agencies or the special pamphlets issued by some of the bureaus. The annual yearbooks of various departments are useful compilations of statistics and general information.

Department of Agriculture

Four former agencies of the U.S. Department of Agriculture (USDA) have been merged to form the Science and Education Administration. They are the Agricultural Research Service, the Cooperative State Research Service, the Extension Service, and the National Agricultural Library. The first and last of these are of concern here.

Central Research

Activities Most of the biological, chemical, physical, and engineering research activities in the Department of Agriculture are consolidated in the Science and Education Administration—Federal Research, which has the following research divisions: animal diseases and parasites, agricultural engineering, animal husbandry, crops, entomology, soil and water conservation, clothing and housing, consumer and food economics, human nutrition, utilization, and marketing.

Headquarters for all this research is at the Agricultural Research Center, Beltsville, MD. The Research and Development Division centers are located regionally—at Wyndmoor, PA; New Orleans, LA; Peoria, IL; Albany, CA; Ames, IA; and Orient Point, NY.

Primarily the work deals with the broad, basic scientific and applied problems of national interest. There is much cooperation with state agricultural experiment stations and with agriculture-related industries on specific problems.

Publications The publications include many technical, semipopular, and popular items each year. In addition, articles prepared by administration scientists are published in professional and other periodicals. See USDA List No. 11, "List of Publications of the U.S. Department of Agriculture."

Forest Products Laboratory

Activities The Forest Products Laboratory, located in Madison, WI, is operated by the Forest Service, U.S. Department of Agriculture, in cooperation with the University of Wisconsin. Its object is to develop new and more efficient methods of converting harvested trees into finished products, to increase possibilities for utilizing both used and unused species, and to find ways of utilizing material which otherwise would be wasted. Information of interest to chemists is developed in the following research fields: (1) solid wood products (the chemistry of glues, fire retardants, wood finishes, and wood preservatives, as well as wood-liquid relationships and the modification of wood to render it incapable of supporting fungus growth); (2) wood chemistry (chemical properties and uses of wood, the chemistry of cellulose, lignin, bark, and extractives; modification of wood to improve its physical properties; and wood for energy and chemicals); and (3) wood fiber products (manufacturing methods and suitability of various woods for pulp, paper, and special products).

Publications   Many articles dealing with these subjects appear in technical journals. In addition, printed reports, bulletins, and circulars are issued.

See "Lists of Publications of the Forest Products Laboratory."

National Agricultural Library

Activities   The library contains over 1,500,000 volumes, covering agriculture in the broadest sense. Substantial holdings in the field of chemistry supplement a comprehensive collection in agricultural chemistry, including material on pesticide residues and on pollution by agricultural chemicals. There is an extensive lending service, and on-line access to the leading data bases is provided. The library adds some 12,000 citations a month to its own on-line file (AGRICOLA), which has grown to over a million citations.

Publications   Two major publications derived from the AGRICOLA files are issued by commercial publishers: the "Bibliography of Agriculture" (Oryx Press) and the "National Agricultural Library Catalog" (Rowman and Littlefield).

Department of Commerce

Bureau of the Census

Activities   The first Census of Population was taken in 1790. In 1810 the first Census of Manufactures was taken. In later years other censuses were initiated by the Bureau. As the present major agency of the Federal Government for gathering statistical facts, the Bureau provides information in the broad fields of population, housing, agriculture, governments, manufactures, business, construction firms, some areas of transportation, and foreign trade.

The present schedule of "regular" censuses is as follows: Population and Housing Censuses have been taken every 10 years in the years ending in 0, but starting in 1985 they will be taken at 5-year intervals pursuant to a law passed by Congress in 1976. Agriculture Censuses are taken at 5-year intervals with one covering 1978, but they are to be phased in with the Economic Censuses. Economic Censuses—manufactures, mineral industries (mining), retail trade, wholesale trade, service industries, construction firms, and unregulated transportation; and governments (state and local)—are taken every 5 years, covering the years ending in 2 and 7. For selected subjects, current data are collected weekly, monthly, quarterly, and annually. The censuses provide data in the form of reports for the United States as a whole, for states, for metropolitan areas, counties, cities, and the like; the current reports generally have national data only.

Publications Information amassed by the Census Bureau is issued as reports. In addition to reports of the censuses, the Bureau issues weekly, monthly, quarterly, and annual reports—the so-called current reports. Among the Bureau's Current Industrial Reports (CIR's) are several in the field of chemicals. Other reports with data on chemicals issued by the Bureau on a regular basis include the annual *Statistical Abstract of the United States* and monthly and annual reports on foreign trade.

One series of current reports, the Annual Survey of Manufactures (ASM), produces statistics on major lines of manufacturing activity including chemicals and allied products, petroleum and coal, primary metals, and instruments, in each of the years in which a Census of Manufactures is not taken.

Data from the 1982 Economic Censuses will be available in late 1983 and in 1984.

Information about all publications of the Census Bureau may be obtained from the Bureau's *Catalog* which is issued on a current basis each quarter and cumulated to an annual volume. A monthly supplement, which lists new publications other than regular monthly and quarterly reports, enables users to be informed more currently on publications as they appear. The *Catalog* may be ordered from the Superintendent of Documents, U.S. Government Printing Office, Washington, DC 20402.

National Bureau of Standards

Activities As stated in its *Journal of Research*, "The Bureau's overall goal is to strengthen and advance the Nation's science and technology and facilitate their effective application for public benefit." To this end, the Bureau conducts research and provides (1) a basis for the Nation's physical measurement system, (2) scientific and technological services for industry and government, (3) a technical basis for equity in trade, and (4) technical services to promote public safety.

The technical work of the National Bureau of Standards (NBS) is performed in three divisions in the general areas indicated:

1 National Measurement Laboratory: absolute physical quantities, radiation research, thermodynamics and molecular science, analytical chemistry, and materials science

2 National Engineering Laboratory: applied mathematics, electronics and electrical engineering, mechanical engineering and process technology, building technology, fire research, consumer product technology, and field methods

3 Institute for Computer Sciences and Technology: systems and software, computer systems engineering, and information technology

Known internationally, this Bureau is one of the most important of the Federal laboratories. The new quarters are located at Gaithersburgh, MD.

Publications The most important periodical is the *Journal of Research of the*

National Bureau of Standards. Papers appear on chemistry, physics, engineering, mathematics, and computer sciences.

Another periodical (monthly) is *Dimensions/NBS* which is issued to inform scientists, engineers, businessmen, industrialists, teachers, and consumers of the latest advances in science and industry, with primary emphasis on the work of the Bureau.

Nonperiodicals A number of nonperiodicals are issued by the National Bureau of Standards as needed or as studies are completed. Included are the following categories:

Monographs. Major contributions on various subjects related to the Bureau's scientific and technical activities are covered.

Handbooks. These comprise recommended codes of engineering and industrial practice developed in cooperation with industry, professional organizations, and regulatory bodies.

Special Publications. These include mainly proceedings of conferences sponsored by the Bureau, annual reports, and bibliographies.

Applied Mathematics Series. Included are mathematical tables, manuals, and studies of special interest to a number of kinds of scientists and engineers engaged in scientific and technical work.

National Standard Reference Data Series. This series provides quantitative data on the physical and chemical properties of materials, compiled from the world's literature and critically evaluated. The worldwide program is coordinated by the Bureau.

Journal of Physical and Chemical Reference Data. Published quarterly for the Bureau by the American Chemical Society and the American Institute of Physics, this is the principal publication outlet now.

Building Science Series. This series disseminates information developed at the Bureau on building materials, components, systems, and whole structures.

Technical Notes. These studies or reports are complete in themselves but not as comprehensive as a monograph.

Voluntary Product Standards. The purpose of the standards is to establish nationally recognized requirements for products, and to provide all concerned with a basis for common understanding of the characteristics of the products.

Consumer Information Series. The series presents practical information, based on NBS research and experience, covering areas of interest to the consumer.

Federal Information Processing Standards Publications. Publications in this series collectively constitute the Federal Information Processing Standards Register. It serves as the official source of information in the Federal Government regarding standards issued by the National Bureau of Standards persuant to the Federal Property and Administrative Service Act.

NBS Interagency Reports. This special series of interim or final reports covers work performed by the Bureau for outside sponsors (both government and nongovernment).

Bibliographic Subscription Series. Three current-awareness and literature-survey bibliographies are issued: Cryogenic Data Center Awareness Service (bimonthly), Liquified Natural Gas (quarterly), and Superconducting Devices and Materials (quarterly).

See "Publications of the National Bureau of Standards," 1977.

National Technical Information Service

Activities   Established in 1970, the National Technical Information Service (NTIS) provides public access to publications of the Department of Commerce, to scientific and technical reports sponsored by Federal agencies, and to data files of these sources.[9] As a clearinghouse, it reproduces and sells reports of research and development, along with other analyses prepared by Federal agencies and their contractors or grantees. These are the *Current Published Searches.*

The NTIS collection of information exceeds 1 million titles, of which about 150,000 are of foreign origin. The NTISearch bibliographies cover some 500 categories.

Publications   Abstracts of new reports are published as *Weekly Government Abstracts* in 26 newsletters, Section 7 being *Chemistry*. The reports originate in a variety of academic, governmental, and industrial laboratories. For example, the 73 abstracts in one issue of *Chemistry* came from 54 laboratories, including 7 in foreign countries. Among these sources are the great national research centers, such as those at Oak Ridge, TN, Brookhaven, NY, and Los Alamos, NM. Several other sections have some chemical interest.

The abstracts are grouped as follows: (1) chemical engineering, (2) inorganic chemistry, (3) organic chemistry, (4) physical chemistry, and (5) radio and radiochemical chemistry. The following example is for a Federally funded project in organic chemistry. It appeared in *NTIS Chemistry,* January 10, p. 28 (1978).

Facile Reduction of Alkyl Tosylates with Lithium Triethylborohydride. An Advantageous Procedure for Deoxygenation of Cyclic and Acyclic Alcohols
S. Krishnamurthy and Herbert C. Brown.

> Purdue Univ., Lafayette, Ind., Richard B. Wetherill Lab. 24 May 76. 5 p. ARO-13810.2-C Availability: Pub. in Jnl of Organic Chemistry, v41, n18, p3064-3066, 1976.
> **AD-A045 846/3WK**   Price Code: PC A01/MF A01
> Lithium triethylborohydride rapidly reduces p-toluenesulfonate esters of both cyclic and acyclic alcohols to the corresponding alkanes in excellent yields and is applicable even to tosylates derived from hindered alcohols.

Government Reports Announcements and Index   In the biweekly all-inclusive journal for the 26 sections, the following indexes are issued: Subject; Personal

---

[9]See Chapter 15.

Author; Corporate Author; Contract Number; and Accession/Report Number.

In addition, there is the SRIM service (Selected Research in Microfiche), which provides complete research reports, carried out on request.

A 1980 catalog, "Special Technology Groups," lists the publications available on data for more than a score of specific areas of technology.

See "NTIS Information Services," a general catalogue of the diverse and extensive information products available from this service. Included are instructions for ordering and payment.

U.S. Patent and Trademark Office   Although this office is in the Department of Commerce, the activities and publications are so specialized that they are treated separately in Chapter 4. There is much current interest in making this office an independent agency.

Department of Defense

Although this department is one of the oldest, the information in the *Government Manual* is completely uninformative on chemically related agencies. As with the Department of Energy, the *Whitten Report* names specific laboratories. The numbers of these for the three main services follow, with the figures in parentheses showing the number conducting chemical research: Army 39 (5); Air Force 23 (6); and Navy 32 (6).

In the Air Force only two of these are not at Air Force bases—The Massachusetts Institute of Technology (MIT) Lincoln Laboratory and the Aerospace Corporation.

In the Army two well-known laboratories are the Chemical Systems Laboratory (the Chemical Warfare Laboratory) and the Picatinny Arsenal Laboratories.

In the Navy a similar pair are the Naval Research Laboratory and the Naval Weapons Center.

Activities   As no information was obtained from individual agencies, only mission statements in the *Whitten Report* can be cited.

Publications   No periodical research publications are reported. Abstracts of specific published reports appear in *Government Reports Announcements and Index* (see the discussion of NTIS above).

Department of Education

In 1979 the new Department of Education was created to take over the educational activities previously a part of the Department of Health, Education, and Welfare. To be transferred also are certain additional education programs now in other agencies. Any prospective publication programs have not been announced.

Department of Energy

The formation of the Department of Energy (DOE) in August, 1977, resulted in considerable government reorganization, with the transfer and consolidation of energy-related agencies. "The DOE Facilities," published in April, 1978, lists the laboratories involved as a national resource for resolving energy problems.

There are 32 research and development laboratories described, 17 for energy technology and 15 program-dedicated.[10] Included also are five nuclear materials production facilities.

Of major interest to chemists and chemical engineers are the lists for each laboratory of the principal program activities and the major facilities.

Activities The Department is concerned with the resources, production, utilization, research and development, demonstration, and commercialization of energy. Most of the Department's efforts are carried forth by industrial concerns and by private and public institutions under contract with the Department.

Publications The Department publishes *Energy Research Abstracts, Energy Abstracts for Policy Analysis, Solar Energy Update, Fossil Energy Update, Geothermal Energy Update, Fusion Energy Update,* and *Energy Conservation Update.* These periodicals are published from the Energy Data Base prepared and maintained by the DOE Technical Information Center. The coverage of the DOE Energy Data Base includes the following areas: all chemistry applied to energy, all combustion chemistry, energy consumption and conservation in chemical processing plants, radiochemistry and radiation chemistry, isotope effects and exchange in chemical processes, selected basic analytical and separation chemistry for certain elements considered of significance to energy-related studies, basic chemistry of aqueous solutions and certain rocks and minerals of interest in geothermal studies, and basic biochemistry referring to yeast and fermentation appropriate to biomass conversion.

Many books pertaining to chemistry, prepared and published under the sponsorship of the Department, are reported in "Technical Books and Monographs." Information on DOE publications may be obtained by writing to the DOE Technical Information Center, P.O. Box 62, Oak Ridge, TN 37850.

The Center cooperates with the International Nuclear Information System of the International Atomic Energy Agency in its publication *Atomindex,* and the International Energy Agency of the Organization for Economic Cooperation and Development in its publication *Coal Abstracts.*

---

[10]The following eight of these laboratories were established to concentrate on nuclear science, nuclear energy, and nuclear weapons: Argonne National Laboratory, Brookhaven National Laboratory, Lawrence Berkeley Laboratory, Lawrence Livermore Laboratory, Los Alamos Scientific Laboratory, Oak Ridge National Laboratory, Pacific Northwest Laboratory, and Sandia Laboratories.

Later the roles of these laboratories were expanded by Congressional action to include environmental and safety research and nonnuclear energy research and development. A new report from the General Accounting Office, EMD-78-62, recommends extension of the roles to include socioeconomic areas of expertise. (See *Chem. Eng. News,* **56** (23), 5 (1978).)

Department of Health and Human Services

Food and Drug Administration

Activities The Food and Drug Administration (FDA) is concerned primarily with enforcement of the Federal Food, Drug, and Cosmetic Act, the Tea Importation Act, the Import Milk Act, the Caustic Poison Act, the Filled Milk Act, the Fair Packaging and Labeling Act, the Radiation Control for Health and Safety Act, and parts of the Public Health Service Act.

In addition to sampling and testing products, staff laboratories conduct research to form groundwork for enforcement policy. Included are evaluations of the safety and efficacy of medicine and medical devices; the toxicity of ingredients in foods, drugs, and cosmetics; the detection and measurement of pesticides and processing chemical residues in foods; and the potency of vaccines and other biological products, drugs, and vitamins. New drugs and medical devices must be approved before marketing. In addition, insulin, antibiotics, and coal-tar colors are tested and certified before distribution.

Publications Scientific papers are published in professional journals. The agency publishes weekly the *FDA Enforcement Report* containing information about prosecutions, seizures, and injunctions, as well as "New Drug Approvals." Also published are calendars of events; Annual Reports; and *FDA Consumer* (a monthly magazine describing the Agency's activities and interests); several newsletters, over 150 consumer pamphlets, brochures, and booklets; and hundreds of technical reports, manuals, and documents for industry and businesses. The *FDA Drug Bulletin* is issued as needed six to eight times yearly to all physicians and health professionals in the United States. Also published are *FDA By-Lines* and laboratory information bulletins which review advances in technical and analytical procedures of interest to the Agency.

See "Catalogue of FDA Publications."

National Library of Medicine

Activities The National Library of Medicine comprises the greatest collection of medical literature in the world, the total exceeding a million items. It was established to assist the advancement of medical and related sciences, and to aid the dissemination and exchange of scientific and other information important to the progress of medicine and to the public health. The library acquires and preserves books, periodicals, and other library materials; organizes these materials by appropriate cataloguing and indexing; and provides reference and research assistance.

The library has a number of computerized on-line bibliographic data bases available over a nationwide network. Two such bases of interest to chemists are TOXLINE (Toxicology Information On-Line) and CHEMLINE (Chemical Dictionary On-Line). See Chapter 15.

Publications The chief publications are bibliographic guides to medical literature.

Public Health Service

Activities   The Public Health Service has the responsibility, under the direction of the Surgeon General, of protecting and improving the health of the people of the nation. Two subdivisions are of most concern to those in various kinds of chemical work.

The National Institute for Occupational Safety and Health is the principal Federal agency engaged in research to eliminate on-the-job hazards to the health and safety of workers. The objectives are to identify hazards and to recommend occupational standards.

The Institute's main research laboratories are in Cincinnati, OH. Much of the work centers on specific hazards, such as asbestos and other fibers, beryllium, volatiles of coal tar pitch, and silica. A laboratory at Morgantown, WV, studies respiratory disease, agricultural and noncoal mining health, and energy.

The National Institutes of Health (NIH) are concerned primarily with the extension of basic knowledge regarding the health problems of people and how to cope with them. The areas of the research programs of the 11 institutes cover cancer, heart disease, allergies and infectious diseases, dental diseases, mental illnesses, neurological diseases, and others. The famous facilities at Bethesda, MD, include the Clinical Center, a 500-bed hospital for clinical-laboratory research.

Publications   Monthly publications include *Public Health Reports, Journal of the National Cancer Institute, Public Health Engineering Abstracts,* and *Index Medicus.* Miscellaneous publications include directories, bibliographies, technical reports, and recommendations on a wide variety of topics. Many articles appear in nongovernmental technical periodicals.

See "List of Publications of the United States Public Health Service"; also "The Public Health Service Today."

The NIH/EPA (National Institutes of Health/Environmental Protection Agency) chemical information system has been developed since 1973 with the cooperation of various Government agencies. The network comprises interactively searchable chemical data bases.[11]

Department of the Interior

Bureau of Mines

Activities   The Bureau was created by Congress in 1910 to conduct scientific and technological investigations concerning mining and the preparation, treatment, and use of minerals and fuels; to increase safety, efficiency, and economy in the development of mineral resources and promote their conservation; to study economic factors affecting the mineral industries; and to disseminate findings on these subjects to the public.[12]

[11]G. W. A. Milne and S. R. Heller, *J. Chem. Inf. Comput. Sci.,* **20,** 204 (1980).

[12]In 1977 all fuels programs, including fuel statistics work and research in fuel extraction and use, became a part of the new Department of Energy.

The Bureau performs research and investigations aimed at improving safety and health conditions in mining, and at minimizing pollution and other harmful environmental effects associated with mineral extraction, processing, and use. Metallurgical investigations (including work on nonmetallics) are conducted to improve efficiency and economy in beneficiation, smelting, and other phases of mineral processing and to find economic ways of recovering mineral values from consumer and industrial wastes. Much of the Bureau's work concerns analytical, environmental, industrial, and organic chemists. The Bureau contributes substantially to published data on thermodynamics and theoretical metallurgy. Results of research on mine safety and health are the basis for a number of mandatory Federal standards.

Headquarters are in Washington, DC, with investigations carried out at mining or metallurgical research centers in Albany, OR, Avondale, MD, Boulder City, NV, Bruceton, PA, Denver, CO, Minneapolis, MN, Pittsburgh, PA, Reno, NV, Rolla, MO, Salt Lake City, UT, Spokane, WA, and Tuscaloosa, AL.

Publications The following technical series are of interest to chemists: *Bulletins* (results of major scientific investigations), *Reports of Investigations* (principal findings and results of minor investigations and of significant phases of major studies), *Information Circulars* (digests, reviews, abstracts, and discussions of activities and developments related to the mineral industries), and *Technical Progress Reports* (newsworthy findings from current research). The *Minerals Yearbook,* an annual reference, contains comprehensive statistics on production, consumption, imports, exports, and shipments of mineral commodities. Periodic releases in the *Mineral Industry Survey* series give more recent preliminary statistics and estimates. *Mineral Facts and Problems,* issued every 5 years, covers all the commercially important minerals in more detail and with more background. Updates on particular minerals are issued in the Bureau's new *Mineral Commodity Profile* series.

See also "List of Publications of the U.S. Bureau of Mines."

U.S. Geological Survey

Activities The principal objectives of the U.S. Geological Survey are the determination and appraisal of the nation's mineral and water resources; the delineation of its physical and cultural features through topographic surveys; and the supervision of development and production of minerals, oil, and gas from leases on Federal and Indian lands. Significant scientific and engineering functions are carried out under four operating divisions: Conservation, Geologic, Topographic, and Water Resources. The Water Resources and Geologic Divisions carry out the most important chemical activities.

The Geologic Division investigates the geology and mineral deposits of the United States and performs research in geology and related chemical and physical problems. Geologic Division laboratories are located in Reston, VA, Denver, CO, Menlo Park, CA, and Casper, WY. The staff analyze and identify rocks and minerals, and conduct special researches on chemical and physical processes that affect rocks and govern geologic processes.

The Water Resources Division provides the hydrologic information needed for the continuing evaluation of the quantity, quality, and use of the Nation's water resources. Major activities are concerned with conducting interpretive water-resource appraisals describing the occurrence, the availability, and the physical, chemical, and biological characteristics of streams, lakes, reservoirs, estuaries, and ground waters.

To support its water-quality programs, the Water Resources Division operates a Central Laboratories System which provides the bulk of its water analyses. The system consists of two highly automated laboratories located in Atlanta, GA, and Lakewood, CO. These laboratories produce an average of 7,000 determinations per day. Water-quality data from the laboratory system are stored in computerized data systems which can be accessed by Federal, state, and local agencies.

Publications The most important publications may be grouped as follows: professional papers (each year one volume in the Professional Paper series summarizes the scientific and economic results of current work of the Geological Survey and includes many chapters of short papers on geology, geochemistry, geophysics, and hydrology); bulletins (reports on economic geology, applied geology, and theoretical geology); water supply papers (reports dealing with the general problems of location, quantity, and quality of the Nation's water resources and their uses); circulars (short papers highlighting studies, methods, programs in geology and hydrology, including river-quality assessment, chemical and mineralogical studies, and many other earth science subjects); water resources data reports, issued annually for each State (include records of stream flow, groundwater levels, and quality of water, as measured at representative locations, nationwide); map series reports, with or without brief texts (covering bedrock, surficial, and general geology of economic and noneconomic areas); maps of the National Topographic series.

See U.S. Geological Survey announcement lists, *New Publications of the Geological Survey,* issued monthly, and the annual catalogs, *Publications of the Geological Survey.*

Office of Water Research and Technology

Activities This agency (OWRT) carries on research and development of water resources (through contracts and grants) and related functions. The fundamental purposes are to develop new or improved technology and methods for solving or mitigating existing and projected state, regional, and nationwide problems of water resources; to train water scientists and engineers through on-the-job participation in research work; and to coordinate and disseminate the results of activities in water research. To do this OWRT performs the following general functions:

Administration of a cooperative program with university Water Resources

Research Institutes, designated by the states, for research, as well as the training of scientists through such research, directed toward solving water and water-related problems of the states and hydrological regions of the Nation.

Conducting research and development activities to develop methods, equipment, and processes for the economical production, from sea and other saline or chemically contaminated water, of water suitable for agricultural, industrial, municipal, and other beneficial uses.

Provision of summary information to the Nation's water resources community about ongoing research projects and results of completed projects, and the interpretation of such information to potential users.

Publications The Office publishes *Selected Water Resources Abstracts,* a semimonthly journal of bibliographic references, including abstracts and availability information for articles, monographs, technical reports, and other documents relating to research results of water resources; completion reports of research projects resulting from support programs; a quarterly listing by states of research reports supported by OWRT; nonscheduled topical bibliographies produced and updated in response to indicated user needs; and an Annual Report of progress and accomplishments.

A series of manuals describe the methods developed and used in the Central Water Quality Laboratories System of the U.S. Geological Survey.

The Office operates an on-line bibliographic and information retrieval system for research-in-progress which is accessible from five terminals around the country.

Justice Department

Drug Enforcement Administration

Activities This agency operates laboratories in seven cities in the United States. Chemists there perform tests and analyses of controlled and other drug substances submitted by Drug Enforcement Administration (DEA) special agents and police officers at all levels of government. The purpose of these tests is to determine and conclusively prove or disprove the existence of controlled substances in contraband materials, and to determine potency levels of any such substances found.

Publications The monthly newsletter, *Microgram,* is a compilation of drug intelligence items worthy of note, as well as analytical methodology for the analysis of drugs. In addition, DEA has published manuals entitled "Analysis of Drugs" and "A Basic Training Program for Forensic Drug Chemists." These publications, as well as *Microgram,* are intended for the exclusive use of laboratories serving law enforcement agencies, and are not available to the public on a general basis.

Other Departments

In several other departments there is considerable research and development work of biological, chemical, physical, and technological interest. Examples are the chemical activities of the Bureau of Customs, the Bureau of Engraving and Printing, the Bureau of the Mint, and the Bureau of Narcotics (all in the Treasury Department), and the Federal Bureau of Investigation (in the Department of Justice). Since much of the work of these laboratories is routine, ordinarily no publications are issued. Consequently, the primary concern with them is to know the kind of work carried on so that personal communication may be employed when knowledge is desired on items of interest.

In still other directions information on the chemical industry may be found. Two examples are the records of congressional hearings[13] and of antitrust cases.

Independent Agencies

There are many agencies not organized in any of the major divisions of the United States government. The few which seem of interest as sources of chemical information are considered briefly.

Consumer Product Safety Commission

Activities The objectives of the Consumer Product Safety Commission (CPSC) are to protect the public against unreasonable risk of injury associated with consumer products, to assist consumers in evaluating the comparative safety of consumer products, to develop uniform safety standards for consumer products, to minimize conflicting state and local regulations, and to promote research and investigation into causes and prevention of product-related deaths, illnesses, and injuries.

Scientific and engineering functions are carried out in the Directorate for Engineering and Science. It develops and evaluates performance criteria, design specifications, and quality control standards for consumer products. It conducts and evaluates engineering tests and test methods. Also it performs or monitors research in the engineering sciences and provides technical supervision to Commission field laboratories and other engineering test facilities.

The Directorate collects scientific and technical data, and reviews and evaluates scientific, technical, and medical reports to determine the physiological effects of injury and potential injury treatment. It conducts tests, evaluates toxicological and chemical hazards, and provides technical supervision to agency field chemical laboratories and other chemical and toxicological testing facilities.

The laboratories are located in Washington, DC, New York, NY, Chicago, IL, and San Francisco, CA.

Publications There is no regular series of publications, but there are "briefing packages" and papers in the scientific literature. The Annual Report, summarizing the Commission's activities, contains the Index of Products Regulated by CPSC.

---

[13]See, for example, "Documentation, Indexing, and Retrieval of Scientific Information," a study of Federal and non-Federal information processing and retrieval programs, Senate Document No. 113, 86th Cong., 1960, and Senate Document No. 15, 87th Cong., 1961.

Environmental Protection Agency

Activities The general mission of the Environmental Protection Agency (EPA), established in 1970, is to control and abate pollution of the environment in the areas of air, water, solid waste, pesticides, and radiation. The programs include air and waste management, water and hazardous materials, and toxic substances.

The *Whitten Report* lists 46 agency laboratories, scattered from Massachusetts to California and from Minnesota to Mississippi. Four are in North Carolina in the region of the Research Triangle Park, and three are in Cincinnati.

Publications There is no general publication covering all the chemical work in these laboratories. The investigations are varied because of the numerous pollutants and the many possibilities of pollution.

Pesticides may be taken as an example. One of the 46 agency laboratories is the EPA Pesticide Laboratory at Beltsville, MD. Specific publications involving pesticides are *Pesticide Abstracts* (PESTAB), *Pesticide Monitoring Journal,* and starting in 1979, "Generic Standards for Pesticide Chemicals."

The "Toxic Substances Control Act Chemical Substance Inventory" was published June 1, 1979. Included in the listing are 43,278 chemical substances manufactured or imported for commercial purposes. The listing is in CAS Registry Number order (see Chapter 7).

The six volumes are as follows: I, "Initial Inventory"; II and III, "Substance Name Index"; IV, "Molecular Formula and UVCB [composition uncertain] Index"; V and VI, "Trademark and Product Names; Reporting Companies."

Since 1973 the NIH/EPA chemical information system has been developed, in cooperation with various other U.S. Government agencies, including the National Institutes of Health. The network comprises interactively searchable data bases.[14]

Federal Trade Commission

Activities The Federal Trade Commission (FTC) was organized as an administrative agency pursuant to the Federal Trade Commission Act of 1914. The basic objective of the Commission is the maintenance of free competitive enterprise as the keystone of the United States economic system.

Although the duties of the Commission are many and varied under the statutes, the foundation of public policy underlying all these duties is essentially to prevent the free enterprise system from being stifled or fettered by monopoly or corrupted by unfair or deceptive trade practices.

Publications The principal publications are "Decisions of the Federal Trade Commission," *Annual Reports,* and reports on individual commodities and industries.

See "List of Publications," 1978; also "Your FTC: What It Is and What It Does."

---

[14]G. W. A. Milne and S. R. Heller, *J. Chem. Inf. Comput. Sci.,* **20,** 204 (1980).

General Services Administration

Activities   The General Services Administration (GSA) provides a wide variety of management and related services for the Government. There are four services in the General Services Administration: the Public Buildings Service, the National Archives and Records Service, the Automated Data Management and Telecommunications Service, and the Federal Supply Service.

The Office of Standards and Quality Control administers the Federal Specifications and Standards Program. Federal specifications cover materials, products, or services used by, or for potential use of, two or more Federal agencies. Some examples are chemicals, paints, office supplies, and building materials.

Federal Standards establish engineering or technical limitations and applications for materials, processes, methods, and designs, including any related criteria considered essential to achieve the highest practicable degree of uniformity in materials or products or interchangeability of parts used in those products. Standards are used primarily as references in specifications.

Publications   Federal specifications and standards are listed in the *Index to Federal Specifications and Standards*. It provides alphabetical, numerical, and Federal Supply Classification listings of *Federal and Interim Specifications, Federal and Interim Federal Standards,* Federal Handbooks, and Qualified Products Lists in general use throughout the Federal Government. The Federal Procurement Regulations is an irregular publication of new or revised regulations.

The "Guide to Specifications and Standards of the Federal Government" is an aid to business people doing business with the Government.

U.S. International Trade Commission

Activities   The U.S. International Trade Commission (ITC) conducts investigations on tariff and international trade issues and provides related information to the Congress and the Executive Branch. In the performance of its duties, the Commission compiles and analyzes technical and economic data on the chemical industry, with emphasis on the interactions of imports, exports, and domestic and foreign production.

Publications   Two annual statistical reports are *Synthetic Organic Chemicals, United States Production and Sales,* and *Imports of Benzenoid Chemicals and Products.* A monthly release is the *Preliminary Report on U.S. Production of Selected Synthetic Organic Chemicals.*

Among the Commission's many reports are several on selected products, such as organic and inorganic chemicals, plastics, and petroleum. Others are in preparation.

See "List of Publications" (1977) and *Annual Report of the United States International Trade Commission.*

National Aeronautics and Space Administration

Activities   The research of the National Aeronautics and Space Administration (NASA) is conducted as programmatic activities, such as space sciences, applications, and aeronautics and space technology. Chemical research exists, to some extent, in each of these activities. There are National Space Technology Laboratories and field centers where the research and applications are conducted. In addition, the agency sponsors an extensive program of research grants to educational institutions and private laboratories.

Publications Specific general publications include "NASA Directory of Services for the Public" and "NASA Publications List."

Abstracts of declassified research reports of the laboratories, such as the Ames Research Center and the Jet Propulsion Laboratory, appear in Section 7, Chemistry, of *Weekly Government Abstracts,* the NTIS publication.

There are three other publications of interest:

1 "Research and Technology—Objectives and Plans" (RTOP). This compilation is a summary of the RTOP programs for the fiscal year 1978.

2 "NASA's University Program." This list covers the active grants and research contracts for fiscal 1977.

3 "Facilities Data." For each of the 16 major and component installations the information includes a statement on the location and a description of the mission. There are interesting photographs.

National Academy of Sciences: National Research Council, National Academy of Engineering, and Institute of Medicine

Activities According to its charter, the National Academy of Sciences (NAS) "shall, whenever called upon by any department of the Government, investigate, examine, experiment, and report upon any subject of science or art."

The National Research Council (NRC) was organized in wartime (1916) to stimulate research in the sciences and in the application of these sciences to agriculture, engineering, medicine, and other useful arts; to increase knowledge; to strengthen the national defense; and to contribute to the public welfare. The Council now comprises eight major units, the following four of which most concern chemistry: the Assemblies of Engineering, Life Sciences, and Mathematical and Physical Sciences, and the Commission on Natural Resources.

The National Academy of Engineering (NAE) was established under the charter of the National Academy of Sciences in 1964, and shares the responsibility of serving the Federal Government through the National Research Council.

The Institute of Medicine was chartered by the National Academy of Sciences in 1970 to deal with problems associated with the adequacy of health services for all sectors of society.

Publications Several general and various specialized publications are issued. Examples of general publications are the monthly *News Report,* the monthly *Proceedings of the National Academy of Sciences,* and the *Annual Report.* Examples of newsletters are *ALS Lifelines* (life sciences), *The Bridge* (engineering), and *Institute of Medicine Newsletter.*

See the brochures "NAS," "NAE," "Institute of Medicine," and "NRC." Also see "Publications Listing," 1978.

National Science Foundation

Activities   In 1950 the Congress established the National Science Foundation (NSF) to promote the progress of science; to advance the national health, prosperity, and welfare; and to secure the national welfare. Some aid is provided for publication of reference aids and for the study of processing, storing, and retrieving scientific information.

Primarily the Foundation helps to initiate and coordinate research in governmental and academic laboratories through its various organizational units. Biological, engineering, mathematical, medical, and physical sciences are included. There is a Science Resources Planning Office and a Division of Information Sciences and Technology.

Publications   Activities are summarized in the pamphlet "Publications of the National Science Foundation." They are classified as follows: (I) Annual Reports, Publications of the National Science Board, Descriptive Brochures, and Announcements; (II) Science Resources Studies; (III) Special Studies; and (IV) Periodicals (of which only the bimonthly *Mosaic* and the monthly *NSF Bulletin* are of chemical interest.

The National Science Board (of NSF) issues a biennial report to the President, for transmission to Congress, designed to summarize the state of science in the United States. Its "Science Indicators 1976" includes, among much else, important details on Federal support of basic research, research and development, and gathering and disseminating information.

In the budget recommendations for 1978[15] more than 80 percent of the funds proposed for basic research were allocated to five Federal departments or entities: Department of Defense, Department of Energy, National Science Foundation, National Institutes of Health, and National Aeronautical and Space Administration.

Tennessee Valley Authority

Activities   The Tennessee Valley Authority (TVA) operates the country's most complete fertilizer development center. It conducts basic and applied research designed to improve the use of natural resources in the production of fertilizer and improve the fertilizer materials themselves to make them cheaper to the farmer. Its research in chemical engineering includes the development of improved manufacturing processes for both new and conventional fertilizers. Soil-plant-fertilizer relationships are studied in greenhouse laboratories at Muscle Shoals, AL, and experimental fertilizers are tested and demonstrated in field plots of state agricultural colleges in many parts of the country. Research also includes the development of process technology for removing sulfur oxides from the stack gases of coal-fired electric power generating plants.

Publications   Research results are published as they occur in current technical publications and may be obtained from the TVA National Fertilizer Development Center, Muscle Shoals, AL 35660.

[15]*Sci. Amer.*, **238**, No. 3, 69 (1978).

Other Government Laboratories

Various laboratories considered thus far are operated as centers of research for specific agencies of the Federal Government. Two very important examples of these are the National Institutes of Health (NIH) and the National Bureau of Standards (NBS).

A different group of laboratories, not so connected administratively, are owned but not operated by the Federal Government. For some the responsibility is a major research university, such as the University of California or the University of Chicago. For others it is a large chemical company, such as the Union Carbide Company or E. I. du Pont de Nemours and Company.

Activities   Probably all these laboratories were established for specific purposes. Thus, the two at Los Alamos, NM, and Livermore, CA, are primarily weapons laboratories of the Department of Defense, and the three at Oak Ridge, TN, Paducah, KY, and Miamisburgh, OH, are uranium-enrichment laboratories. Nuclear reactors, nuclear wastes, effects of atomic irradiation, and related subjects concern others. Basic studies on atomic structure are fundamental at the great Fermi Laboratory at Batavia, IL.

Judging by abstracts of the technical reports submitted for distribution, the interests of at least some of these centers are now much broader than originally planned.

Publications   Predictably, only unclassified and cleared material is published. There are annual reports of directors which outline the general nature of the programs and summarize the year's accomplishments.

The specific items made public are submitted as technical reports to the National Technical Information Service. Abstracts of these reports then appear in the *Government Reports Announcement and Index.* If desired, copies of the reports may be obtained, as stated in the description of NTIS earlier in this chapter.

Legislative Branch

Library of Congress

Activities   The Library of Congress is not the kind of organization discussed in the preceding sections. It is a library, said to be the largest in the world in its combined holdings. Because of its vast collections and its extensive activities, it is one of the great centers of the world for bibliographic and documentation research. In the sciences this is in the Science and Technology Division.

Publications   Probably most of the 1,500 publications of the Library of Congress have little chemical interest. However, the *Monthly Checklist of State Publications,* the *Monthly Index of Russian Accessions,* the *New Serial Titles,* and bibliographies on a wide range of subjects may be useful.[16]

---

[16]A complete listing of Library of Congress *Publications in Print* is available free of charge from the Central Services Division, Library of Congress, Washington, DC 20540.

The National Referral Center for Science and Technology issues the following two directories of chemical interest: "Directory of Information Resources in the United States on Physical Sciences, Biological Sciences, and Engineering" and "Directory of Information Resources in the United States on Water."

PUBLICATIONS OF FOREIGN GOVERNMENTS

In various foreign countries scientific work comparable to that in the United States is being done under government supervision. In general, such activities are not as extensive as our own. The present discussion is limited to sources of further information.

Great Britain[17]

"Government Publications." See sectional lists, such as Department of Scientific and Industrial Research, Chemical Research Board, Forest Products Research Board, National Physical Laboratory, and Scientific Research in British Universities. See also reports of the Medical Research Council.

Canada

"Canadian Government Publications"
"Publications of the National Research Council of Canada"
"Technical Information Service Reports"
"Information Notes"
"Sources of Information on Canada"
Consult Technical Information Service, National Research Council, Ottawa, Canada.

France

"Bibliographie de la France," Supplement F, Publications officielles

Germany

"Monatliches Verzeichnis der reichdeutschen amtlichen Druckschriften"

THE UNITED NATIONS

Activities The United Nations is primarily a political organization, but it performs various other important activities. Some of these have significant scientific and technological interest.

---

[17]C. Oppenheim, chap. 14 in R. T. Bottle (ed.), "Use of Chemical Literature," Butterworths, London, 1979.

Publications   Most of the publications of the United Nations deal with the official records, but about one-third are surveys, yearbooks, and other periodicals.

Among the yearbooks of interest here are the Yearbook of the United Nations, Statistical Yearbook, and Yearbook of International Trade Statistics. Examples of other publications of value are "Commodity Survey"; estimated "World Requirements of Narcotic Drugs and of World Production of Opium in 1976"; "Mineral Resources Development Series"; "International Conference on the Peaceful Uses of Atomic Energy, 1971"; and "World Energy Supplies, 1950–1974."

There have been two comprehensive publications: Proceedings of the United Nations Conference on Conservation and Utilization of Resources (8 v.), and Science and Technology for Development (8 v.).

UNDEX is the United Nations Document Index: Series A, Subject Index; Series B, Country Index; and Series C, List of Documents Issued.[18]

PUBLICATIONS OF STATE INSTITUTIONS

In addition to the extensive list of public documents issued by the several Federal departments, we have, in the aggregate, a large number of publications coming from similar institutions which are under the control of the state governments. As intimated, these publications are not essentially different in character from those already considered. Frequently the subjects treated have a distinctly local significance; but probably just as often the subject matter is of more general interest than many productions of the Federal bureaus.

Two of the main sources of these publications are the state engineering experiment stations and agricultural experiment stations, which are located in many cases at the state universities. The general function of the first of these stations is to conduct investigations along various lines of engineering and to cooperate with engineering societies in pursuing industrial investigations, particularly for the engineering interests of the states. The agricultural investigations bear a similar relationship to the agricultural interests of the states.

One of the functions of these institutions is to publish the results of their work. This usually takes place in the form of bulletins. One may secure lists of available publications by writing the directors of the various stations, and ordinarily copies of the bulletins may be secured in a similar manner, without charge.

Certain other state institutions issue publications which are less numerous but often of value, such as the annual reports of the state boards of health and the bulletins of the state geological surveys.

Many publications of state institutions, such as experiment stations, geological surveys, boards of health, and similar organizations, are listed in the *Monthly*

---

[18]A UNISIST guide for teachers is A. J. Evans, R. G. Rhodes, and S. Keenan, "Education and Training of Users of Scientific and Technical Information," UNESCO, Paris, 1977.

Check List of State Publications, Superintendent of Documents, Washington, DC.

REPORTS OF NONGOVERNMENTAL INSTITUTIONS

Publications are also issued from several institutions whose financial status is such that they should hardly be included among those already discussed as government projects. These institutions are maintained, or at least were started, by private endowments and may be under the general supervision of some government agency. The nature of the publications themselves does not differ from that of public documents. Two typical institutions of this kind will be considered briefly.

Carnegie Institution of Washington

Activities   The Carnegie Institution of Washington was organized by Andrew Carnegie "to encourage in the broadest and most liberal manner investigation, research, and discovery, and the application of knowledge to the improvement of mankind." Accordingly, means are provided to undertake large research programs within the institution itself, giving the scientific staff the freedom to pursue long-term goals without pressure for scheduled results.

The institution aims to lend its work, whenever possible, to the advancement of fundamental research in fields of study not yet opened by other agencies, concentrating its attention upon specific problems and shifting the direction of attack as more pressing needs develop out of the increase in knowledge.

The research programs of the institution are carried on in several centers throughout the United States. The ones of chief chemical interest are the Department of Plant Biology at Stanford, CA, the Department of Embryology in Baltimore, MD, and the Geophysical Laboratory in Washington, DC.

Publications   Many of the contributions from these laboratories are reported in scientific journals. Others are issued under the title *Carnegie Institution of Washington Publication.* The annual *Year Book* summarizes the research work. Publications are distributed by Academic Press, Inc., New York.

See "Price List of Publications."

Battelle Memorial Institute

Activities   The Institute was established by the will of Gordon Battelle "for the purpose of education . . . the encouragement of research . . . and the making of discoveries and inventions. . . ." It attempts the charge of its founder and the challenge of today through a broad range of research, education, and invention- and technology-development activities. Its sponsored research embraces the physical, life, social, and behavioral sciences.

Publications   Many technical articles appear in appropriate journals. There is available an annual list of these publications, *Published Papers and Articles.*

GENERAL REFERENCES

Andriot, J.: "Checklist of Major United States Government Series," Documents Index, McLean, VA, 1973.

———: "Guide to United States Government Publications," Documents Index, McLean, VA, 1973.

Auger, C. P. (ed.): "Use of Report Literature," Shoestring Press, Hamden, CT, 1975.

Morehead, J.: "Introduction to United States Public Documents," Libraries Unlimited, Littleton, CO, 1975.

Parish, D.: "State Government Reference Publications," Libraries Unlimited, Littleton, CO, 1974.

Pemberton, J. E.: "British Official Publications," Pergamon Press, New York, 1973.

PRIMARY SOURCES—
PATENT DOCUMENTS

For words, like Nature, half reveal and half conceal the Soul within.

Alfred, Lord Tennyson

The subject of patents is so specialized, and sources of information pertaining thereto are so varied, that this separate chapter is devoted to them. Publications of patent offices could be grouped with public documents, of course, as these offices are governmental agencies.

By origin the word "patent" is an adjective, meaning open in certain legal senses. "Letters patent" are legal documents. The term has uses far older than the grant of letters patent to protect inventions. So the word, as used here, is simply a short term for "patent of invention," and has no concern with grants of land or titles of nobility.

The patent literature began in 1617 with the grant of the first British patent.[1] Great Britain averaged only about 60 patents per year during the first two centuries, with around a million in the third century. In 1816 the present numbering system began. It is now past 1.5 million (beginning with 100,001 to avoid duplication of the older annual numbers).

The total number of patents granted by all patent-granting nations runs far into the millions. The United States alone reached 4,131,951 on December 26, 1978, using the numbering system started July 28, 1836.

---

[1] For historical aspects of patent systems, see H. Skolnik, *J. Chem. Inf. Comput. Sci.*, **17**, 119 (1977).

As part of the chemical literature, patents have become very important. During 1978 the United States issued 66,245 patents, of which 21,751 were classified as chemical in the *Official Gazette* (Patents) of the U.S. Patent and Trademark Office. Many of the documents contained structural formulas of new organic compounds.

According to Dr. R. J. Rowlett, Jr., editor of *Chemical Abstracts,* 428,342 documents were abstracted in 1978, of which 13 percent were patents. The percentage distribution in the five areas involved follows: 37.6, applied chemistry and chemical engineering; 27.9, macromolecular chemistry; 16.0, organic chemistry; 11.9, physical and analytical chemistry; and 6.6, biochemistry.

Because many nations, with wide variation in their patent practices, are involved, patents of the United States receive the chief attention here. Consideration of this patent literature involves three principal questions: (1) What value has it for chemists? (2) What are the significant publications? (3) What services are available for using these publications?

CHEMISTS' INTEREST IN PATENTS

The first British patent (1617) covered an oil for prevention of rust. The first U.S. patent (July 31, 1790), entitled "Making Potash and Pearl Ash," was also a chemical patent. It bears the signatures of George Washington, then President; Edmund Randolph, then Attorney General; and Thomas Jefferson, then Secretary of State.

Alert and competent chemists or chemical engineers should be potential inventors. If they have imagination and analytical minds, they are capable of detecting deficiencies in old procedures and instruments or machines, and/or devising improvements. If they have creative and inventive talents and wish to develop a novel process or product, or improve on what is known in the art, automatically they have an interest in their patentability. Theoretical chemists are unlikely to be concerned.

Many chemists, particularly in educational institutions, seem relatively unaware of the existence, let alone the significance, of patent documents as primary sources of chemical information. As the first requirement of patentability is novelty, patents must contain new information. Many hundreds of new compounds, and/or methods of making or using them, are patented annually.

Some reasons why the patent literature is overlooked as a source of basic information are the use of uninformative titles, turgid language, inadequate review of anything but the claim(s) of the patent, and lack of easy access to patent files, which are the only source of details of description and claims.

Information about patents which have been issued or are under litigation in the courts, or in a patent office, is often highly important, especially to industrial chemists.[2] They deal not only with industrial activities, but also with laboratory developments having potential value.

---

[2] J. Fleischer, *Adv. Chem. Ser.,* No. 30, pp. 97, 208 (1961).

The following statement by F. E. Barrows[3] has lost nothing with time:

Patents, from their very nature, require consideration from various aspects other than as a part of the chemical literature. Thus, the patentability of inventions, the filing and prosecution of applications for patents, the construction, validity, and scope of issued patents, questions of infringement of unexpired patents, patent litigation, property rights in patents, the rights and obligations of patentees, etc., are matters primarily involving patents as patents rather than as publications, and are matters requiring consideration from the standpoint of the relevant principles of the patent law applicable thereto, as well as from the standpoint of the chemical principles that are involved.

Patent searches or investigations may thus be of a special character. In considering questions of infringement, for example, the primary search extends only through the United States patents granted during the last seventeen years and requires consideration of the invention claimed rather than, or in addition to, the invention described; but inasmuch as the claims of a patent may require to be construed by the accompanying description and in the light of the Patent Office proceedings leading up to the grant of the patent, as well as in view of the prior state of the art disclosed by prior patents and publications, and in accordance with relevant principles of the patent law, a more extended search to include expired United States patents and other patents and publications may be important or even essential.

Investigations of the patentability of inventions, such as are made to determine the advisability of making application for a patent, or by the Patent Office examiners in determining the patentability of inventions set forth in patent applications, as well as investigating the scope and validity of issued patents, may likewise require an extended search of the prior patents and publications, to determine whether the invention is new and whether it is a patentable invention or discovery, within the meaning of the patent law.

Patent investigations thus include both investigations of patents as patents, and investigations or searches of patents as publications and as a part of the chemical literature.

Considered as a part of the chemical literature, the patent literature furnishes one of the most important fields of search, inasmuch as it records the inventions and improvements, and hence the progress, made in almost all fields of chemical industry. Not infrequently inventors have patented their inventions without having published any descriptions of them elsewhere and without any abstracts or digests of their inventions appearing in the periodical literature. The patent literature, therefore, contains much that is not available elsewhere.

An industrial chemist has stated that patents reveal most of the important technical developments. Therefore, the examination of others' patents (and especially their applications) may (1) avoid repeating their work, (2) prevent infringement of their patents, (3) suggest ideas for research, and (4) reveal competitors' progress.

The applications may or may not result in the grant of a patent.

Relevant papers of interest cover the following subjects: obsolescence of

[3]*Chem. Met. Eng.,* **24,** 517 (1921); see also T. S. Reid, *Chem. Eng. News,* **40** (3), 90 (1962).

patent literature;[4] scientific and technical information contained in patent specifications;[5] patents as a source of technical information;[6] patents: another way to publish;[7] a unique source of information;[8] and the significance of a patent.[9]

Much information, still current, is contained in 11 papers included in a report of two symposia, "Patents for Chemical Inventions."[10] One covered the broad range of problems encountered in obtaining a patent, and the other covered the general requirements of patentability.

A recent symposium of nine papers dealt with the subject, "Meeting the Challenges of the Changing Patent Literature."[11]

In order to make the most of their opportunities, potential chemical inventors should understand what patents are and something of what can and cannot be done with them. Three matters of primary importance are the legal character of a patent, what kinds of inventions are patentable, and the requirements for patentability.

Character of Patents

The statement that an inventor has a patent means that letters patent have been issued to him or her by a government for an invention. The document comprises the certificate of grant, the description and claim(s) with full, complete disclosure of the invention, and drawing(s), if any. For United States patents the certificate of grant sets forth that there is granted "to the patentee, his heirs, and assigns, for a period of 17 years from the date of issue, the exclusive right to make, vend, or use the invention throughout the United States and the territories thereof." What is really granted is the right to prevent others from *making, using,* or *selling* the invention.

The specification is the information furnished by the patentee regarding the object or process patented. This section of the letters patent consists, in addition to any drawings, of two parts, the description and the claim(s).

> The *description* is a disclosure of the invention, and of the manner or process of making, constructing or compounding, and using it, in such full, clear, concise, and exact terms as to enable any person skilled in the art or science to which it appertains, or with which it is most nearly connected, to make, construct, or compound, and to use the same. . . . The function of the *claims* is to define the exact limits of the

[4]C. V. Clark, *J. Documentation,* **32,** 32 (1976).
[5]F. Liebesny et al., *Inf. Scientist,* **8,** 165 (1974).
[6]K. M. Sanderson, *ASLIB Proc.,* **24,** 244 (1972).
[7]C. M. Dann, *Phys. Today,* **31,** 23 (1978).
[8]J. F. Terapane, *CHEMTECH.,* **8,** 272 (1978); see also S. M. Kaback, *CHEMTECH,* **10,** 172 (1980).
[9]E. J. Saxi, *Am. Lab.,* **10,** No. 9, 31 (1978).
[10]R. F. Gould (ed.), *Adv. Chem. Ser.,* No. 46 (1964).
[11]*J. Chem. Inf. Comput. Sci.,* **17,** 119–157 (1977).

invention, and no matter what has been described in the body of the specification, or illustrated in the drawings, the invention patented is the invention set forth in the claims—nothing more. The patentee is bound by his claims, and these will not ordinarily be enlarged by reference to the specification. Failure to claim described matter dedicates it to the public use, unless claimed in other applications which should be properly referred to.

The present practice of the Office is to insist that each claim must be the object of a sentence starting with "I (or We) claim (or the equivalent). . . . Each claim begins with a capital letter and ends with a period."

Patentability

There is frequent misunderstanding of just what is patentable. Any chemist who wishes to exploit her or his inventive opportunities should have the possibilities clearly in mind. Thus, a new analytical balance might be patentable, but a new law for the absorption of radiant energy would not. Philosophical concepts and principles are not patentable.

The rules governing patentability recognize six patent types:

1 *Machines.* For chemists, machines are usually mechanical, electrical, or optical devices for laboratory or industrial control, operations, analysis, or the like. Centrifuges, recording instruments, and devices for automatic process control are examples.

2 *Processes.* Any means or procedure for effecting a result involves or is a process. In chemistry a process may be synthesis of a compound, or degradation of large molecules as in hydrolyzing starch or cracking petroleum, or an analytical procedure. The chemical process industries cover a tremendous range of commercial operations and their trail of progress is marked by many thousands of patents.

3 *Compositions of matter.* Novel individual chemical compounds are compositions of matter; so also are combinations of different substances. These, if they have properties not predictable from the sum of the properties known for the original substances, are patentable. Many alloys have been patented on this basis. Medicinal compositions are barred by the patent laws of some nations but not in the United States.

Two recent decisions of the U.S. Supreme Court [*Chem. Times & Trends,* **3,** No. 5, 16 (1980)], seem to broaden the interpretation of the coverage of this class. In 48 U.S.L.W. 4714 new life forms created by humans are considered patentable. The specific new form patented, a human-made bacterium, qualified as a composition of matter.

The second decision, 48 U.S.L.W. 4908, gives patent protection for the invention of new uses for known chemicals, i.e., compositions of matter.

Developments from these two decisions promise considerable interest.

4 *Manufactures.* Any useful article made by imparting to raw or processed materials novel forms, qualities, properties, or combinations of these is patentable. A classic example is Glidden's U.S. Patent 157,124 for barbed wire. More modern examples include plastic foams and aerosol sprays.

5 *Plants.* Plant hybrids were not patentable until revision of the patent law in 1930. They have acquired much importance in agriculture, including horticulture and floriculture. They are the only patents of invention which are permitted to have illustrations in color.

6 *Designs.* Not patents of invention, design patents cover novel, original ornamental designs for commercial products. They are much sought for protecting legal rights to jewelry, housewares, automotive parts, and the myriad containers brought out for packaging.

All types of patents of invention are likely to have some chemical interest, but the first three overshadow the others in chemical aspects.

Another patent office function, in the United States and many other nations, is administration of the trademark laws. There are many trademarks for chemicals, e.g., "Gesarol" for DDT and "Gammexane" for lindane, and for apparatus, e.g., the "Quantometer" spectrometer. Respect for rights of trademark owners is important. Owners have the responsibility of diligence in protecting their marks. The trademark "Cellophane" became public property for lack of such diligence.

Copyrights are important for chemical publications, but are not covered by the patent laws nor administered by the patent offices in many countries. Copyright law in the United States is independent of patent law and is administered by the Library of Congress.

Requirements

There are three general requirements which must be met to secure a patent. Court interpretation at times may have left uncertainty about these requirements, but they may be summarized as indicated here.

1 *Novelty.* The invention must be new. The item must not have been patented anywhere previously nor the idea for it published anywhere more than 1 year before application is made for a patent. Obviously, a decision on novelty puts great responsibility upon a patent examiner in the Patent and Trademark Office.

This requirement also makes it imperative that a prospective patentee keep adequate and reliable dated records to support the application for a patent. Issuance of a patent, and also defense of one issued, may depend upon proof of priority of novelty of the idea.

In connection with utility, the patent examiner must decide that the subject matter sought to be patented, in view of the prior art, would not have been obvious, at the time the invention was made, to a person having ordinary skill in the art to which said subject matter pertains.

2 *Utility.* In theory, at least, to be patentable the item must be useful. Otherwise presumably there would be no point in patenting it. Actually, there seems to have been some laxity in adhering to this requirement, and probably with some justification. For example, it seems likely that some new chemical compound may not at once be recognized as useful for anything but later may be found very useful in some particular application.

3 *Completeness of disclosure.* The courts have recognized a broad, general test here. In brief, the description must be adequate to enable "one skilled in the art" to make and/or use the item. That is, an organic chemist, competent in synthetic work, should be able to apply in the laboratory a process patent for making some new organic compound. Decisions in famous patent cases have hinged upon demonstration of this requirement.

THE PATENT LITERATURE

Although over 100 countries grant patent protection and a considerable number issue official publications from their patent offices, this discussion is confined for the most part to the practice followed in the United States and to the publications of the U.S. Patent and Trademark Office. These publications are considered under three divisions.

Letters Patent

A printed patent is a copy of the beribboned document issued to the patentee, omitting the proclamation of the grant and the official seals.

In current United States patents the heading includes the following items: patent number; date of issue, title, name of inventor(s), name of assignee (if any), date of filing, application number, related United States application data (if any), United States classification number, international classification number, field of search (i.e., classes and subclasses examined), references cited (patents and any other publications), names of examiners and attorney (if any), and an abstract.

Then follows the specification, that is, the description and the claim(s).

Patent documents vary widely in length. That of D. Isenberg, U.S. patent 1,650,071, for a paint, has one-third of a page of description and one claim. In contrast, that of W. S. Gubelmann, U.S. patent 1,817,451, for a calculator, has 975 claims filling, with the description, 205 pages, not counting 40 sheets of drawings. Longer, but with only 209 claims, is U.S. patent 2,320,548 to J. H. Voss for telephony, with 220 printed pages and 174 sheets of drawings. U.S. patent 2,991,451, granted to E. W. Flint and A. E. Hague for a device to sort recorded data, has only 96 printed pages, but leads in drawings with 264 sheets. U.S. patent 2,925,957, issued to A. E. Joel, Jr., for an accounting system, contains 354 sheets, with 243 claims.

To illustrate the general character of chemical patents, a specimen U.S. patent is reproduced here. It was selected because it has drawings. Usually product patents do not have drawings, but many chemical patents show structural formulas for the new compounds involved.

United States Patent [19]

Bock

[11] **3,992,109**

[45] **Nov. 16, 1976**

[54] **CYCLIC COLORIMETRY METHOD AND APPARATUS**

[75] Inventor: **Ditmar H. Bock,** Boston, N.Y.

[73] Assignee: **Calspan Corporation,** Buffalo, N.Y.

[22] Filed: **Sept. 15, 1975**

[21] Appl. No.: **613,647**

Related U.S. Application Data

[63] Continuation of Ser. No. 341,438, March 15, 1973, abandoned.

[52] **U.S. Cl. . . . . . 356/181; 23/230 R; 23/253 R; 250/565; 250/573; 356/205**

[51] **Int. Cl.^{2} G01J 3/50**

[58] **Field of Search** 23/230 R, 232 R, 253 R, 23/254 R; 250/564, 565, 573, 574, 576; 356/180, 181, 184–186, 195, 201, 205

[56] **References Cited**

UNITED STATES PATENTS

1,977,359	10/1934	Styer	23/232 R X
2,389,046	11/1945	Hare	23/232
3,089,382	5/1963	Hecht et al.	356/181
3,549,262	12/1970	Hozumi	250/576
3,643,102	2/1972	Harper et al.	356/180
3,708,265	1/1973	Lyshkow	356/181
3,739,263	4/1973	Engholdt	356/184
3,748,044	7/1973	Liston	356/180

Primary Examiner—John K. Corbin
Assistant Examiner—F. L. Evans
Attorney, Agent, or Firm—Allen J. Jaffe

[57] **ABSTRACT**

A process of and apparatus for colorimetry, whereby an indicator is added cyclically to a fluid stream to indicate the presence or absence of a condition in the stream. The presence of the condition may be indicated by the formation of a precipitate, a change of color, etc. A light source and photocell are located downstream of the point of addition of the indicator and by the difference in an optical characteristic such as the transmissivity or scattering of the fluid, at one or more wavelengths, due to the introduction of the indicator, a measure is obtained of the degree to which the tested-for condition is present in the fluid.

2 Claims, 3 Drawing Figures

CYCLIC COLORIMETRY METHOD AND APPARATUS

This is a continuation of application Ser. No. 341,438, filed Mar. 15, 1973, now abandoned.

The present invention relates to the broad field of colorimetry and, more specifically to the use of colorimetry for detecting hazardous chemicals in water or air.

It is an object of this invention to detect the presence of a chemical in a liquid or gas by the use of colorimetry.

It is an additional object of this invention to provide a reliable colorimetric device which can be emplaced where the fluid to be tested varies slowly in color, turbidity or transmissivity.

It is a further object of this invention to provide a system whereby a minimum of indicator is added to the fluid to provide lower operating costs and to minimize the polluting effects of the indicator. These objects, and others as will become apparent hereinafter, are accomplished by the present invention.

Hazardous chemical spills should be detected quickly, both to limit damage to biotae and to facilitate treatment which is most effective when the spill is still concentrated. Although many spills cannot be anticipated, in certain probable locations such as industrial rivers, detector arrays could be effective at a reasonable cost. The detectors should be suitable for unattended, long-term use where a spill may reasonably be expected.

Among applicable methods for detecting pollutants in water, electrical conductivity is effective in detecting the presence of ionic solutes; pH and certain other specific

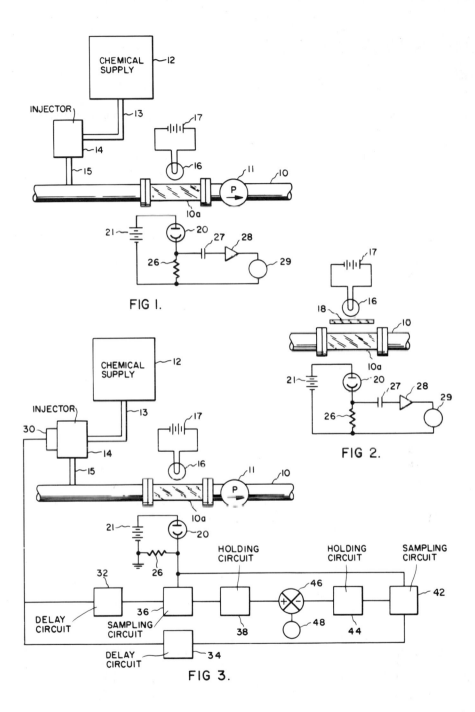

FIG 1.

FIG 2.

FIG 3.

1

ion probes are useful for indicating the presence of acids, bases and metallic pollutants. A variety of other compounds, especially heavy metal pollutants such as manganese, iron, cobalt, nickel, copper, zinc, silver, cadmium, antimony, mercury and lead, are detectable with a cyclic colorimeter.

In prior art devices the use of colorimetry is unsatisfactory where the fluid to be tested varies slowly in color, turbidity or transmissivity so as to mask the condition to be tested for due to the similar optical characteristics of the contaminant and the final indication. For example, silt which has an ochre appearance would mask or give an erroneous indication of the presence of manganese in a sulfide test because manganese sulfide is also ochre in appearance.

Heretofore this problem has been overcome by a careful sample preparation process such as centrifuging, filtering, etc. in order to remove the interfering contaminant such as silt. These processes are costly, time consuming and may even be inaccurate due to the removal of the constituent being tested for. Moreover, such existing processes are impractical for in situ measurements which are currently essential in pollution control efforts directed to industrial effluents.

Basically the present invention provides a method of and apparatus for cyclic colorimetry suitable for use in the presence of interfering contaminants which comprises introducing an indicator into a fluid then measuring and comparing the transmissivity or scattering of the fluid with and without the indicator whereby the presence of a condition of the fluid may be determined by the difference in transmissivity or scattering.

BRIEF DESCRIPTION OF THE DRAWINGS

For a fuller understainding of the present invention reference should now be had to the following detailed description thereof taken in conjunction with the accompanying drawings wherein:

FIG. **1** is a schematic diagram showing

2

the use of the invention for detecting the presence of a chemical in the fluid;

FIG. **2** is a partial schematic diagram showing the use of the invention for determining a condition of the fluid; and

FIG. **3** is a schematic diagram showing a modified form of the invention.

DESCRIPTION OF THE PREFERRED EMBODIMENTS

In FIG. **1, 10** represents a line carrying fluid to be tested and may, in practice, be the main line, a branch line, a sampling bypass line or even a freely falling stream. Line **10** may be made of any suitable material but a section **10a**, thereof, must be made of a transparent material such as glass or plastic where line **10** is in fact a pipe. If necessary, a pump **11** is located in line **10** to provide a sufficient and/or regulated flow of fluid in line **10**. A regulated supply of chemical is supplied to line **10** via line **15** by injector mechanism **14** which may take the form of a dosing pump, syringe, needle valves or other conventional structure. The chemical is furnished to the injector mechanism **14** via line **13** from a chemical reservoir or storage container **12**. A light source **16** powered by a power source **17** transmits light through section **10a** of line **10**. A photocell **20** is located opposite light source **16** with section **10a** of line **10** therebetween. A power source **21** and resistor **26** are located in series with photocell **20** and one of the terminals of resistor **26** is connected by means of an AC coupling capacitor **27** to the amplifier **28** and AC meter **29**.

In the cyclic colorimetry system of FIG. **2**, the numerals used in FIG. 1 have been used to designate similar structure. The system of FIG. 2 differs from that of FIG. 1 in the use of a filter **18** located intermediate light source **16** and section **10a** of line **10**.

In the cyclic colorimetry system of FIG. **3**, the numerals used in FIG. 1 have been used to designate similar structure. The system of FIG. 3 differs from that in FIG. 1 in the use of a synchronous demodulating system to discriminate against random fluctuations in transmissivity as might occur if a fish, debris, etc. should pass through line

3

section 10*a*. An injection sensor **30** senses the injection of the chemical into line **10** by the injector mechansim **14** and furnishes a signal to delay circuits **32** and **34**. A sampling circuit **36** is connected to delay circuit **32**, photocell **20** and holding circuit **38**. A sampling circuit **42** is connected to delay circuit **34**, photocell **20** and holding circuit **44**. A subtracting circuit **46** is connected to holding circuits **38** and **44** and to meter **48**.

The operation of the process of FIG. 1 is as follows: fluid is pumped through line **10** by pump **11** and passes between light source **16** and photocell **20**. Fluctuations in the transmissivity or scattering of the fluid passing through section 10*a* of line **10** are indicated on meter **29** which is connected to photocell **20** by means of AC coupling capacitor **27**. A source of chemicals **12** and an injector mechanism **14** are located upstream of section 10*a* to inject a slug of indicator into line **10**. If the tested-for condition is present in line **10**, the addition of the indicator will change the transmissivity or scattering of the fluid and the difference will be indicated on meter **29**. This process is particularly applicable for the testing of a fluid for the presence of heavy metals especially those forming a sulfide precipitate.

EXAMPLE 1

To detect the presence of iron sulfate (clear) in a dilute water solution, concentrated sodium sulfide solution (clear) is sdded to the dilute water solution to be tested. If iron sulfate is present, a black precipitate, iron sulfide will result which can be detected by the method and apparatus of this invention.

The operation of the process of FIG. 2 is as follows: fluid is pumped through line **10** by pump **11** and passes between light source **16** and photocell **20**. Fluctuations in the transmissivity or scattering of the fluid passing through section 10*a* of line **10** are indicated on meter **29** which is connected to photocell **20** by AC coupling capacitor **27**. A filter **18** of an appropriate color is located intermediate light source **16** and line section 10*a*. A source of chemicals **12** and an injector mechanism **14** are located

4

upstream of section 10*a* to inject a slug of indicator into line **10**. If the tested for condition is present in line **10**, the addition of indicator will change the color of the fluid and hence its transmissivity or scattering and the difference will be indicated on meter **29**. This process is particularly applicable for the testing for pH where there is a change of color by the indicator.

EXAMPLE II

To detect the presence of any acid, bromothymol blue containing a trace of base (blue) turns red in the presence of any acid. The use of a blue filter in the instant invention would permit the transmittance of less light through line 10*a* when the fluid has been turned red by the introduction of bromothymol blue in the presence of an acid in the fluid in line **10**. This change in the transmittance of the light can be detected by photocell **20** and indicated on meter **29**.

The operation of the process of FIG. 3 is as follows: fluid is pumped through line **10** by pump **11** and passes between light source **16** and photocell **20**. Fluctuations in the transmissivity or scattering of the fluid passing through section 10*a* of line **10** are indicated on meter **48** if they occur within a prescribed time period following injection. A source of chemicals **12** and an injector mechanism **14** are located upstream of section 10*a* to inject a slug of indicator into line **10**. If the tested-for condition is present in line **10**, the addition of indicator will change the transmissivity or scattering of the fluid in section 10*a* after a short delay and hence the output of photocell **20** will be changed. An injection sensor **30** senses the injection of a slug of indicator into line **10** by injector mechanism **14** and gives a signal to delay circuits **32** and **34**. The delay circuit **32** after a first delay sufficient for the injected chemical to have reached line section 10*a* activates sampling circuit **36** which is connected to photocell **20** and transmits a signal indicative of the transmissivity or scattering of the fluid to holding circuit **38**. The signal can represent the value of the transmissivity or scattering averaged over a short time period, the

5

integral of the value over such a period, or any other suitable form. The delay circuit **34** after a second delay sufficient for the injected chemical to have been purged from line section **10**a activates sampling circuit **42** which is connected to photocell **20** and transmits a signal indicative of the transmissivity or scattering of the fluid to holding circuit **44**. The holding circuits **38** and **44** thereafter transmit their signals to the subtracting circuit **46** which compares the two signals and indicates any difference therebetween on meter **48**.

In each of the above descriptions of the operation of the embodiments of the invention, a single cycle has been described and it is obvious that this cycle will be repeated. The frequency and regularity of the periods of addition of indicator will be a function of the injector mechanism chosen. Likewise, it will be apparent to those skilled in the art that a control function and/or a sampling function can be had in addition to or in substitution for the indicating function.

In summary, this invention provides a method and apparatus for testing for the presence of a chemical, such as a heavy metal, or a condition, such as pH, where a precipitate or color change may be produced upon the introduction of a chemical, even in the presence of a masking contaminant, since the masking effect of the contaminant is reduced by determining fluctuations of, or the differences in, the transmissivity or scattering of light passing through the fluid rather than the value of the transmissivity or scattering after the addition of the indicator.

This invention is applicable to standard tests other than those specifically described above such as may be found in any reference on colorimetric analysis such as Bruno Lange, "Kolorimetrische Analyse," Berlin 1944.

Although preferred embodiments of the present invention have been illustrated and described, other changes will occur to those skilled in the art. It is therefore intended that the scope of the present invention is to be limited only by the scope of the appended claims.

6

I claim:

1. A method for eliminating the effects of optical system deterioration and the presence of interfering contaminants when determining the presence of a condition in a flowing fluid and including the steps of:

flowing the fluid to be tested through the light path of an optical testing means;

using said optical testing means to periodically determine an optical characteristic of the flowing fluid and producing a signal indicative of said optical characteristic;

storing said signal indicative of said optical characteristic of the fluid;

periodically introducing a predetermined amount of an indicator indicative of the presence of the tested-for condition into the fluid to be tested at a point upstream of said optical testing means to cause a chemical reaction and to change said optical characteristic of the fluid to be tested only in the presence of the tested-for condition;

flowing the fluid to which the indicator has been added through the identical light path of said optical testing means as said fluid to be tested;

using said optical testing means for synchronously determining said optical characteristic of the flowing fluid to which indicator has been added and producing a signal indicative of said optical characteristic;

storing said signal indicative of said optical characteristic of the fluid to which said indicator has been added; and

comparing said stored signals indicative of said optical characteristic whereby the optical effects produced by the presence of interfering contaminants in the fluid, by nonsynchronous changes in the fluid and by the deterioration of said optical testing means will cancel out and the presence of the tested-for condition will be indicated by a difference in said optical characteristic determined periodically.

2. A cyclic colorimetry system for eliminating the effects of optical system deterioration and the presence if interfering contaminants when determining and indicating

7

the presence of a condition in a flowing
fluid to be tested and comprising;
 light source means;
 photocell means for producing an output
 signal proportional to the amount of 5
 light from said light source means
 which is incident thereon;
 means for causing the fluid to flow intermediate said light source means and
 said photocell means and including a 10
 transparent portion defining a light
 path between said light source means
 and said photocell means;
 indicator means for producing a chemical reaction in the fluid in the presence 15
 of the tested-for condition and to
 thereby produce a change in an optical
 characteristic of the fluid only in the
 presence of the tested-for condition;
 means for periodically causing the intro- 20
 duction of a predetermined quantity
 of said indicator means into the fluid
 at a point upstream of said light source
 means and said photocell means;
 means for sensing the introduction of 25
 said indicator means into the fluid and
 for producing a synchronizing actuating signal;
 first means responsive to said actuating
 signal for determining the output sig- 30
 nal of said photocell means after a first

8

 predetermined time period and for
 producing a first signal output representative of the tested-for optical characteristic of the fluid to which said
 indicator means has been added;
second means responsive to said actuating signal for determing the output
 signal of said photocell means after a
 second predetermined time period
 which is longer than said first predetermined time period to permit the
 purging of fluid to which said indicator
 means has been added to produce a
 second signal output representative of
 the tested-for optical characteristic of
 the untreated fluid;
means for storing said first and second
 signal outputs; and
means for comparing said stored first
 and second signal outputs whereby
 optical effects produced by the presence of interfering contaminants, by
 nonsynchronous changes in the fluid
 and by deterioration of the optical
 system comprising the light source
 means, the photocell means and said
 light path are canceled and the presence of the tested-for condition is indicated by any difference in said stored
 first and second signal outputs.

*   *   *   *   *

Numbering of early United States patents was at the discretion of the
Superintendent of Patents (the title was changed to Commissioner in 1836).
Numbering was not more systematic than annual numbered series until the
present sequence was adopted, July 28, 1836. Earlier patent records are in the
Patent Office publications "List of Patents Granted by the U.S., April 10, 1790
to December 31, 1836" and *Annual Reports* of the Superintendent of Patents.
Numbers in the present sequence passed 4,242,756 in December, 1980; cf
statistics in Table II.

The proportion of chemical patents to the total number granted has varied,
ranging upward from less than 5 percent to more than 20 percent. This is easy to
trace since July, 1952, when the *Official Gazette* (Patents) began to list general
and mechanical, chemical, and electrical patents in separate groups.

The total number of U.S. patents is not quite that shown in Table II at a
specific date. Over the years about 3,000 prospective patents had to be

TABLE II
NUMBER OF UNITED STATES PATENTS, BY DECADES*

Decade	Years	Number issued		Decade issue all
		Chemical	All	
1	1836–1845	128	4,338	4,338
2	1846–1855	369	13,999	9,661
3	1856–1865	1,588	51,774	37,775
4	1866–1875	5,896	171,631	119,857
5	1876–1885	10,547	333,484	161,853
6	1886–1895	16,991	552,492	219,008
7	1896–1905	26,578	808,608	256,116
8	1906–1915	39,788	1,166,409	357,801
9	1916–1925	58,870	1,568,030	401,621
10	1926–1935	95,906	2,026,506	458,476
11	1936–1945	144,563	2,391,846	365,340
12	1946–1955	195,247	2,728,903	337,057
13	1956–1965	290,339	3,226,729	497,826
14	1966–1975	199,841	3,930,271	700,335

*E. A. Hurd, *J. Chem. Doc.,* **10,** 167 (1970).

withdrawn from issue at a time in the issuance process when it was too late to enter substitutes for them.

Patent Classification

Subject classifications have been used by patent offices for more than a century to guide searchers in locating pertinent patents or determining that none exist.

The present United States classification may tempt exasperated chemists to vow that it has no basis in logic. Actually, it does have a logical basis, and familiarity with the system, backed by trained use of the alphabetical index, makes the "Manual of Classification," now in loose-leaf form with issues of new sheets after each revision, an effective searching tool. Developing familiarity with the system requires some effort. It is greatly aided by the "Definitions of Revised Classes and Subclasses," and the occasional "Classification Bulletins" all available from the U.S. Patent and Trademark Office.

The "Manual" lists some 350 broad classes, along with about 78,000 official subclasses and 17,000 unofficial subclasses. All are subject to revision with changing technology. The class number is an essential aid in searching.

The "Index to United States Patent Classification," December, 1980, contains an alphabetical list of subject headings referring to specific classes and subclasses.

For the purpose of segregating certain patents as chemical in the *Official Gazette* (Patents), the 36 classes listed in Table III have been designated as covering the "chemical arts." As one would predict, Class 260 is very large. It requires 26 two-column pages in the "Manual."[12]

An unusual citing of class and subclass numbers appears in *Adv. Chem. Ser.*, No 78 (1968). In 33 of the 40 chapters the bibliographical lists of references include the relevant numbers.

A fourth edition (1980) of "Concordance," from the U.S. Patent and Trademark Office, links the U.S. Patent Classification to the International Patent Classification.

Obtaining Copies of Patent Documents

The introductory pages of issues of *Chemical Abstracts* carry the addresses of the patent offices of countries whose patents are abstracted, along with the prices for copies. Photocopies of most foreign patents are available from the U.S. Patent and Trademark Office, Washington, DC 20231, at a cost of 30 cents per page. The total number of pages is given in the heading of the abstract.

From the same Washington office copies of United States patents are available at 50 cents per copy, with payment in advance. Payment must be by United States specie, treasury notes, national bank notes, U.S. Postal Service money orders, or certified checks. Money orders and checks are payable to the Commissioner of Patents and Trademarks. Coupons for use in the purchase of patents are available as follows: 50-cent coupons are sold in pads of 10 for $5 and 50 (with stub for records) for $25.

If one is concerned with a particular class or subclass (see Table III), all patents in such classification, from the beginning or within the range of any two dates specified, may be obtained. Then, to keep up to date, one may set up an account to have each new patent in this classification mailed as issued.

Patents as Puzzles

Among the various types of chemical literature, patents are unique in being legal documents. The basic reason for publication in this form is to enable patentees to protect their inventions in court, if necessary. Presumably, patentees and their attorneys try to make the documents defensible in all respects.

How well does a patent accomplish this primary purpose? To find an answer to this question is to solve a kind of puzzle. Questions of priority, novelty, and utility, although always important, are relatively simple. The difficult questions involve the specifications.

Is the description perfectly clear, generally understandable, or not adequately informative? For example, could "one skilled in the art" construct and successfully operate the apparatus described in the patent included above? Or could a chemist, of recognized competency in synthetic work, synthesize polyphenylene

[12]See E. A. Hurd, *J. Chem. Doc.*, **10**, 170 (1970).

TABLE III
CHEMICAL ART AS DEFINED BY THE *OFFICIAL GAZETTE* (PATENTS)

Class	Title
8 (subs. 1–142)	Bleaching and Dyeing; Fluid Treatment and Chemical Modification of Textiles and Fibers
21	Preserving, Disinfecting and Sterilizing
23	Chemistry
29 (subs. 180–199)	Metal Working
44	Fuel and Igniting Devices
48	Gas, Heating and Illuminating
51 (subs. 293–309)	Abrading
65	Glass Manufacturing
71	Chemistry, Fertilizers
75	Metallurgy
96	Photographic Chemistry, Processes and Materials
99 (subs. 1–233)	Foods and Beverages
106	Compositions, Coating or Plastic
117	Coating: Processes and Miscellaneous Products
127	Sugar, Starch and Carbohydrates
134 (subs. 1–42)	Cleaning and Liquid Contact with Solids
136	Batteries
148	Metal Treatment
149	Explosives and Thermic Compositions
156	Adhesive Bonding and Miscellaneous Chemical Manufacture
161	Stock Material and Miscellaneous Articles
162	Paper Making and Fiber Liberation
176	Nuclear Reactions and Systems
195	Chemistry, Fermentation
196	Mineral Oils: Apparatus
201	Distillation: Processes, Thermolytic
202	Distillation: Apparatus
203	Distillation: Processes, Separatory
204	Chemistry, Electrical and Wave Energy
208	Mineral Oils: Processes and Products
210 (subs. 1–64)	Liquid Purification or Separation
252	Compositions
260	Chemistry, Carbon Compounds
263 (subs. 52, 53)	Heating
264	Plastic and Non-Metallic Article Shaping or Treating: Processes
424	Drug, Bio-Affecting and Body Treating Compositions

oxide by the procedure described by W. M. Kruse in his U.S. patent 4,120,824?

Turning to the claims, has a patentee, in a given case, claimed more than the description justifies, or failed to claim all to which he or she is entitled? Probably only an expert in the field covered could solve this puzzle for patents with many intricate drawings and lengthy descriptions, followed by scores, or even hundreds, of separate claims. The patents of Gubelmann and of Joel, cited above, are very complicated examples.

Chemists do encounter these puzzles wherever their work involves patentable items. Students headed in this direction should be concerned about becoming interested in the challenges.

Periodicals and Patents

Many periodicals carry lists of abstracts of patents. In general, three types of periodicals contain information relating to chemical patents, namely, patent journals, abstracting journals, and journals devoted to industrial chemistry.

Special Journals on Patents These periodicals are devoted exclusively to patents, designs, trademarks, and matters relating to them. Sources of such information include the official publications of the various patent offices around the world.

A prime example is the *Official Gazette* (Patents) of the U.S. Patent and Trademark Office, 1872 to date. It is dated on Tuesday (the day for announcing patent grants) of each week. For each patent granted, it gives the number, filing date, serial number of the application, class and subclass numbers, inventor(s), assignee (if any), at least one claim, and a selected drawing (if any). The *Gazette* also carries official notices of patent actions such as interferences and litigation, the state of the applications pending, and often the full text of court decisions or the Commissioner's ruling in hearings of the Patent and Trademark Office. Design patents and trademarks have separate sections. Plant patents have been reported separately ever since the amended patent law allowed them in 1930. Since 1952 the previously strictly numerical sequence of patents has been divided into three numerical sequences: general and mechanical, chemical, and electrical.

Patents are indexed monthly in alphabetical order of patentees, and by class and subclass number. In the Introduction, class titles are arranged in order of class numbers, and also in alphabetical order.

Beginning in 1975, the *Official Gazette* has been issued in two parts, *Official Gazette* (Patents) and *Official Gazette* (Trademarks).

The Commissioner's *Annual Report,* not a part of the *Official Gazette* (Patents), includes (Part I) an alphabetical index of inventors and assignees, and (Part II) an alphabetical list (by courtesy called an index) of subjects, based only on patent titles. This list has some utility as a checklist, but is by no means a searching aid.

In Chapter 6A, Indexing Journals, attention is directed to the use of uninformative titles for publications by chemists. Titles of many chemical patents are especially open to this criticism. Some recent examples follow: analytical apparatus; catalyst; composition of matter; crate; composite materials; novel compositions; organic compounds; resistive material; and support member.

For the illustrative patent document included here, the following abridgment appeared in the *Official Gazette* (Patents), **952,** 1154 (1976), in the issue containing lists of patents granted Tuesday, November 16, 1976:

3,992,109
CYCLIC COLORIMETRY METHOD AND APPARATUS

Ditmar H. Bock, Boston, N.Y., assignor to Calspan Corporation, Buffalo, N.Y.
Continuation of Ser. No. 341,438, March 15, 1973, abandoned. This application Sept. 15, 1975, Ser. No. 613,647

Int. Cl.^{2} G01J 3/50

U.S. Cl. 356–181　　　　　　　　　　　　2 Claims

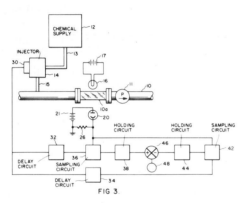

FIG 3

2. A cyclic colorimetry system for eliminating the effects of optical system deterioration and the presence of interfering contaminants when determining and indicating the presence of a condition in a flowing fluid to be tested and comprising;

light source means;

photocell means for producing an output signal proportional to the amount of light from said light source means which is incident thereon;

means for causing the fluid to flow intermediate said light source means and said photocell means and including a transparent portion defining a light path between said light source means and said photocell means;

indicator means for producing a chemical reaction in the fluid in the presence of the tested-for condition and to thereby produce a change in an optical characteristic of the fluid only in the presence of the tested-for conditions.

means for periodically causing the introduction of a predetermined quantity of said indicator means into the fluid at a point upstream of said light source means and said photocell means;

means for sensing the introduction of said indicator means into the fluid and for producing a synchronizing actuating signal;

first means responsive to said actuating signal for determining the output signal of said photocell means after a first predetermined time period and for producing a first signal output representative of the tested-for optical characteristic of the fluid to which said indicator means has been added;

second means responsive to said actuating signal for determining the output signal of said photocell means after a second predetermined time period which is longer than said first predetermined time period to permit the purging of fluid to which said indicator means has been added to produce a second signal output representative of the tested-for optical characteristic of the untreated fluid;

means for storing said first and second signal outputs; and

means for comparing said stored first and second signal outputs whereby optical effects produced by the presence of interfering contaminants, by nonsynchromous changes in the fluid and by deterioration of the optical system comprising the light source means, the photocell means and said light path are canceled and the presence of the tested-for condition is indicated by any difference in said stored first and second signal outputs.

Official journals of foreign patent offices either include abridgments or, especially in the smaller nations, only number-and-title entries. The Canadian *Patent Office Record* and the West German *Auszüge aus den Patentschriften,* like the *Official Gazette* (Patents), use claims and drawings. The British *Official Journal* is limited to brief entries but is accompanied by the *Illustrated Abridgments* which are carefully condensed abstracts of the specifications. In France the *Bulletin officiel de la propriété industrielle* was limited to brief entries until 1958 but now, as in Great Britain, it has a separate publication, *Abreges descriptifs des brevets d'invention,* using true abridgments.

With the opening June 1, 1978, of the European Patent Office, the situation with foreign patents will change.[13] The purpose of the Euorpean patent is to grant in each country in which protection is desired the same patent rights as are conferred by a national patent in that country. With this arrangement, the patentee does not have to secure a separate patent in each country. The first article cited lists the countries already participating, along with those expected to follow later.

Unfortunately, these European patents may grant protection only in certain countries. Consequently, Chemical Abstracts Service aims to list the countries in which the patent is operative.

This new patent office is located in Rijswijk, The Netherlands. Several types of publications are under way.

General Abstracting Journals   These periodicals carry abstracts of chemical and other patents, unfortunately never with complete coverage. Some include an abstract to show the technical advance made by the invention; some merely state the patent number, date, country, patentee(s), and general character of the patent. Searchers needing details must consult the original document.

Chemical Abstracts   A general description of this periodical is contained in Chapter 7. In the first issue of each volume the Introduction contains the Patent Abstract Heading to show the various data elements involved in a patent citation.

Included in this description is a list of countries whose patents are covered, with the abbreviations and codes used in *Chemical Abstracts*. There is also a general statement of the nature of the coverage of chemical patent documents.

Through the period 1963–1980 references for duplicate patents in different countries for the same invention were handled in a Patent Concordance.[14] It has been replaced by the new Patent Index (1981–   ).[15]

Patent dates are important for establishing priority of inventive discoveries and for fixing the effective time at which patent protection begins. Dates cited in *Chemical Abstracts* are, as nearly as can be determined from information in patent documents, the date cited by the U.S. Patent and Trademark Office in patent references. Date of issue and date of publication are identical, or nearly so, for most countries.

For some countries, such as Germany, Austria, and Switzerland, the class designation is helpful in locating patent specifications in libraries.

Chemical Abstracts contains the following abstract for the patent document quoted earlier in this chapter:

---

[13]Jean-Francois Mézières, *CHEMTECH*, **8**, 658 (1978); R. Lawrence, *Chemist*, **56** (9), 4 (1979).
[14]Patentrückzitate in *Chemisches Zentralblatt*.
[15]Cumulated semiannually.

86: 83256n Cyclic colorimetry method and apparatus

Bock, Ditmar H. (Calspan Corp.) U.S. 3,992,109 (Cl. 356-181. GO1J 3/50), 16 Nov 1976, Appl. 341,438, 15 Mar 1973; 5 pp. To detect an anolyte (e.g., a heavy metal) or a condition (e.g., pH) in a fluid stream in the presence of interfering contaminants an indicator is added cyclically to the stream and the difference in transmissivity or scattering of the fluid with and without the indicator is obtained with a colorimeter. The app. consists of a radiation source, a photocell, an optical filter, and an elec. synchronous demodulating system to discriminate against random fluctuations in transmissivity, which can arise, e.g., from debris passing through the sensed section of the analyzed fluid. Thus, acids or Fe were detected in aq. solns. by using bromothymol blue or Na_2S as indicators, resp.

It will be noted that, following the short descriptive title, there follow, in sequence, the name(s) of the inventor(s), the assignee (if any), the patent number, the U.S. classification number, the dates of issue and application, and the number of pages. Most patent abstracts were brief in the early years of *Chemical Abstracts.*

Since the demise of *Chemisches Zentralblatt* in 1969, *Chemical Abstracts* is the most available, comprehensive, and informative source of abstracts of chemical patents. It is the major permanent reference source for retrospective searching of the patent literature.[16]

The Introduction in each volume of *Chemical Abstracts* contains information relating to abstracts of patents. Included are a list of countries whose patents are covered, the nature of the coverage, and directions for procuring copies of the patents and (in some cases) the applications.

Derwent Patent Abstracts and Indexes A number of patent services are available from Derwent Publications, Ltd., London. Several brochures, with accompanying subscription price lists, outline what may be secured. Further details are contained in these brochures and instruction manuals for the specific services. In addition, a fourth brochure is *On-Line Search Service 1978* (see Chapter 15).

1 *World Patents Abstracts 1978.* Issued in the form of weekly booklets, the chemical editions comprise 13 divisions.

2 *Central Patents Index 1978.* This abstracting and retrieval service covers chemical patents under 12 categories. An article by M. Hyams[17] describes the importance of a *Basic Abstracts Journal* and an *Alerting Bulletin.* This service began in 1970.

3 *World Patents Index 1978.* Started in 1974, one of the four editions of

---

[16]J. T. Maynard, *J. Chem. Inf. Comput. Sci.,* **17,** 136 (1977), R. J. Rowlett, Jr., *CHEMTECH,* **9,** 348 (1979).
[17]*Role Pat. Inf. Res. Dev., Coll. Lect. Moscow Symp.,* **1975,** 275.

weekly gazettes covers chemical patents. There are 12 divisions. To follow the flow of chemical patents, and especially the submission of applications for patents, this index is recommended.[18] Supplementing it is an internal *WPI Alert*.

Judging by the various divisions and categories, these services seem likely to be of most value to those in chemical industries.

Uniterm Index to Chemical Patents   Beginning with 1950, this index includes the du Pont files through 1976. It is stated to cover more than 350,000 patents. The printed format consists of two volumes per year. One is a dictionary of terms, and the other contains reprints from the *Official Gazette* (Patents). There are a microfilm edition and a computer-readable version.

Details and prices are available from the IFI/Plenum Data Company, Arlington, VA.

International Index of Patents   This compilation lists patents. Published by the Interdox Corporation of New York, in 1964, a six-volume set covers patents for the period 1790–1960. The references are arranged numerically under classes (48) and subclasses (some 9,500), with about 737,000 cross references. There is a yearly volume, *National Catalogue of Patents,* with a supplement to the index every 5 years.

Journals on Industrial Chemistry   Some periodicals contain notices, abstracts, or reviews of the more important patents, together with occasional discussions of general interest in patent practice. In this case the chemical patents included are those considered most important. The information given for each patent is correspondingly fuller than the usual abstract. Such periodicals are not numerous.

The following journals are examples: *Adhesives Age, Journal of the American Petroleum Institute, Journal of Cellular Plastics, Metal Finishing,* and *Textile Manufacturer.*

Court Records

Patent litigation often introduces into court records the literature and patent citations from prior art searches which have been made with utmost skill and care. Cost of searching is no barrier to meticulous thoroughness when large sums depend on a court decision. Infringement suits are often defended by exhaustive prior art searches seeking to prove that the invention was not novel and that the patent is invalid. The defense of patentability seeks to prove that there is no pertinent prior publication. The parties in an interference hearing make

[18]W. J. Mayer, Jacqueline A. Angus, and Pauline Mariucci, *J. Chem. Inf. Comput. Sci.,* **16,** 76 (1976); also W. J. Mayer, *Prog. Res., 29,* No. 2, 1 (1976).

strenuous efforts to prove priority over any known prior art. So, whether in court decisions or in transcripts of testimony, valuable assistance in searching can often be had for the effort of examining the records.

Interference proceedings contain the records involving issued patents. In some cases they contain lists of prior patents and other publications relating to the same subject matter as the patent, and in others testimony of interest in connection with the patented invention.

MAKING PATENT SEARCHES

The general problem of searching the chemical literature is discussed later (see Chapter 14 on printed sources and Chapter 15 on computer-readable data bases). Dealing with patents, particularly the patent documents themselves, is sufficiently specialized that separate consideration seems justified. Only the printed sources are considered here.

Patent Documents

Of primary importance are actual copies of the patent documents, for only in them are all the details revealed. In nearly all cases abstracting services can indicate only bibliographic data, with a brief statement of the general nature of the invention. The omitted details may be crucial in determining just what is, or is not, described and claimed.

Where are the copies to be found? Of utmost importance are the Search Room and the Scientific Library of the United States Patent and Trademark Office. They are located in an office complex known as Crystal Plaza, Buildings 2, 3, 3-4, and 4, at 2021 Jefferson Davis Highway, Arlington, VA.[19] A very useful brochure is "Guide to the Public Patent Search Facilities of the United States Patent and Trademark Office." Included are many details about the sources available and conditions for their use. There is a helpful sketch of the overall arrangement of the various service areas.

Some important items are a complete collection of copies of U.S. patent documents, a numerical file of U.S. patents, a classified file, a patent index, and a file of the *Official Gazette* (Patents).

In addition to the patents of the United States, patent files of all major foreign countries are available in the Scientific Library, located immediately above the Search Room for U.S. patents. This library also contains an extensive collection of scientific books and journals, especially works on applied science, technology, and patents. Is the proposed invention really novel? In addition to checking patents of the United States and foreign countries, may the idea have been

---

[19]Correspondence should be addressed to Commissioner of Patents and Trademarks, Washington, DC 20231.

revealed in some related journal or book? Ultimately, the patent examiners pass upon such questions, but the searcher or an attorney might try to decide.

A second brochure, "Obtaining Information from Patents," contains valuable suggestions for using the facilities in Arlington.

In view of the facilities available at Arlington, it would seem best, if possible, to make patent searches there. Besides the resources at hand, skilled professional searchers are available, many of whom are patent attorneys having chemical training and experience in locating domestic and foreign patents.

In addition to the complete file of U.S. patents at Arlington, there are at least partial files of bound volumes of these patents in the libraries listed as follows:

REFERENCE COLLECTIONS OF U.S. PATENTS AVAILABLE FOR PUBLIC USE IN PATENT DEPOSITORY LIBRARIES

The libraries listed herein, designated as patent depository libraries, receive current issues of U.S. Patents and maintain collections of earlier issued patents. The scope of these collections varies from library to library, ranging from patents of only recent months or years in some libraries to all or most of the patents issued since 1870, or earlier, in other libraries.

These patent collections are open to public use and each of the patent depository libraries, in addition, offers the publications of the patent classification system (e.g. The Manual of Classification, Index to the U.S. Patent Classification, Classification Definitions, etc.) and provides technical staff assistance in their use to aid the public in gaining effective access to information contained in patents. With one exception, as noted in the table following, the collections are organized in patent number sequence.

Depending upon the library, the patent may be available in microfilm, in bound volumes of paper copies, or in some combination of both. Facilities for making paper copies from either microfilm in reader-printers or from the bound volumes in paper-to-paper copies are generally provided for a fee.

Owing to variations in the scope of patent collections among the patent depository libraries and in their hours of service to the public, anyone contemplating use of the patents at a particular library is advised to contact that library, in advance, about its collection and hours, so as to avert possible inconvenience.

State	Name of Library	Telephone Contact
Alabama	Birmingham Public Library	(205) 254-2555
California	Los Angeles Public Library	(213) 626-7555 Ext. 274
	Sacramento: California State Library	(916) 322-4572
	Sunnyvale: Patent Information Clearinghouse*	(408) 738-5580
Colorado	Denver Public Library	(303) 573-5152 Ext. 223
Georgia	Atlanta: Price Gilbert Memorial Library,	
	Georgia Institute of Technology	(404) 894-4519
Illinois	Chicago Public Library	(312) 269-2814
Massachusetts	Boston Public Library	(617) 536-5400 Ext. 265
Michigan	Detroit Public Library	(313) 833-1458
Minnesota	Minneapolis Public Library & Information	
	Center	(612) 372-6552

*Collection organized by subject matter.

State	Name of Library	Telephone Contact
Missouri	Kansas City: Linda Hall Library	(816) 363-4600
	St. Louis Public Library.	(314) 241-2288 Ext. 214
Nebraska	Lincoln: University of Nebraska-Lincoln,	
	Love Library	(402) 472-3411
New Hampshire	Durham: University of New Hampshire	
	Library	(603) 862-1777
New Jersey	Newark Public Library.	(201) 733-7740
New York	Albany: New York State Library.	(518) 474-5125
	Buffalo and Erie County Public Library.	(716) 856-7525 Ext. 267
	New York Public Library (The Research	
	Libraries)	(212) 790-6291
North Carolina	Raleigh: D. H. Hill Library, N.C. State	
	University	(919) 737-3280
Ohio	Public Library of Cincinnati & Hamilton	
	County	(513) 369-6936
	Cleveland Public Library	(216) 623-2932
	Columbus: Ohio State University Libraries	(614) 422-6286
	Toledo/Lucas County Public Library.	(419) 255-7055 Ext. 212
Oklahoma	Stillwater: Oklahoma State University Library . . .	(405) 624-6546
Pennsylvania	Philadelphia: Franklin Institute Library	(215) 448-1224**
	Pittsburgh: Carnegie Library of Pittsburgh	(412) 622-3138
	University Park: Pattee Library,	
	Pennsylvania State University	(814) 865-4861
Rhode Island	Providence Public Library	(401) 521-7722 Ext. 224
Tennessee	Memphis & Shelby County Public Library	
	and Information Center	(901) 528-2957
Texas	Dallas Public Library.	(214) 748-9071
	Houston: The Fondren Library, Rice	
	University	(713) 527-8101 Ext. 2587
Washington	Seattle: Engineering Library, University	
	of Washington.	(206) 543-0740
Wisconsin	Madison: Kurt F. Wendt Engineering Library,	
	University of Wisconsin	(608) 262-6845
	Milwaukee Public Library	(414) 278-3043

**Call only between the hours of 12 o'clock noon and 5:00 p.m.

Another article[20] deals with the subject of searching foreign patents. Presumably it reflects the practice and experience in the central library of E. I. du Pont de Nemours and Company.

The advice given by F. E. Barrows[21] to searchers in 1921 is still valid:

In making searches at the Patent Office at Washington, not only does the searcher have available the classified patents (U.S.), but he has access to the files of such patents containing the various Patent Office proceedings leading up to the grant, including the original application papers and amendments thereto, and the official

[20]Martha M. Duffy, *J. Chem. Inf. Comput. Sci.*, **17**, 126 (1977).
[21]*Chem. Met. Eng.*, **24**, 518 (1921).

communications of the examiners who examined the application, in which are references to the prior patents and publications which the examiners found on their search and considered relevant to the invention sought to be patented. By examining the files of U.S. patents which are of the nearest subject matter to the investigation in question, it may be possible to obtain reference to prior patents or publications which are of particular interest and which would not be readily available elsewhere.

A further value of the patent files is the indication therein of the classification of the respective patents and the indication in the various official actions of the classification of the prior patents which are referred to therein. Attention may thus be directed not only to prior patents of relevant subject matter, but even to other classes of the Patent Office which might otherwise have been overlooked but which may contain patents of interest. The number of classes and subclasses is so great and the classification of patents is in many instances so complex that relevant classes or subclasses may well be overlooked unless special precautions are taken to insure their proper consideration.

In all such searching of descriptions and claims it is well to keep in mind the urge among inventors and patent attorneys to cover all the ground that the patent law allows. Hence, the use of generalized terms; a filament lamp becomes a substantially evacuated rigid transparent envelope containing an electrical device, a jointed doll an articulated toy, and hydrogen gas a gaseous element having a positive valence of one. This means that inventors, in pitting their skill against that of the patent examiners, have also challenged searchers to sharpen their detective skills.

The Hill Formula Index

In order to answer quickly the question of novelty for patent applications covering new chemical compounds, E. A. Hill[22] devised a combined inorganic-organic index for use in the Classification Division of the U.S. Patent and Trademark Office. The Hill system, with slight modifications,[23] has been used since 1920 in the formula indexes of *Chemical Abstracts* (Chapter 7).

The arrangement of symbols in the formulas is alphabetical except that in carbon-containing compounds C always comes first, followed immediately by H, if hydrogen is present. The arrangement of formulas is also alphabetical except that the number of atoms of any specific kind influences the order of compounds; e.g., all formulas with C (one carbon atom only) come before those with C_2; thus, CCl_2O, CCl_4, $CHCl_3$, CHN, $CHNO$, CH_2Br_2, CH_2O, CH_3Cl, CO, C_2Ca, $C_2H_3O_2$. The arrangement of entries under any heading is alphabetical according to the names of the isomers.

To use the index, first reject any water of crystallization and rewrite the empirical formula in the alphabetical order of the chemical symbols, except that

---

[22]*J. Am. Chem. Soc.*, **22**, 478 (1900); **29**, 936 (1907); **34**, 416 (1912).
[23]E. J. Crane and E. Hockett, *Chem. Abstr.*, **14**, 4557 (1920).

in carbon compounds put carbon first and hydrogen second; then look in the proper alphabetical location in the index, noting the first symbol of the rewritten formula, together with the number of times it occurs. According to this arrangement, one would find the following compounds listed under the respective formulas: rubidium permanganate, MnO_4Rb; ammonium sulfate, $H_8N_2O_4S$; and acetyl bromide, C_2H_3BrO.

The Hill system is used in the latest formula indexes of the Beilstein treatise (see Chapter 10).

Computer Checking

A portion of Chapter 15 is devoted to the application of computers in searching data bases for information on patents, and the useful data bases are listed there. The objective here is to show the information provided by Search Check, Inc., for the sample patent document included in this chapter.

This computer service is designed to identify all U.S. patents issued since February 4, 1947, that cite a specified pertinent patent. Validity, infringement, and novelty searches may be the objective of such a search.

To illustrate the nature of the data yielded by this service, a search was made November 12, 1979, for the illustrative patent, U.S. 3,992,109. The following two groups of numbers were reported:

1 1,977,359;   2,389,046;   3,089,382;   3,549,262;   3,643,102;   3,708,265; 3,729,263; 3,748,044. This group represents a *back search* of the patents cited against U.S. 3,992,109. They are listed in the patent document under "References Cited," United States Patents. Presumably, they were considered relevant patents by the examiner(s).

2 4,029,416; 4,054,384; 4,111,560. This group represents a *forward search* for the patents in which the examiner(s) cited U.S. 3,992,109 as a relevant reference. In this respect, a forward search is comparable to what is done in compiling the data for *Science Citation Index* (see Chapter 6A).

A brochure, "Use of Search Check," is available from Search Check, Inc., Arlington, VA.

Microform Copies

The distribution of microform copies of patent documents, and the *Official Gazette,* should extend the general availability of such information. An example of what is now available is described in a pamphlet, "U.S. Patent Publications on Microfilm." Issued by Research Publications, Inc., Woodbridge, CT, it includes prices for back files and current subscriptions.

The following kinds of copies are available currently:

1 U.S. Patents, 1790–   . Microfilm. Reissues, 1838–

2 CDR File, 1973–   . Collected changes in patents, such as certificates of correction, adverse decisions, disclaimers, deductions, and withdrawals. Since 1973, reissues are included.

3 *Official Gazette,* 1872–   . Microfilm. 1972–   . Microfiche.

GENERAL REFERENCES

In connection with specific points in the discussion a number of papers have been cited as references. A symposium, "Trends in Handling Patent Information,"[24] included 12 papers dealing with various aspects of the problem.

Perhaps the most condensed summary of relevant information, at the time of its writing, is in Volume 14 (1967) of the "Kirk-Othmer Encyclopedia of Chemical Technology." There are two detailed sections on patents. The one by Robert Calvert, a patent attorney, covers patent practice and management (pp. 552–582). The other, by E. S. Turner of the Bell Telephone Laboratories, covers the patent literature, with a bibliography of 294 references (pp. 583–635).

The following more general references may be helpful:

Brink, R. E., D. C. Gipple, and H. Hughesdon: "An Outline of United States Patent Law," Interscience Publishers, Inc., New York, 1959.

Burge, D. A.: "Patent and Trademark Tactics and Practice," John Wiley & Sons, New York, 1980.

Costas, P. L.: "Introduction to Patents," American Chemical Society, Washington, DC, 1974.

Houghton, B.: "Technical Information Sources: A Guide to Patent Specifications, Standards, and Technical Report Literature," Shoestring Press, Hamden, CT, 1972.

Horwitz, L.: "Patent Office Rules of Practice," 5 v., Matthew Bender, New York, 1975.

Kase, F. J.: "Foreign Patents: A Guide to Official Patent Literature," Oceana Press, Dobbs Ferry, NY, 1973.

Liebnesny, F. (ed.): "Mainly on Patents: The Use of Literature of Industrial Property and Its Literature," Archon Press, Hamden, CT, 1973.

Maizell, R. E.: "How to Find Chemical Information," chap. 11, John Wiley & Sons, New York, 1979.

March, W. (ed.): "Patent Policy: Government, Academic, and Industry Concepts," ACS Symposium Series, No. 81, American Chemical Society, Washington, DC, 1978.

Maynard, J. T.: "Understanding Chemical Patents," American Chemical Society, Washington, DC, 1978; also How to Read a Patent, *CHEMTECH,* p. 91, 1978.

Oppenheim, C.: chap. 14 in R. T. Bottle (ed.), "Use of Chemical Literature," Butterworths, London, 1979.

Rossman, J.: "The Law of Patents for Chemists," Williams and Wilkins, Baltimore, MD, 1934.

Thomas, E.: "Chemical Inventions and Chemical Patents," Clark Boardman Company, Ltd., New York, 1964.

"General Information Concerning Patents," U.S. Patent and Trademark Office, Washington, DC, 1975.

[24]*J. Chem. Inf. Comput. Sci.,* **18,** 61–87, 121–129 (1978).

IEEE Trans. Prof. Commun., PC-22, 45–127 (1979). This special issue on patents includes 18 articles by 17 authors. Although published by the Institute of Electrical and Electronic Engineers, the general coverage of a broad range of subjects is very pertinent to the interests of chemists and chemical engineers.

"Consolidated Listing of Recent Gazette Notices—Re Patent and Trademark Office Practices and Procedures, U.S. Government Printing Office, Washington, DC, 1981.

Concordance: United States Patent Classification to International Patent Classification, U.S. Department of Commerce, Patent and Trademark Office, Washington, DC, 1980.

"Manual of Patent Examining Procedure," Patent and Trademark Office, Washington, DC. Loose-leaf, revised as needed.

PRIMARY SOURCES— MISCELLANEOUS CONTRIBUTIONS

Those who cannot remember the past are condemned to repeat it.

George Santayana

There are two other types of publications hardly belonging in the previous chapters. Their total number is not large, and often the material in an individual publication is not extensive; but the information may be important, and they should not be overlooked.

DISSERTATIONS

For many years candidates for the doctoral degree in philosophy or science were usually required to publish their dissertations and to deposit a specified number of copies in the library of the institution granting the degree. Most of the copies were distributed to important libraries. In some cases an entire dissertation was published in a journal. Then reprints of the article were accepted as fulfilling the deposition requirement. In these instances the information may be found in the appropriate journal.

Very often, however, a detailed account of the work is so long that only lengthy abstracts or abridgments, giving the most important contributions, are published in journals. Then one must obtain a copy of the dissertation to consult details of the work. Occasionally nothing is published.

Until recently dissertations have not been easy to locate. No general list of all scientific and technical dissertations is known, even by title. Since 1938 *Chemical Abstracts* has been including titles of dissertations issued by Xerox University Microfilms (Ann Arbor, MI).

Xerox University Microfilms began its abstracting service in 1938. Now most American graduate schools cooperate. The author of a dissertation prepares a summary of not over 600 words. This summary is printed in *Dissertation Abstracts International (Dissertations Abstracts* prior to 1969), published monthly. The complete dissertation is microfilmed. This provides for reproduction of positive copies, which may be purchased.

Locating dissertations written before the establishment of this service may be a problem. If a copy is in a given library, the card catalogue will list it. Frequently one learns of the existence of a dissertation through references to it in articles or bibliographies.

The Library of Congress accumulated thousands of dissertations from European universities. These publications are now in the Center for Research Libraries, Chicago, which tries to provide any foreign doctoral dissertation requested by a member institution.

There are some compilations of dissertation titles. There is now an annual listing in *American Doctoral Dissertations* (University Microfilms). The same service has issued a *Comprehensive Dissertation Index,* 1861–1972, for North America only. Volumes 1 to 4 cover chemistry, with author and keyword indexes.

Usually such lists give only the title. Every experienced searcher knows how uninformative dissertation titles may be. Some dissertations are catalogued under the name of the institution granting the degree.

Lists of theses by title were published by Irlene R. Stephens in *Adv. Chem. Ser.,* **30,** 110 (1961). The listings are under 14 countries. For the United States 18 national sources are included. Twelve states are mentioned under individual institutions, in most cases with one institution per state. Most of the lists are limited in the time covered.

An annual compilation of titles for the United States and Canada is *Master's Theses in the Pure and Applied Sciences.* Started in 1956 by University Microfilms, since 1974 (Vol. 19) it has been edited by W. H. Shafer at the Center for Information and Numerical Data Analysis and Synthesis (CINDAS) at Purdue University. It is published by the Plenum Publishing Corporation, New York.

M. M. Reynolds (ed.) has published a "Guide to Theses and Dissertations", 1975, Gale Research Company, Detroit.

Some sources for foreign dissertations and theses follow:

Great Britain and Ireland: *Index to Theses,* ASLIB, London, 1950–
France: *Catalogue des Theses et Ecrits Academiques,* 1960–
Germany: *Jahresverzeichnis der Hochschulschriften der DDR, der BDR, und Westberlins,* 1880–

MANUFACTURERS' TECHNICAL PUBLICATIONS

Many manufacturers of materials and equipment used in scientific and technical work issue pamphlets or circulars dealing with the construction and use of new apparatus, appliances, materials, and processes. These bulletins serve partly as advertising but more as technical information. Often there is included a summary of the theory of the construction and operation of scientific apparatus, together with instructions for using the apparatus for its intended purpose. Accompanying the description may be a bibliography of references relating to the apparatus and its applications. Along with current advertising, these publications constitute the latest published announcements of producers.

Many of these publications are quite technical in nature. Many more are designed to acquaint the public with the product described. They are of particular value to teachers of undergraduate students as an authoritative statement concerning the nature and applications of new products.[1]

Various names are used for this kind of publication. Examples are "trade literature," "house organs," "data sheets," "package circulars," and "sales-development brochures." "Customer magazines" is another name employed.

Few libraries file such works. If they do, the collection is usually small. Ordinarily the best way to obtain late copies is to write the concerns advertising in periodicals or taking space in trade catalogues. Some companies maintain a permanent mailing list upon which interested people may have their names placed, thus ensuring the receipt of the latest bulletins. For convenience some periodicals include with their advertising cutout or tear-out mailing cards for use in sending for such industrial literature. Each notice or advertisement is numbered so that one has only to encircle the appropriate numbers and mail the card. Thus, *Chemical & Engineering News* maintains a readers' information service for such purposes.

This kind of publication is infrequently cited as a reference. However, a striking example of such citing occurs in "The Literature of Chemical Technology," *Adv. Chem. Ser.,* No. 78 (1968). A number of the extensive bibliographies in the 40 chapters include as references many trade publications and/or industrial advertising literature.

Ella M. Baer and H. Skolnik have discussed house organs and trade publications as sources of information.[2] A systematic scheme has been suggested for cataloguing such engineering trade literature according to unit operations.[3]

For convenience the trade literature is considered in two groups: (1) journals or serials and (2) brochures or pamphlets. The former are continuing publications, many appearing regularly. The latter appear irregularly and usually are not interrelated.

[1]See W. G. Kessel, *J. Chem. Educ.,* **31,** 255 (1954), for a list of such literature.
[2]*Adv. Chem. Ser.,* **30,** 127 (1961).
[3]F. C. Vilbrandt, *J. Chem. Educ.,* **10,** 354 (1933).

1. House Organs

Although the names attached to various kinds of trade publications vary and may be inconsistently used, house organ is used here for serials. That is, they are somewhat analogous to periodicals of societies.

Most of them do not have the quality, nor do they enjoy the status, of the best technological periodicals. They are issued by industrial organizations. Consequently, they provide background information for the companies' products. There are descriptions of new and improved apparatus, industrial equipment, and chemicals. Also, very good, rather general articles frequently appear.

Several publications discuss these scientific journals of industry. One considers the characteristics of technically oriented house journals.[4] Another reports on the industrial house journals of the United States, the United Kingdom, and France. It is concluded that the coverage of these publications is low in indexing and abstracting journals.[5]

House journals are considered in the discussion of commercial technical publications available in the Scientific Reference Library of England.[6] R. T. Bottle[7] mentions British trade literature. "Internal Publications Directory,"[8] lists many house publications.

Examples of Trade Journals

Aldrichimica Acta, Aldrich Chemical Company
Arcs and Sparks, Ultra Carbon Corporation
Bell Laboratories Record, Bell Laboratories
Chemist-Analyst, J. T. Baker Chemical Company
Chromatography Review, Spectra Physics
Chemalog Hi-Lites, Chemical Dynamics Corporation
Dupont Magazine, E. I. du Pont de Nemours and Company
Eastman Organic Chemicals Bulletin, Eastman Kodak Company
FMC Progress, FMC Corporation
Foote Prints, Foote Mineral Company
Gulf—The Orange Disc, Gulf Oil Corporation
Hewlett-Packard Journal, Hewlett-Packard Company
IBM Journal of Research and Development, IBM Corporation
Kodak Laboratory Chemical Bulletin, Eastman Kodak Company
Perkin-Elmer Instrument News, Perkin-Elmer Corporation
PPG Products, PPG Industries
Progress thru Research, General Mills
Rohm and Haas Reporter, Rohm and Haas

[4]M. C. Drott and B. C. Griffith, *IEEE Trans. Prof. Comm.,* PC-18, 45 (1975).
[5]M. C. Drott, T. C. Bearman, and B. C. Griffith, *ASLIB Proc.,* **27,** 376 (1976).
[6]M. J. Thomson, *ASLIB Proc.,* **28,** 186 (1976).
[7]R. T. Bottle (ed.), "Use of Chemical Literature," Butterworths, London, 1979, p. 212.
[8]National Research Bureau, Inc., Chicago, 1981.

Technical Journal, Leeds and Northrup Company
The Spex Speaker, Spex Industries
VIA (Varian Instrument Applications), Varian Instrument Division

2. Brochures and Pamphlets

A large number of industrial publications may be referred to as pamphlets, booklets, brochures, or bulletins. They deal with a wide range of subjects and vary considerably in nature and quality.

For some years the Committee on Pharmaceutical Nonserial Industrial Publications of the Pharmaceutical Section of the Special Libraries Association has published a quarterly *COPNIP List.* This publication lists and briefly describes such pamphlet materials, which may be obtained from manufacturing organizations and some other agencies. The *COPNIP* committee obtains the items, evaluates them, and selects those considered worthwhile.

Technical brochures have been valuable sales tools in the chemical industry. They may be aimed at the research and development person or the purchasing agent.

Technical bulletins range in size from a page or two to an occasional hardbound book. The average lifetime is estimated at from 2 to 3 years. New developments and products necessitate new publications.

Some journals, especially those carrying advertising, have been including a special section for announcements of trade brochures and pamphlets. Thus, in *Analytical Chemistry,* the section is "Manufacturers' Literature," and in the *Journal of Chemical Education* it is "Editor's Basket." Other section names may be "New Products" and "Catalogues."

Sometimes there is a series of numbered bulletins from a particular company.

Examples of Industrial Brochures

"Eastman Reagents for Protein Chemistry," Eastman Kodak Company
"Fatty Amides: Their Properties and Applications," National Dairy Products Corporation
"Intermediate Chemicals," E. I. du Pont de Nemours and Company
"Laser Micro Raman Spectrometry," Walter C. Crone Associates
"Orion Research: Analytical Methods Guide," Orion Research, Inc.
"Preparation, Properties, Reactions, and Uses of Organic Peracids and Their Salts," FMC Corporation
"Products for the Analyst," United Mineral and Chemical Corporation
"4,4'-Sulfonyldiphenol," Crown Zellerbach
"Sodium Borohydride," Thiokol, Danvers, MA
"Sulfuric Acid," Allied Chemical Corporation
"Tetrahydrofuran as a Reaction Solvent," E. I. du Pont de Nemours and Company
"Water Treatment," Bull. M-1050, Carns Chemical Company

SECONDARY SOURCES— PERIODICALS

All that mankind has done, thought, gained or been—it is lying, as in magic preservation, in the pages of books.

Thomas Carlyle

The various types of publications already considered constitute the primary sources of chemical information. This does not imply that they are of greater importance than those which remain to be discussed. Such publications, as a rule, contain material that is new, or at least previously unpublished. From our consideration of the nature, number, and variety of these primary sources, it is evident that the facts relating to some given subject will, in all probability, be widely scattered and, as a consequence of this, be difficult to locate when needed. The material included in them is necessarily unorganized. The separate and distinct published chemical items of some significance—that is, papers, bulletins, patents, and others—must number many millions.

The publications considered in this chapter, together with several more following, have been termed "secondary sources," since they contain information compiled from one or more of the primary sources and arranged according to some definite plan. This material, then, is essentially organized in its nature. The general object in issuing such publications is to provide means for collecting, classifying, arranging, and discussing the scattered myriads of facts already recorded, so that the most efficient use may be made of the results of others' work. These secondary sources consist of periodicals, bibliographies, reference works, monographs, and textbooks. Those appearing periodically are considered in this chapter.

With the rapid increase in the number of primary sources, together with the accompanying extension in the amount included in them, it became very desirable to have some agency for collecting, classifying, and summarizing this material as soon as possible after its appearance. No individual or small group of individuals is capable of performing such a task. Comprehensive efforts in this direction are limited mostly to scientific societies.

Depending upon the information furnished and upon the method of arrangement used, the periodicals supplying this kind of information have been grouped under the headings of indexing, abstracting, and reviewing journals.

A. INDEXING JOURNALS

Indexing journals contain compilations of references only, including author's name, title of the article, and citation to the original. Their general function is to serve as an index by means of which one may easily find scientific papers.

Some individuals may be inclined to put indexing journals and bibliographics in the same class. In the sense that they are both lists of references, this is justified. The two have been separated, however, upon the general basis that the former are periodical publications and are devoted to some general field of scientific endeavor. Bibliographies, on the other hand, usually are not periodical publications, and they relate to some specific subject in one field of science.

The inadequacy of many titles of articles for searching purposes is often a serious defect in indexing periodicals. An example is the title "Metallurgical Analysis." In one case it was used for a book which contained photometric methods for the determination of the minor constituents in white-metal alloys. In another case it was used for a paper which described a method for the simultaneous spectrophotometric determination of chromium and manganese in steel.

This problem of uninformative titles is important enough to emphasize further with the following examples: "Through an Energy Window," "Nonlinear Calibration," "S&B," "Inorganic Compacts," "Green Factories," "96,400 Coulombs," "The Next Mile," and "Chemistry, Curves, and Color."

Use of such titles makes difficult the selection of keywords for an index such that manual or computer searching will yield the desired references. Several papers have reported studies on this difficulty. G. A. Cooke[1] stated that searching titles alone may retrieve only about one-third of the relevant articles. R. T. Bottle[2] reached the same conclusion for alerting services in the chemical and life sciences. More recently J. O'Connor[3] found that only about 30 percent of the titles (or abstracts) of 100 papers indicated the data contained in the original. Much earlier, the author[4] had a similar experience in checking references for the determination of nickel in iron and steel.

[1]*Chem. Can.*, **22,** 17 (1970).
[2]*J. Amer. Soc. Inform. Sci.*, **21,** 16 (1970).
[3]*J. Chem. Inf. Comput. Sci.*, **17,** 181 (1977).
[4]*Adv. Chem. Ser.*, **30,** 73 (1961).

A little thought on the part of writers, reviewers, and editors could largely avoid such difficulties. Ideally, a title should be a one-sentence abstract. Examples of recent indexable titles in analytical chemistry follow: "Triphenyl-methane Dyes as Reagents for the Extraction-Photometric Determination of Elements," "Extraction-Photometric Determination of Cerium (IV) Using α-Naphthylamine," and "Extraction-Spectrographic Determination of Silver in Rocks and Minerals." The author has long advocated analytical titles for quantitative methods which indicate the method of separation (if any) and the method of measurement.[5]

Although no indexing journal covers all current chemistry, *Current Contents* and *Chemical Titles* do index several hundred of the most important periodicals. Other indexing journals cover science in general, or some broad technical field. A brief chronological list of important works follows.

J. D. Ruess, Repertorium commentationum a societatibus litterariis editarum . . . , t. 3. Chemia et res metallica Dieterich (1803)

This work is an index to the publications of the learned societies of various countries from the founding of the society to 1800.

Catalog of Scientific Papers (1800–1900); Name changed in 1900 to *International Catalog of Scientific Literature*

The Royal Society had under way what was termed the "most comprehensive index to general science ever attempted." Although a number of volumes were issued, publication of the work ceased.

Repertorium der technischen Journal-literatur

Appearing first as an annual publication in 1874, the *Repertorium der technischen Journal-literatur* had been issued previously, under the name of *Repertorium der technischen Literatur* prior to 1879, in three volumes covering the period back to 1823. The German Patent Office issued it from 1877 to its discontinuance in 1908. Many references were included. The arrangement is alphabetical by subjects.

Index Medicus (1879–   )

In 1879 the librarian of what became the National Library of Medicine started *Index Medicus* as a current index of periodical publications on medicine. The monthly issues, arranged by subjects, have an author index. They are cumulated annually. Beginning in 1964, the material is coded for searching by computer to retrieve answers to specific bibliographic requests. See Chapter 15 for MEDLARS.

[5] *J. Chem. Educ.*, **50,** 690 (1973).

Abridged Index Medicus (1970–   )

This monthly bibliography, based on articles from some 100 English language journals, is designed for the needs of the individual practioneer, and for libraries of small hospitals and clinics.

Engineering Index (1884–   )[6]

For many years this work was an indexing publication. The various kinds of publications were listed by title. The information dealt with costs, finance, and management of industrial concerns. The technical and engineering subjects include aspects of applied chemistry and chemical engineering.

Four volumes cover the period 1884 to 1905. At this time the publication became an annual. Ultimately, it was issued monthly, with annual cumulation.

In the 1920s the purely indexing function was expanded as the Engineering Index Service. For each publication covered a 3- by 5-inch card was prepared. It contained essentially the same information as an abstract in *Chemical Abstracts*. Subscription could be made to the entire service or to any one of more than 300 subdivisions. This CARD-A-LERT (CAL) service was terminated December 31, 1975, after a period of nearly 50 years.

In 1977 the CAL Classification Codes was issued as an aid in using COMPENDEX, the magnetic tape service of Engineering Index, Inc. (see Chapter 15). The present printed publications of Engineering Index, Inc., are discussed under abstracting journals.

Bioengineering Abstracts (1974–   )

A by-product of the *Engineering Index* data base, this monthly publication is designed for the specialist in bioengineering and related disciplines. Reference is made to more than 2000 periodicals, conference papers, and reports issued through government, industrial, and academic activities.

Reader's Guide (1900–   )

Articles of a popular nature, appearing in nonchemical periodicals, may be found through the *Reader's Guide*.

Industrial Arts Index (1913–1964)

The *Industrial Arts Index* was issued in magazine form as an indexing service to technical journals, mostly American, which selected by the subscribers as the leaders in their respective fields. The entries dealt with technical articles and descriptions of new apparatus and machinery relating to engineering, electrical appliances, chemistry, business, printing, and textiles and were arranged

---

[6]Different titles have been used.

alphabetically by subject under sufficient subject headings to bring out the points of interest.

In 1958 this monthly periodical was divided into *Applied Science and Technology Index* and *Business Periodicals Index.*

Agricultural Index (1916–1964)

The *Agricultural Index* was a monthly subject index to agricultural periodicals, publications of the U.S. Department of Agriculture and of American state and foreign agricultural experiment stations, and occasionally, other literature.

Current Chemical Papers (1954–1969)

This periodical, published by the Royal Society of Chemistry, was intended to inform chemists of new work more quickly than an abstract journal. The titles, classified in 13 sections, were obtained by scanning some 300 periodicals.

Current Contents (1958–   )

This weekly publication from the Institute for Scientific Information contains reproduced tables of contents from journals selected from six fields of research. It is issued in the following editions: *Agriculture, Biology & Environmental Sciences; Arts & Humanities; Clinical Practice; Engineering, Technology & Applied Sciences; Life Sciences, Physical, Chemical & Earth Sciences;* and *Social & Behavioral Sciences.*

The table of contents of each journal title covered in these editions is enhanced, rearranged, and reproduced in a format easy to scan. Each issue includes a variety of features, such as editorials by Eugene Garfield, a selection of commentaries from the professional and lay press, and announcements of new scientific and technical books. In addition an author address index, a subject index, and a journal index are included.

Chemical Titles (1961–   )

A biweekly product of Chemical Abstracts Service, this current-awareness journal reports the titles of recently published papers of chemical interest. It covers approximately 700 of the world's most important chemical journals. Some 150,000 titles appeared in 1978. Following publication of *Chemical Titles,* abstracts of all of the articles appear in *Chemical Abstracts.* One survey showed a lapse of 6 to 7 weeks before publication of the abstracts.[7]

An issue of *Chemical Titles* consists of three parts: a Keyword-in-Context Index; a Bibliography; and an Author Index. To illustrate each part, reference is made to the listing of a paper by T. Lloyd and J. Weisz, entitled "Direct Inhibition of Tyrosine Hydroxylase Activity by Catechol Estrogens" and pub-

[7]Martha E. Williams, *J. Chem. Doc.,* **12,** 217 (1972).

lished in the *Journal of Biological Chemistry* [**233,** 4841 (1978)]. (See *Chemical Titles,* No. 17, Aug. 28, 1978).

1 *Keyword-in-Context Index.* Keywords selected from the title are sorted alphabetically and printed in the center of two columns on each page of the index. In addition to the keyword, each line of the index contains as many additional words from the title, in context, as space permits. An equal sign (=) is printed at the end of the title. A plus sign (+) is printed to indicate that not all of the title could be printed in the index. In addition to whole keywords from the title, many title words are segmented into important word fragments and these fragments also are sorted alphabetically in the index. To the right of each title is the reference code which provides a highly condensed bibliographic citation for the source of the title. The reference code consists of the five-letter CODEN and its check character and the volume and/or issue of the journal, along with the beginning page number of the article.

An effective way to use this index is to scan the alphabetical keyword list vertically, pausing at words of interest to examine the horizontal context. If the context indicates relevancy, use the reference code at the right of the entry to locate the full title, author(s), and journal citation in the Bibliography. Inside each issue cover is a list of periodical titles and their CODEN.

For the article selected to illustrate the index format, the keyword entries follow, with the keywords italicized (not done in *Chemical Titles*):

sine hydroxylase activity by	*cathechol* estrogens= +inhibition of tyro	JBCHA3-0253-4841
roxylase activity by catechol	*estrogens=* +inhibition of tyrosine hyd	JBCHA3-0253-4841
Direct inhibition of tyrosine	*hydroxylase* activity by catechol estrogen	JBCHA3-0253-4841
ity by catechol estro+ Direct	*inhibition* of tyrosine hydroxylase activ	JBCHA3-0253-4841
l estro+ Direct inhibition of	*tyrosine* hydroxylase activity by catecho	JBCHA3-0253-4841

The words "direct" and "activity" are not considered keywords. They are in the list of words prevented from indexing, near the front of the issue cover.

2 *Bibliography.* The Bibliography organizes the titles indexed in the Keyword-in-Context Index into a table of contents for each of the journals covered in the issue. The arrangement is alphabetical by CODEN of the journals covered. In addition to the complete journal citation, the Bibliography provides the complete title, all author names, and the inclusive pagination for the papers included.

For the illustrative paper, the Bibliography entries follow:

1 JBCHA3 J. Biol. Chem., 253, No. 14 (1978). This is the issue heading.

2 Next come the individual paper entries for issue 14 of the *Journal of Biological Chemistry*. That for the illustrative paper is

4841   Lloyd, T., Weisz, J.
    Direct inhibition of tyrosine hydroxylase activity by catchol estrogens   4841-3

3 *Author Index*. This index lists all author names for the papers indexed in the issue, the arrangement being alphabetical according to the names. Given names are shortened to the initial(s). Following each name, Lloyd, T., and Weisz, J., in the illustrative case, is the code entry JBCHA3-0253-4841.[8]

Science Citation Index

Since 1961 the Institute for Scientific Information (ISI) of Philadelphia has been publishing a different kind of index. In the usual papers, the authors cite prior publications which they consider relevant to their contributions.

In a citation index the references are documents published after the one cited, say paper A. Some 3,500 scientific and technical journals are searched regularly. In case of paper A, a record is noted of every document which cites this paper. Then the computer printout shows under the name of the author of paper A the names and references of all papers citing paper A.

The number of citations to a given paper is one measure of the importance others attach to it. From 1964 through 1978 the publication schedule was quarterly. Now it is bimonthly.

Some study and practice are needed to familiarize oneself with the resources of this publication, and with the method of handling the information included. Perhaps the best advice to the uninitiated is to study the current "Science Citation Index Guide," a supplementary publication. More detailed and informative is a 274-page book by the editor,[9] Eugene Garfield, entitled "Citation Indexing—Its Theory and Application in Science, Technology, and Humanities."

The sample display on page 119 is reproduced from the "Guide" to show how citations are issued. Although this example includes only journals, other kinds of publications covered include patents, proceedings, symposia, monographic series, and books.

Note should be made that only initials of cited authors are given. Thus, the entry "Smith, HJ" (for Harry John) may list citations not only to Harry John Smith, but also to any other Smiths whose given names have the initials H. J.

Other sections of the 1977 "Guide" to be understood for effective use include the following items, with pages:

Source Display (15)
Corporate Index (17)
Permuterm Subject Index (19)
Lists of Source Publications and Publishers' Guide (43)
Publishers' Address Directory (61)
Source Publications

---

[8]See E. Kiehlmann, *J. Chem. Doc.,* **12,** 157 (1972), for the journal coverage by the major title and abstracting periodicals.
[9]John Wiley & Sons, Inc., New York, 1979.

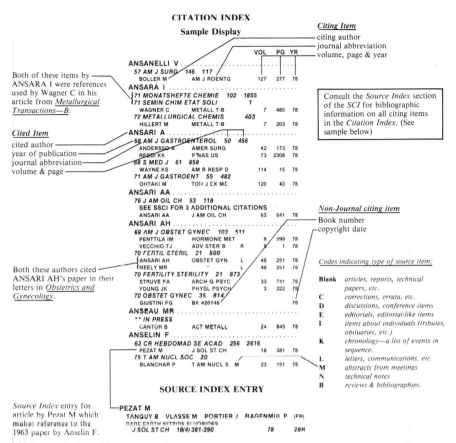

CITATION INDEX

Sample Display

(With permission of Institute for Scientific Information.)

Arranged by Full Title (74)
Arranged by Subject Category (88)
Arranged by Country of Origin (93)
Lists of Books Covered (99)

A new 1980 "Guide" differs in details.

Chemical-Biological Activities (1965–1971)

Another biweekly from Chemical Abstracts Service covered current literature on the biological activity of organic compounds. Some 300 journals were scanned for data. The computerized format resembles that shown for *Chemical Titles*.

In this case a Digest Section provides the key information. Then the indexing is by (1) keyword-in-context, (2) molecular formulas, and (3) authors.

This coverage was continued in 1972 as a computer-readable file from Chemical Abstracts Service.

Current Index to Conference Papers in Chemistry (1969–   )

This monthly listing is issued in three subject areas: chemistry, life sciences, and engineering. Subject and author indexes are provided.

Current Programs (1973–   )

Issued monthly, this listing contains the printed programs (including titles of the papers to be presented) of regional, national, and international scientific and technical meetings. As the announcement of titles of the papers precedes their publication (if any) later in *Current Contents* and/or *Chemical Titles,* there is an advantage for current-awareness information. In 1978 the journal title was changed to *Conference Papers Index.*

The quality of the titles submitted for programs may be even less informative than those approved later by reviewers and editors for publication of the papers.[10]

C&EN Calendar of Meetings and Events

In January and again in July *Chemical and Engineering News* lists events of chemical interest for the year. Both international and domestic meetings are included. Each entry gives date, title of meeting, location, and someone to write for details.

Other Indexing Publications

There are a number of other similar publications. Several examples follow which are more specialized and/or less comprehensive than *Chemical Titles.*

Biweekly List of Papers on Radiation Chemistry and Photochemistry (1968–   )
British Technology Index (1962–   )
Current Papers in Physics (1966–   )
Current Physics Titles (1976–   )
Current Titles in Electrochemistry (1965–   )

B. ABSTRACTING JOURNALS

Abstracting journals contain contemporaneous, concise summaries of the various articles, reports, patents, and other primary publications. Each abstract usually provides the following information: title of the original document, author's name, original reference, and generally a brief extract of the principal points or results in the publication.

Ideally an abstracting journal should provide prompt service, cover the field

---

[10]See H. Baum, *J. Chem. Doc.,* **13,** 187 (1973).

intended, present abstracts of high quality, and publish good annual and collective indexes as promptly as possible.

There are two general types of abstracts. Indicative abstracts may either elaborate the title or describe the scope of a publication and its significant contribution. Informative abstracts give the information itself.

In *Chemical Abstracts* (see Chapter 7) the abstracts are designed to be concise, informative statements of the major disclosures reported in the original documents. Thus, the abstracts are "findings-oriented." They are not critical or evaluative summaries. Their primary purpose is to give, accurately and quickly, sufficient detail about the information reported in the abstracted document to allow readers to determine whether they need to consult the original publication. Abstracts do not always include all the new information in the source documents, such as a list of newly synthesized compounds or the details of an experimental method. In *Chemical Abstracts* the semiannual volume indexes do attempt to reference such new information, particularly that related to chemical substances.

The text of the abstract follows the bibliographic heading or citation. In general, the first sentence of the abstract highlights the new findings reported in the original document. The rest of the abstract elaborates upon these findings and emphasizes the following types of information: (1) the purpose and scope of the reported work; (2) new reactions, substances, techniques, procedures, apparatus, uses, and properties; (3) new applications; and (4) the results of the work, together with the author's interpretations and conclusions.

Publications are not only a means of communicating ideas but also a barrier between a discoverer and the potential user of the discovery. Generally the discoverer, as author, bears the responsibility of describing adequately his or her contribution. At this point the abstractor begins. Assuming a full comprehension of the significance of the publication, the quality or value of the abstract depends very largely upon the person's skill in preparing an adequate summary. In some cases the title may be considered sufficient to indicate the probable value of the publication. Papers in foreign languages, and especially those in rare and inaccessible periodicals, are usually abstracted in some detail.

Three authors, with a wealth of experience with abstracts, published a book on abstracting.[11] It is a general description of the process of preparing abstracts of primary publications. Many examples are included from different abstracting services, along with specific suggestions for the preparation of abstracts.

Following the abstractor, the indexer has a very important function to perform, for the compilation of an adequate and accurate index for the abstract is second only in importance to the preparation of the abstract itself. As one

---

[11]R. E. Maizell, J. F. Smith, and T. E. R. Singer, "Abstracting Scientific and Technical Literature," Wiley-Interscience Co., New York, 1971. See also R. L. Collison, "Abstracts and Abstracting Services," Clio Press, Santa Barbara, CA, 1971; and B. M. Manzer, "The Abstract Journal, 1790–1920," Scarecrow Press, Metuchen, NJ, 1977.

depends so much upon indexes in looking for material, an index must be dependable if nothing is to be missed.

One very serious, and possibly insurmountable, difficulty connected with all abstracting journals is the failure to list or index what may be termed hidden facts. Unfortunately, titles and indexes often give no adequate indication of the nature of the material in a publication, and one is dependent upon the abstractor's judgment about what should be mentioned in the abstract. Dependence upon indexes alone may cause one to miss important points. This problem is discussed further in Chapter 14, under searching.

As far as their indexes are adequate and accurate, abstracting journals do enable one to ascertain what general references are available upon a given subject and where the original may be found, if needed. Frequently, the information contained in the abstract is enough.

The following three abstracts for the same article are examples of variability in abstracting:

Absolute Method of Turbidimetric Analysis

E. J. Meehan and W. H. Beattie (Univ. of Minnesota, Minneapolis, Minn., U.S.A.) *Anal. Chem.*, 1961, **33**:632–635,—The method described, based on the measurement of turbidity at an experimentally determined wavelength at which the turbidity is proportional to the reciprocal of the wavelength, eliminates empirical comparison of known and unknown suspensions. The method can be applied to heterodisperse suspensions of spherical, non-absorbing particles of known refractive index, for which accurate Mie theory scattering coefficients are available. An example is given of application to a silver bromide sol. (From *Analytical Abstracts*, **8**: Abs. 4908, 1961.)

An Absolute Method for Turbidimetric Analysis

E. J. Meehan and W. H. Beattie (Univ. of Minnesota, Minneapolis). *Anal. Chem.* **33**: 632–5 (1961),—An abs. method of turbidimetric analysis is described which eliminates comparison of known and unknown suspensions and is applicable to heterodisperse suspensions. The method is based on measurement of turbidity at an exptl. detd. wavelength at which turbidity is proportional to the reciprocal of the wavelength. The observed turbidity is then related to weight concentration through the Mie scattering coeff. [From *Chemical Abstracts*, **55**:12,141 (1961).]

Eine Absolutmethode der Turbidimetrischen Analyse

E. J. Meehan and W. H. Beattie. (*Analytic. Chem.* **33**:632–35. April, 1961. Minneapolis, Minn., Univ. of Minnesota, School of Chem.) (Engl.)

Es wird eine absol. Meth. der turbidimetr. Analyse beschreiben, die ohne empir. Vgl. von bekannter u. unbekannter Suspension auskommt u. auf heterodisperse Suspension anwendbar ist. Die Meth. beruht auf der Messung der Trübung bei einer

experimentell bestimmten Wellenlänge. bei der die Trübung hängt mit der einge-
wogenen Menge bzw. der Konz. über den Mieschen Zerstreuungs-Koeff. zusammen.
Das beschreiben Verf. wird auf ein AgBr-Sol angewandt. Die Genauigkeit des Verf.
beträgt einige Prozent. [From *Chemisches Zentralblatt*, **132:**16,213 (1961).]

Borderline publications present at least two problems. The first concerns the
editor of the abstracting periodical, who must decide whether the publication
has sufficient chemical interest to warrant abstracting and indexing. The second
concerns the searcher, who obviously cannot find the item in a given abstracting
journal if no abstract was prepared. As an example we may consider analytical
balances and spectrophotometers, both instruments being the product of the
physicist and of primary theoretical interest to the physicist from the standpoint
of mechanics and optics, respectively. Probably all new items on balances will be
abstracted, since every chemist is familiar with the instrument. Certain analysts
may be just as much interested in a new spectrophotometer, but until around
1930 there was no assurance a publication on it would be included. The worker
in this field had to watch suitable sources in physics as well as chemistry.

Abstracting all publications of chemical interest, or even most special fields,
has become an undertaking of such magnitude that only an organization, such as
a chemical society, can do it.[12] Some of the larger chemical manufacturing firms
have their libraries operate private abstracting services. Generally such internal
abstracts include only those thought (usually by the librarian) to be of interest to
the research staff, and they are likely to be more detailed than those otherwise
available.

Between the time of publication of an item in an original source and the
appearance of an abstract of it there is an inevitable time lag, usually not less
than 2 to 3 months or more than 1 year. Most delay is likely to be encountered in
the rarer foreign periodicals. Abstract editors strive for promptness, but the
distribution, writing, and printing processes take time.

Figure 1 (Chapter 1) shows how the number of abstract journals have
paralleled the number of scientific periodicals for many years.

Abstracting Abstracts

To increase efficiency in retrieving the essence of papers, C. L. Bernier[13]
proposed condensing the papers to terse conclusions. To illustrate the possibility, he suggested, as a terse conclusion for an announcement by W. Mertz et al,[14]
"Chromium deficiency may cause the body's failure to use carbohydrates." Such
conclusions are designated by Bernier as terse literature.[15]

[12]Hans Vogel, in *Chemiker-Ztg.*, **87:**187 (1963), suggested a supranational organization to issue
Chemical Abstracts, Chemisches Zentralblatt, and *Referativnyi Zhurnal Khimiya* in English, German, and Russian to avoid extensive duplication and achieve better accessibility of abstracts.
[13]*J. Am. Soc. Inf. Sci.*, **21,** 316 (1970).
[14]*Chem. Eng. News*, **48** (52), 35 (1970).
[15]*J. Chem. Inf. Comput. Sci.*, **15,** 189 (1975).

Ultraterse literature would be an overall condensed statement of such conclusions for a number of papers on the same general subject. A proposed computer program could handle such condensations.[16]

A later paper presents a proposal for managing the problem of reading overload.[17]

Journals Publishing Abstracts

Abstracts appear in two kinds of journals: those devoted exclusively to abstracts and those whose contents consist partly of abstracts and partly of other items which were mentioned in Chapter 2. The second type is found particularly in publications devoted to special fields; here, as one might expect, the abstracts are selected upon the basis of their importance to those engaged in the special work.

Publication of abstracts goes back almost to the beginning of modern chemistry. Thus, Manzer's book[18] covers the period 1790–1920. For the more than 300 journals considered, there are three lists: an alphabetical list of titles, a chronological list of titles, and a classified subject list.

The present emphasis is on the more important general abstracting journals, but a later list does include examples of specialized journals. To show the historical development of the general journals discussed, chronological listing is followed.

Chemisches Zentralblatt (1830–1969)

Because of the length of time covered, the *Zentralblatt* is probably the most important general abstracting journal. It started in 1830 as *Pharmaceutisches Zentralblatt,* changed to *Chemisch-Pharmaceutisches Centralblatt* in 1849, to *Chemisches Centralblatt* in 1856, and finally to *Chemisches Zentralblatt* in 1897. Publication lapsed in 1945 because of the war. Various efforts were later made to reestablish the journal, including a German and a Russian edition. Normal publication was resumed in 1952. The breaks are now covered by supplementary volumes.[19]

The journal was long published by the Deutsche Chemische Gesellschaft. In 1950 it became a joint project of the Deutsche Akademie der Wissenschaften (Berlin), the Chemische Gesellschaft (Soviet East Zone), the Akademie der Wissenschaften (Göttingen), and the Gesellschaft Deutscher Chemiker.

The abstracts are classified as follows:

[16]T. L. Glazener and C. L. Bernier, *J. Chem. Inf. Comput. Sci.,* **18,** 16 (1978).
[17]C. L. Bernier, *Inf. Proc. Manag.,* **14,** 445 (1978); see also C. L. Bernier and A. N. Yerkey, "Cogent Communication: Overcoming Reading Overload," Greenwood Press, Westport, CT, 1979.
[18]Loc. cit.
[19]See C. Weiske, *Chem. Ber.,* **1973,** 106, I–XVI, for a summary of the history and development of this periodical from 1830 to 1969.

A General; physical; inorganic chemistry (9 subdivisions)
B General and theoretical organic chemistry
C Preparative organic chemistry; naturally occurring substances (8 subdivisions)
D Macromolecular chemistry
E Biological chemistry; physiology; medicine (6 subdivisions)
F Pharmaceutical chemistry; disinfection
G Analysis; laboratories
H Applied chemistry (24 subdivisions)

The abstracts as a whole are likely to be more detailed than those in *Chemical Abstracts*. Special features are the use of italics for the names of new compounds, a brief outline at the beginning of the volume for each division of chemistry covered, a list of numbers of patents mentioned, and statistical data on the abstracts included in the various divisions. A valuable feature is the list of equivalent patents for different countries (Patentrückzitate).

References to the *Zentralblatt* usually have the form "*Chem. Zentr.* 1928 II, 789," which indicates year, part, and page (there being no volume number). Simply "*C*" may replace "*Chem. Zentr.*" "*D.R.P.*" means "*Deutsches Reichs-Patent.*"

Before 1919 no attempt was made to cover applied chemistry thoroughly, because early industrial abstracts appeared in *Angewandte Chemie* and the *Chemisch-technische Übersicht* of *Chemiker-Zeitung.*

Collective indexes were issued from 1897 through 1954 for subjects and authors, with ones for formulas and patents covering only 1922–1954. Until 1950 the Richter formula index system was used, after which the Hill system of *Chemical Abstracts* was adopted.

Following the discontinuance of *Chemisches Zentralblatt* in 1969,[20] the renamed society, Gesellschaft Deutscher Chemiker and Farbenfabriken, undertook a new publication, *Chemischer Informationsdienst*. A weekly abstracting service, it is selective in covering the fields of inorganic, organic, and organometallic chemistry. Only a relatively small number of original sources are covered.

The abstracts are arranged and classified according to subject areas. Each abstract is characterized by a code consisting of letter designation, year, issue number, and abstract number, e.g., Cl-1972-2-263. Included are the German title of the paper, author's name, bibliographic data, address, language, text, and name of the abstractor.

The German society has now signed an agreement with Chemical Abstracts Service to cooperate in the production and distribution of *Chemical Abstracts* and its derived services.

[20]*Nach. Chem. Tech.*, **17**, 243 (1969).

Bulletin de la société chimique de France (1863–1945)

This journal first included abstracts in 1863. From 1892 to 1933 even-numbered volumes contained the abstracts. Since that time this volume has been known as the "documentation volume." Since 1918 French abstracts on applied chemistry have appeared in *Chimie & Industrie* (Paris).

British Abstracts (1871–1953)

The two journals just described are the main ones devoted entirely to general abstracts, but many others contain abstracts along with other material. Two of these deserve special attention because of their general nature. They are the *Journal of the Chemical Society* (London) and the *Journal of the Society of Chemical Industry* (London). Abstracts appeared first in the former journal in 1871, and a separate volume was devoted to them first in 1878. From then until 1926 the even-numbered volume for each year was devoted to abstracts on general, organic, and physical chemistry. The latter journal was started in 1882 and from the beginning included abstracts on all aspects of applied chemistry.

Beginning in 1926 the abstracting work of the two journals was combined in *British Chemical Abstracts,* Part *A,* pure chemistry, bound separately, and Part *B,* applied chemistry, bound as part of the *Journal of the Society of Chemical Industry* (name changed to *Chemistry and Industry*). Later the abstracting name was changed to *British Chemical and Physiological Abstracts,* and finally, in 1945, to *British Abstracts.* In 1944 Part *C,* analysis and equipment, was added. All the parts had subparts, generally paged separately. Finally, in 1953, this fine abstracting service was terminated.

The general quality of these English abstracts was high. Together they enable one to follow abstracts in the English language back to the dates given. *Chemical Abstracts* consistently covered more periodicals. Since 1953 what is left of *British Abstracts* appeared in the following two journals:

1 *Journal of Applied Chemistry.* The abstracts are classified under (1) Chemical Engineering and Atomic Energy; (2) Fuel and Fuel Products; (3) Industrial Inorganic Chemistry; (4) Industrial Organic Chemistry; (5) Fats, Waxes, Detergents; (6) Fibres; and (7) Laboratory Apparatus and Technique; Unclassified. The periodical was discontinued in 1971.

2 *Analytical Abstracts.* The abstracts are classified under (1) General Analytical Chemistry, (2) Inorganic Analysis, (3) Organic Analysis, (4) Biochemistry (five subdivisions), (5) General Technique and Laboratory Apparatus (five subdivisions). Several new subdivisions have been added.

Cumulative subject and author indexes were issued from 1954 through 1963 for *Analytical Abstracts.*

Chemical Abstracts (1907–   )

As *Chemical Abstracts* is now the only general abstracting journal remaining of the big three, separate consideration of it is contained in Chapter 7.

Prior to 1910, one needed the British and/or German abstracting journals. Subsequently, they were valuable checks on *Chemical Abstracts* when the original sources could not be consulted.

Bulletin Signaletique (1940–   )

Issued by the Centre National de la Recherche Scientifique in Paris, this abstracting journal appeared from 1940 to 1955 as the *Bulletin Analytique*. An international publication, this journal attempts to cover the world's scientific and technical literature. Titles are translated into French; abstracts are in French. In 1969 this title was issued as separate abstracting journals. In 1978 fifty sections were published; six section groupings were also made available. Sections of chemical interest include: 170: *Chimie;* 320: *Biochimie;* 780: *Polymeres;* and 800: *Genie chimique.*

D. Pelissier described manual methods for using the data base of this abstracting journal.[21]

Referativnyi Zhurnal [22]

In 1952 Russia established the All-Union Institute for Scientific and Technical Information (VINITI) of the Academy of Sciences of the U.S.S.R. It publishes abstract journals from a central office for the following subjects: astronomy and geodesy, biochemistry, biology, chemistry (established in 1953), electrical engineering, geology and geography, mathematics, mechanical engineering, mechanics, metallurgy, and physics. The one for chemistry is entitled *Referativnyi Zhurnal, Khimiya,* for which the abbreviated name is *RZhKhim.*

Practices of *Chemical Abstracts* have been followed somewhat, including use of the Hill system for formula indexes. The indexes include author and patent index for each issue; annual indexes for authors, subjects, patents, and formulas; and an annual index of organic reactions.

Current Abstracts of Chemistry (1960–   )[23]

This weekly periodical appeared first with the title *Index Chemicus*, a designation now reserved for the cumulative indexes.

Structural diagrams are featured. Only journals describing new chemical compounds or reactions are covered. These include especially journals on biological, medicinal, organic, and pharmaceutical chemistry.

Figure 2 shows an example of one of the shorter abstracts. In addition to the usual bibliographical data, the following information is included:

The ISI index number for locating the abstract.

[21] *Pure Appl Chem.*, **49**, 1797 (1977).
[22] See E. J. Copley, "A Guide to *Referativnyi Zhurnal*," British Library/Science Reference Library, London, 1970.
[23] Published by Institute for Scientific Information, Philadelphia, PA.

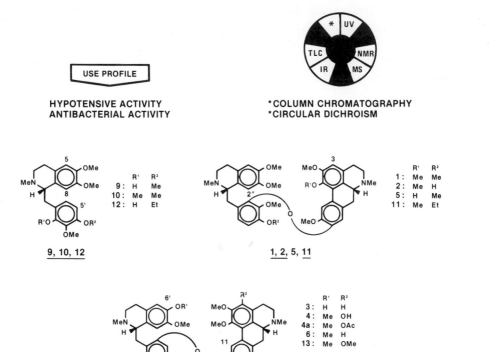

Figure 2
Abstract from *Current Abstracts of Chemistry* (with permission).

An abstract by the author(s).

Suggested applications of the compounds (if any).

An analytical wheel showing the techniques used to identify the compounds.

Structural diagrams of the compounds.

Compound numbers to speed examination of the abstract and the original article. New compounds are underlined.

Issue Indexes

The following items are indexed in each issue: molecular formulas; keywords; biological activities alert; alert of new reactions, new syntheses, and labeled compounds; authors; corporate entities; and list of journals covered in the issue.

Annual Indexes

The cumulated issue indexes are published as *Index Chemicus*. In addition to including the items listed for issue indexes, there is a Rotoform and a Registry Number Index.

The semiannual Rotaform formula index aids in searching molecular formulas for compounds containing elements in addition to carbon, hydrogen, nitrogen, and/or oxygen. The formula is rearranged to bring the element of interest to the front position. If there are several such elements in the compound, a rearranged form is shown for each one, as illustrated in the following example:

$C_{12}H_{18}BF_4IrN_2$ [not boldface in the journal]
    $BF_4IrC_{12}H_{18}N_2$
    $F_4BIrC_{12}H_{18}N_2$
    $IrBF_4C_{12}H_{18}N_2$

A molecular "Chemical Substructure Index" provides for manual searching for partial chemical structures. It employs the Wiswesser line notation (see Chapter 15) to list new compounds in permuted alphabetical order of the significant structural elements. Reference should be made to the index for details of use.

The Registry Number System covers new organic compounds, reactions, and syntheses.

International Pharmaceutical Abstracts (1964–   )

This semimonthly abstracting journal covers more than 1,000 primary journals in pharmacy, medicine, and related fields.

Based on this publication, the American Society of Hospital Pharmacists issues the *IPA Information System,* a bimonthly publication dealing with pharmacy practice and technology, a trade-name cross-reference list, magnetic tape services, microfilm services, cumulative indexes, special bibliographies, and certain other services.

INIS Atomindex (1970–   )

INIS (International Nuclear Information System) is a cooperative, decentralized information system, set up by the International Atomic Energy Agency and its member states. Its purpose is to construct a data base identifying publications relating to nuclear science and its practical applications. Member states and the cooperating international organizations scan the scientific and technical literature published within their boundaries or by them, select from it the items which fall within the subject scope of INIS, and process the data according to the agreed rules and standards. By 1978 more than 50 countries were participating in the project.

The following services are maintained: (1) the INIS magnetic tape service; (2) *INIS Atomindex,* a semimonthly abstracting journal; and (3) INIS nonconventional literature on microfiche.

The abstracts are arranged as follows: Physical Sciences; Chemistry, Materials, and Earth Sciences; Life Sciences; Isotopes, Isotope and Radiation Applications; Engineering and Technology; Other Aspects of Nuclear Energy.

There are several kinds of indexes for each issue. Users of the periodical should consult the issue introduction for details of format and data elements.

Energy Abstracts (1974–   )

In this transdisciplinary publication the energy-related abstracts are listed according to 39 categories established by the former Energy Research and Development Administration, now a part of the Department of Energy. The abstracts are sequentially numbered for each 12-month period.

Bioengineering Abstracts (1974–   )

A by-product of the Engineering Index data base, this monthly publication is designed for the specialist in bioengineering and related disciplines. Reference is made to more than 2,000 periodicals and conference papers and reports issued through governmental, academic, and industrial activities.

Engineering Index Monthly (1976–   )

The indexing period of *Engineering Index* was discussed in the previous section. With the termination of the CARD-A-LERT Service in 1975, an abstracting journal has been issued in printed form. It is cumulated annually, with indexes of subjects, authors, and affiliations of authors. There is an annual microform available.

The organization is by subject headings to reflect the technology of the world today. Considerable information is included on chemical engineering, the chemical industry, the chemical profession, and chemical aspects of the other areas of engineering.

Weekly Government Abstracts (1976–   )

In Chapter 3 there is a discussion of the general activities of the National Technical Information Service (NTIS). Included is mention of the published summaries of most unclassified Federally funded research.

As these summaries are abstracts, the following example is included to illustrate the general format:

PB-273 080/2GA
National Bureau of Standards, Washington, DC
Luminescence in Metal Flames Irradiated with CO_2 Laser
J. C. Mulder and A. F. Clark. 1977, 4 p.
Sponsored in part by Air Force Office of Scientific Research, Bolling AFB, DC.
Pub. in *Chemical Physics Letters* **49,** 471 (1977).
Descriptors: Calcium, Magnesium, Flames, Luminescence, Carbon dioxide lasers.
Identifiers: Reprints, Laser enhanced fluorescence, Heterogeneous reactions.
Rapid-scan spectrometry of calcium and magnesium flames in air with a CO_2 laser reveals luminescence excited by heterogeneous reactions on condensed

metal oxide particles. The luminescence is stimulated by laser radiation. An explanation for this effect is proposed.

Other Sources of Abstracts

The publications mentioned above are considered best for general work. Abstracts are widely scattered, however, since many periodicals have contained such information for at least part of the period covered by the journals. In case the periodical was not restricted in its field, its abstracts are general in nature; but if it pertained to some special field, the abstracts are confined to this and closely related fields. The latter arrangement is the more common and forms a desirable combination for the individual who wishes to watch developments only in his or her own limited field.

The sources go back to the first continuous chemical journal, *Annales de chimie,* which carried some abstracts from 1789 to 1870. A previous edition of this book[24] included a selected list of such periodicals with their starting dates. For searches before 1875 these sources may be of some value.[25]

Of more importance today are the journals currently carrying such abstracts. The latest compilation of sources is "Ulrich's International Periodicals Directory," by the R. R. Bowker Company, New York, 1981. From more than 60,000 titles listed, 1,059 were selected as belonging in the indexing and abstracting group. Various helpful details are included to facilitate use of the directory.

The following selected list of specialized abstracting sources includes well-known examples:

Specialized Abstracting Journals

Analytical Chemistry

Analytical Abstracts, 1954–
Atomic Absorption and Flame Emission Spectroscopy Abstracts, 1969–
Electroanalytical Abstracts, 1963–
Gas and Liquid Chromatography Abstracts, 1958–
Mass Spectrometry Bulletin, 1966–

---

[24]2d ed., pp. 85–87, 1940.
[25]For a more extensive list of sources of early and specialized abstracts, see the following publications:

Periodical Bibliographies and Abstracts for the Scientific and Technological Journals of the World, *Bull. Natl. Research Council (U.S.),* vol. 1, no. 3, 1920.

E. J. Crane, A. M. Patterson, and E. B. Marr, "A Guide to the Literature of Chemistry," John Wiley & Sons, New York, 1957, pp. 139–150.

E. E. Reid, "Introduction to Organic Research," p. 92, D. Van Nostrand Company, Inc., Princeton, NJ, 1924.

C. J. West and D. D. Berolzheimer: "Bibliography of Bibliographies on Chemistry and Chemical Technology," p. 12, National Research Council, Washington, 1925.

Biological Chemistry

Amino Acid, Peptide, and Protein Abstracts, 1972–
Biological Abstracts,[26] *1927–*   *; and Biological Abstracts/RRM*, 1980–
Carbohydrate Chemistry and Metabolism Abstracts, 1973–
Excerpta Medica, 1947–   ; various sections began later
Nucleic Acid Abstracts, 1971–
Tissue Culture Abstracts, 1963–

Chemical Engineering

Chemical Market Abstracts, 1950–
Petroleum Abstracts, 1961–
Theoretical Chemical Engineering Abstracts, 1964–

Organic Chemistry

Synthetic Methods of Organic Chemistry (Theilheimer), 1942–

Physical Chemistry

Laser-Raman Spectroscopy Abstracts, 1971–
Nuclear Magnetic Resonance Spectroscopy Abstracts, 1971–
Physics Abstracts, Science Abstracts, Section A, 1898–
Physics Briefs, Physikalische Berichte, 1979–
Solid State Abstracts, 1957–

Miscellaneous Abstracts

Air Pollution Abstracts, 1970–
Alloys Index, 1974–
Energy Research Abstracts, 1976–
Metals Abstracts, 1968–
Plastics Abstracts, 1959–
Water Pollution Abstracts, 1972–
World Aluminum Abstracts, 1970–

---

[26]See H. E. Kennedy and D. A. Fisher, *Fed. Proc., Fed. Amer. Soc. Exp. Biol.*, **33,** 1714 (1974), for other services and products of BioSciences Information Service BIOSIS; also W. C. Steere, "Biological Abstracts BIOSIS," Plenum Press, New York, 1976. In 1980 *Biological Abstracts/RRM* replaced *BioResearch Index*.

C. REVIEWING PERIODICALS AND SERIALS

A third type of survey publication contains brief accounts of the developments in various fields for some given period of time. These accounts are designed to be reviews of progress, and the publications are usually known as review serials.

Various titles are used for such publications, "reviews" and "advances" being the most common English terms. Others are "yearbooks," "reports," "developments," and "conferences." German equivalents are *Jahrbuch, Jahresbericht, Fortschritt,* and *Ergebnis.* If the English term is preceded by the word "annual" (or "biennial"), the publication is a periodical. If not, librarians use the term "serial," although they may include journals with the serials. Titles alone often do not indicate whether the publication is a periodical, with the reviews being published at regular intervals of time.

The use of "advances" here for general reviews should not be confused with the use mentioned in Chapter 2. There the term was equivalent to symposium or conference, and meant a series of more or less related papers presented at a scientific meeting.

A "periodical-type" review is a survey of progress and developments during the time period covered. Thus, in an annual review on light absorption spectrometry, of the 129 papers cited dealing with new information in 1949 only three involved heteropoly chemistry in spectrophotometry.[27]

In contrast, a "serial-type" review is not a report of annual developments, and so the publication is not a periodical. Rather, it is irregular, and often has the characteristics of a minimonograph. That is, it is a current status survey of the subject. Taking heteropoly compounds again as an example, one survey dealt with their role in absorptimetry.[28] An earlier survey covered their role in all quantitative chemical analysis, dating from the proposal of L. Svanberg and H. Struve in 1848 to use ammonium molybdophosphate in analytical chemistry.[29] Such reviews resemble the longer articles in encyclopedias.

Even the annual reports may not deal with developments in the same fields year after year. For example, the *Annual Reports of the Chemical Society* vary the subjects covered from volume to volume.

That the significance of reviews was realized early in the development of modern chemistry is indicated by the publication, in 1795, of the *Berlinisches Jahrbuch für die Pharmacie.* Some of the succeeding early publications were listed in the second edition of this book, page 89, 1940.

An estimate of the value of, and need for, well-written reviews was published by E. Garfield.[30] His conclusions are based on a study by the Institute for Scientific Information on the role review articles play in the process of scientific communication.

---

[27]M. G. Mellon, *Anal. Chem.,* **22**; (1950).
[28]M. G. Mellon, *Rec. Chem. Prog.,* **11**, 177 (1950).
[29]E. R. Wright and M. G. Mellon, *Proc. Indiana Acad. Sci.,* **50**, 110, (1941).
[30]*J. Chem. Inf. Comput. Sci.,* **18**, 1 (1978).

Many researchers face the question of what is known on a given subject, or what developments were reported in a stated period. Typical subjects are desulfurization of coal, fluoridation of potable water supplies, and disposal of nuclear wastes. A researcher familiar with the resources of indexing and abstracting journals should be able to compile a review, but finding one ready-made saves time and effort.

This raises the question of who does the reviewing. Usually the method of compilation consists in having someone familiar with a given field examine the papers and other contributions relating to this field and then prepare a summary stating the general trend of the developments during the year and noting any particular advances reported. Or in the serial-type of review the product is designed to present an overview of the current status of the subject. In both cases references are included for the publications cited. One hopes that the reviewer can, and will, critically evaluate the literature cited. However, reviewers cannot be expected to try out experimentally a series of new methods.

Many reviews appear in periodicals, but many current reviews are in irregular, serial publications. Some of the periodicals, such as *Chemical Reviews,* contain nothing else. In others, such as *Analytical Chemistry,* the reviews are only a small part of the total "editorial" content for the year. In this journal they are segregated, both with respect to time of appearance and to their nature. Thus, they appear in a special April issue, with "Fundamental" reviews in the even years, and "Application" reviews in the odd years.

Reviewing chemical periodicals go back a century and a half. Selected examples follow, in chronological order:

1821–1849, *Berzelius, Jahresbericht uber die Fortschritte der physischen Wissenschaften* (Later *der Chemie*)

This serial was the first of its kind, except for the *Berlinisches Jahrbuch,* and was very valuable for investigators working during the time covered. Where investigations take one back to the literature of that period, it is still of distinct service. In 1849 the material was classified under *Anorganische Chemie, Pflanzen Chemie,* and *Tier Chemie.* The index is not satisfactory.

1847–1910, *Liebig and Kopp, Jahresbericht uber die Fortschritte der reinen, pharmaceutischen und technischen Chemie, Physik, Mineralogie und Geologie*

Issued by different editors and varying somewhat in nature, this serial was for years the main source of critical summaries on the important advances in the fields covered. In 1847 the division of material was as follows: physics and physical chemistry; inorganic, organic, analytical, and technical chemistry; mineralogy; and chemical geology. The early indexes are not satisfactory, but this defect was later corrected with the publication of collective indexes.

1855–1937, *Wagner, Jahresbericht uber die Leistungen der chemischen Technologie* (Issued under different names for some years)

The contents of about 200 journals were summarized in this serial. A classified arrangement was used with the material on organic and inorganic chemistry in separate volumes.

1891–1918, *Meyer, Jahrbuch der Chemie (Bericht uber die wichtigsten Fortschritte der reinen und angewandten Chemie)*

Seventeen divisions are included (1915), with many subdivisions indicated. Original references and subject and author indexes are given.

1892–1941, *Mineral Industry*

The publication known as *Mineral Industry* was an annual review of the mining and metallurgical industry, including statistics, and an account of the year's technical progress.

1904–   , *Annual Reports of the Chemical Society* (London)

In 1955 the annual reports on the progress of chemistry were divided into the following classes: general and physical chemistry, inorganic chemistry, organic chemistry, analytical chemistry, biological chemistry, and crystallography.

1906–   , *Science Progress*

This English quarterly is devoted to reviews of various fields of science.

1916–   , *Applied Chemistry Reports* (of the Society of Chemical Industry)

An excellent series of annual reports on the progress of applied chemistry is given. The arrangement is about the same as that used in the abstracts section of the *Journal of Applied Chemistry*.

1924–   , *Chemical Reviews*

This periodical is a fine example of a review serial covering a wide range of chemical interests. Usually each issue is limited to from one to three articles.[31]

---

[31]See H. Hart, *Chem. Rev.*, **74**, 125 (1974), for a fiftieth-anniversary account of the periodical.

1947– , *Quarterly Reviews*

Supplementing the earlier British annual reports, this quarterly resembles the *American Chemical Reviews.* in 1972 the title was changed to *Chemical Society Reviews.*

1970– , *CRC Critical Reviews*

Recently a series of "critical" reviews was begun by the CRC Press (Boca Raton, FL). They are designed to assess the reliability and importance of individual publications, and to summarize the present status of the topic. Thus, they are essentially minimonographs, rather than annual surveys of progress, or comprehensive historical overviews.

The areas covered of most interest to chemists follow: Analytical Chemistry, Biochemistry, Clinical Laboratory Sciences, Food Science and Nutrition, Solid State and Materials Sciences, and Toxicology. There is a separate volume series for each area. Two or three reviews, of about 100 total pages, constitute an issue, with usually four issues per year.

Other Sources of Reviews

The following list includes selected reviewing publications. There are many others. In general these publications are more specialized than those in the preceding list. No differentiation is made here between "periodical" and "serial" types.

Agricultural and Food Chemistry

Advances in Food Research, 1948–
Nutrition Reviews, 1942–

Analytical Chemistry

Advances in Analytical Chemistry and Instrumentation, 1960–
Advances in Chromatography, 1965–
Chromatographic Reviews, 1959–
Electroanalytical Chemistry, 1966–

Biological Chemistry

Advances in Clinical Chemistry, 1958–
Advances in Enzymology and Related Subjects of Molecular Biology, 1941–

Advances in Protein Chemistry, 1944–
Annual Review of Biochemistry, 1932–
Quarterly Reviews of Biophysics, 1968–

Chemical Engineering

Advances in Chemical Engineering, 1956–
Advances in Electrochemistry and Electrochemical Engineering, 1961–1973.
Advances in Heat Transfer, 1964–

Inorganic Chemistry

Advances in Inorganic Chemistry and Radiochemistry, 1959–
Advances in Organometallic Chemistry, 1964–
Coordination Chemistry Reviews, 1966–
Progress in Inorganic Chemistry, 1959–

Organic Chemistry[32]

Advances in Alicyclic Chemistry, 1966–
Advances in Carbohydrate Chemistry, 1945–
Advances in Heterocyclic Chemistry, 1963–
Advances in Physical Organic Chemistry, 1963–
Organic Reaction Mechanisms, 1965–
Topics in Stereochemistry, 1967–

Physical Chemistry

Advances in Catalysis, 1948–
Advances in Chemical Physics, 1958–
Annual Review of Nuclear and Particle Science, 1952–
Catalysis Reviews—Science Engineering, 1967–

Indexes to Reviews

The following indexes are of value in locating reviews:

Bibliography of Reviews in Chemistry (1958–1962)   Abstracts of reviews found
in *Chemical Abstracts* are arranged by sections as originally reported in this

---

[32]In the July 21, 1978 issue, the *Journal of Organic Chemistry* began a biennial section entitled
Recent Reviews.

abstracting service. An author and subject index is found following each time period.

CA Reviews Index (CARI) (1975–   )   A comprehensive index to abstracts of reviews found in *Chemical Abstracts* is produced twice yearly by the United Kingdom Chemical Information Services. This computer-produced index uses the same Keyword-In-Context (KWIC) as in *Chemical Titles.*

Index of Reviews in Organic Chemistry (1971–   )   This index was initially compiled by D. A. Lewis for use by organic chemists in the Plastics Division of Imperial Chemical Industries Ltd. in the United Kingdom; assistance for publication was later given by the Chemical Society. Review articles in organic chemistry which have appeared in the primary and secondary literature may be retrieved through the cumulative volumes now available. The index is arranged alphabetically and is subdivided under each letter heading into three sections: (1) references to articles on individual compounds or classes of compounds, (2) references to articles on specific "name reactions," and (3) references to articles on specific processes or phenomena.

FEBS Letters Index of Biochemical Reviews (1971–   )   Published on behalf of the Federation of European Biochemical Societies (FEBS), this annual index gives bibliographical citations to reviews in biochemistry under more than 150 subject headings.

Index to Scientific Reviews (1974–   )   This computer-produced semiannual publication of the Institute for Scientific Information (ISI) is designed to aid in locating more than 20,000 review articles and state-of-the-art surveys published annually in over 100 areas of science. Cumulations are prepared annually; a Permuterm subject index is provided for easy access (in the Permuterm subject index every significant title word is paired with every other significant word in the title to produce word pairs).

SECONDARY SOURCES II— *CHEMICAL ABSTRACTS*

A well-used library is one of the few corrections of premature senility.

Sir William Osler

The major publishing division of the American Chemical Society is Chemical Abstracts Service (CAS).[1] Located for over half a century in quarters provided by the Ohio State University, Columbus, OH, it is now housed in two large nearby buildings of its own.[2] Figure 3 shows a view of the two connected buildings.

The primary product of the service is the world-famous periodical, *Chemical Abstracts* (CA). Along with the abstracts are several kinds of indexes and computer-readable files, all dealing with what is new and significant in chemistry, chemical engineering, and related fields.

From a meager beginning in 1907, as the successor to *Review of American Chemical Research* (1895–1906), *Chemical Abstracts,* has grown to its present position as the dominant general chemical abstracting journal in the world. Currently the staff monitors nearly 14,000 periodicals from more than 150 nations, published in more than 50 languages; patents issued by 26 nations and two industrial property organizations; and new books, proceedings of conferences, dissertations, and technical reports from around the world. CA cited 548,676 documents in 1980.

For some decades the editorial office assigned the various kinds of primary

[1]In its own publications Chemical Abstracts Service uses the abbreviations CAS and CA to save space, a practice followed partly here.
[2]2540 Olentangy River Road (or Box 3012), Columbus, OH 43210.

Figure 3
View of the home of Chemical Abstracts Service. CAS has granted permission to use this view, the two charts in the text, and certain data from its publications.

documents to volunteer abstractors selected for the specific subjects involved. The abstract manuscripts, prepared according to detailed instructions,[3] were edited and sent to the printer for composition. Galley proof was then sent to the section editors and the editorial office for checking. After this double checking, the proof was returned to the printer for printing. Preparation of the volume indexes involved further delay of months.

This familiar publishing process became more and more time-consuming and cumbersome with the exponential increase in size and number of publications in the 1950s. The possibility of at least partial automation became more promising with the development of sophisticated computers. Many steps were tried experimentally. Individual developments cannot be related here. Suffice it to state that a processing system was developed for a data base which is now the foundation of the Chemical Abstracts Service Information System. R. E. O'Dette[4] has described the concept for this data base, and later[5] he described what the data base is and how it is assembled.[6]

Finally, in 1972, conversion of the text of abstracts to computer processing and photocomposition began. This conversion was completed in mid-1975. A special report describes the person-machine partnership being used.[7]

The system of photocomposition so developed is now being used to compose the text of a number of primary journals of the American Chemical Society, as well as several nonchemical journals for other organizations.

For printing, the abstracts so prepared are grouped into 80 sections, each section covering a specific subject in chemistry or chemical engineering. The

---

[3]See "Directions for Abstractors," CAS, 1975.

[4]*J. Chem. Inf. Comput. Sci.*, **15,** 165 (1965).

[5]*Pure Appl. Chem.* **49,** 1781 (1977).

[6]See also "CAS TODAY," special publication recounting the overall development of Chemical Abstracts Service (1980); for a history of abstracting at Chemical Abstracts Service, see D. B. Baker, J. W. Horiszny, and W. V. Metanomski, *J. Chem. Inf. Comput. Sci.,* **20,** 193 (1980).

[7]*Chem. Eng. News,* **53,** (24), 30 (1975); **59,** (22), 29 (1981).

number of sections, and their titles, have changed considerably over several decades. The groups and section titles adopted in 1975 for volume 82 are shown in Table IV.

A special publication summarizes the guidelines used to assign a document to one of the 80 sections.[8] Included in the manual for each section are descriptions of the subjects covered; listings of specific exclusions and related subjects covered in other sections and the policies for cross-referencing of abstracts published in one section to another section; and explanations of the subsection arrangement of abstracts within each section.

As an example of subsection distribution, the recommended arrangement for Section 78, Inorganic Chemistry, follows: (0) reviews, (1) elements, (2) acids, (3) bases and metal oxides, (4) synthetic minerals, (5) salts (including metal borides, carbides, and nitrides), (6) double salts, (7) coordination compounds, (8) nonmetal compounds, (9) reactions, (10) other.

There are no marks in any section to differentiate one subsection from another. Thus, in Section 79, Inorganic/Analytical Chemistry, one will not find spectrophotometric methods (or any other kind) segregated.

Within each section, abstracts of journal articles, proceedings of conferences, dissertations, and technical reports are placed first; announcement of new books second; and abstracts of patents third. Each section ends with a set of cross references (if considered desirable) to direct readers to abstracts of possible interest in one or more other sections of that issue.

Such cross references are either to an entire section of *Chemical Abstracts,* e.g., "**9** Biochemical Methods," or to a specific abstract, e.g., "87: 121,039a, Studies of Oxide Inclusions in Steel."[9]

Abstracts with the notation "Title only translated" are from abstract journals published in countries whose copyrights cover reproduction of the text of an abstract. An example of such a title entry is "87: 145,276a, Determination of Scandium in Metallic Nickel." Obviously, all that is revealed is what was determined and in what. To obtain any further information necessitates consulting the Russian abstract and/or the original document.

At the back of each issue of *Chemical Abstracts* is a keyword index, with accompanying explanation. As the terminology of the author(s) is used, the vocabulary is uncontrolled. It is an alphabetical listing of phrases containing one to five characteristic words (keywords) taken from the title, text, or context of each abstract.

Since 1967, volume 66, the references are to abstract numbers rather than to columns or pages. The numbering is continuous throughout a volume. Beginning with volume 84 (1976), the volume number precedes the abstract number, as shown in the title cited above.

Prior to volume 66 (1967), there are three variations in the references.

---

[8]"Subject Coverage and Arrangement of Abstracts by Sections in *Chemical Abstracts,*" CAS, (1975).

[9]Commas are not used to indicate hundreds, thousands, and millions in multidigit numbers in *Chemical Abstracts.* Commas have been inserted in this chapter to facilitate reading.

TABLE IV
THE 80 SECTIONS OF CHEMICAL ABSTRACTS, DIVIDED INTO SECTION GROUPINGS

Biochemistry	Organic Chemistry	Macromolecular Chemistry	Applied Chemistry and Chemical Engineering	Physical and Analytical Chemistry
1. Pharmacodynamics	21. General Organic Chemistry	35. Synthetic High Polymers	47. Apparatus and Plant Equipment	65. General Physical Chemistry
2. Hormone Pharmacology	22. Physical Organic Chemistry	36. Plastics Manufacture and Processing	48. Unit Operations and Processes	66. Surface Chemistry and Colloids
3. Biochemical Interactions	23. Aliphatic Compounds	37. Plastics Fabrication and Uses	49. Industrial Inorganic Chemicals	67. Catalysis and Reaction Kinetics
4. Toxicology	24. Alicyclic Compounds	38. Elastomers, Including Natural Rubber	50. Propellants and Explosives	68. Phase Equilibriums, Chemical Equilibriums, and Solutions
5. Agrochemicals	25. Noncondensed Aromatic Compounds	39. Textiles	51. Fossil Fuels, Derivatives, and Related Products	69. Thermodynamics, Thermochemistry, and Thermal Properties
6. General Biochemistry	26. Condensed Aromatic Compounds	40. Dyes, Fluorescent Whitening Agents, and Photosensitizers	52. Electrochemical, Radiational, and Thermal Energy Technology	70. Nuclear Phenomena
7. Enzymes	27. Heterocyclic Compounds (One Hetero Atom)	41. Leather and Related Materials	53. Mineralogical and Geological Chemistry	71. Nuclear Technology
8. Radiation Biochemistry	28. Heterocyclic Compounds (More than One Hetero Atom)	42. Coatings, Inks, and Related Products	54. Extractive Metallurgy	72. Electrochemistry
9. Biochemical Methods	29. Organometallic and Organometalloidal Compounds	43. Cellulose, Lignins, Paper, and Other Wood Products	55. Ferrous Metals and Alloys	73. Spectra by Absorption, Emission, Reflection, or Magnetic Resonance, and Other Optical Properties
10. Microbial Biochemistry	30. Terpenoids	44. Industrial Carbohydrates	56. Nonferrous Metals and Alloys	74. Radiation Chemistry, Photochemistry, and Photographic Processes
11. Plant Biochemistry	31. Alkaloids	45. Fats and Waxes	57. Ceramics	75. Crystallization and Crystal Structure
12. Nonmammalian Biochemistry	32. Steroids	46. Surface-Active Agents and Detergents	58. Cement and Concrete Products	76. Electric Phenomena
13. Mammalian Biochemistry	33. Carbohydrates		59. Air Pollution and Industrial Hygiene	77. Magnetic Phenomena
14. Mammalian Pathological Biochemistry	34. Synthesis of Amino Acids, Peptides, and Proteins		60. Sewage and Wastes	78. Inorganic Chemicals and Reactions
15. Immunochemistry			61. Water	79. Inorganic Analytical Chemistry
16. Fermentations			62. Essential Oils and Cosmetics	80. Organic Analytical Chemistry
17. Foods			63. Pharmaceuticals	
18. Animal Nutrition			64. Pharmaceutical Analysis	
19. Fertilizers, Soils, and Plant Nutrition				
20. History, Education, and Documentation				

Through volume 27 (1933), they are to page numbers. During this period there is one column per page.

In volume 28 (1934) a two-column page was adopted and the references changed to column numbers. In volumes 28 through 40 (1946), a small superscript number is used with the column number to indicate the fraction of the page, counting downward, where the abstract appears. These small numerals are between the two columns.

Then in volume 41 (1947) through 65 (1966) small letters (a to i) replace the superscript numerals. Since 1967 letters at the end of abstract numbers have been computer check letters.

The issues of the five groups of abstracts (see Table IV) are bound separately as parts, and provision is made for subscription for a given part. Thus, one concerned with analytical chemistry, Sections 79 and 80, may subscribe to Sections 65 through 80. However, in this case, as no doubt in some others, the grouping may not be adequate because usually there are so many cross references to a large majority of the other sections. Presumably this results from analytical chemistry being the identifying and measuring means employed so widely in chemical industry and research. Very often, then, analytical users may need access to the other parts of *Chemical Abstracts*.

Included with issue 1 of each volume is the important Introduction. It contains abbreviations and symbols used and descriptions of abstract formats and content, with examples of typical headings showing the bibliographical details for each kind of publication. Also included are the patent coverage, the nature of the issue index, and suggestions for procuring copies of the original documents. The latter item includes a list of available codes for deposited documents, such as NTIS (National Technical Information Service).

The following example of a format, with accompanying notes, is for the abstract heading of a journal article:

JOURNAL-ARTICLE ABSTRACT HEADING

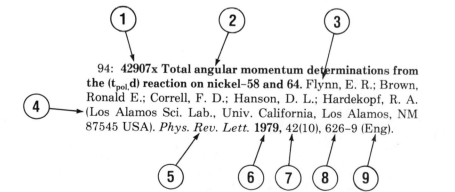

1 The abstract number appears in boldface type at the head of each abstract. The CA volume number in lightface type, followed by a colon, precedes the abstract number. Abstracts are numbered continuously through an entire 6-month volume.

2 The document title in boldface type is an edited reproduction of an English-language title, or as literal a translation of a foreign language title as is consistent with good English.

3 The complete author names are given in inverted order (last name first). Coauthors' names are separated by semicolons. Author names are listed in the order in which they appear on the original document.

4 The address in parentheses which follows the author name(s) is the location at which the reported work was done or where correspondence regarding the document is to be sent. Generally accepted, unambiguous abbreviations are used extensively in these addresses.

5 The journal title, in abbreviated form, is printed in italic type. Title word abbreviations are based on the International Title Word Abbreviations.

6 The year of publication is highlighted by boldface type.

7 The volume number is given in lightface type followed by the issue number in parentheses. Not all journals use both numbers. In these cases the CA citation includes only the number used by that particular journal. Parentheses indicate that it is an issue number. The absence of parentheses indicates that it is a volume number.

8 The inclusive pagination of the article follows.

9 The language of the original document is indicated, usually by an abbreviation enclosed in parentheses.

The Introduction mentioned also includes similar examples of headings for the other kinds of publications, namely, proceedings and symposia, deposited documents, dissertation abstracts, book announcements, and patents. They differ only in details involving the different bibliographic data elements involved.

At the front of issues 1 and 26 of each volume appears a complete masthead for Chemical Abstracts Service and a list of section advisors and abstractors.

Beginning with 1962, the 26 biweekly issues of *Chemical Abstracts* were divided for volume-period publication to make two 13-issue volumes per year. The current practice of publishing weekly issues was initiated in 1967, with Sections 1 to 34 appearing one weak, and Sections 35 to 80 the other week. Comprehensive volume indexes facilitate effective access to these vast semiannual accumulations.

The chart in Fig. 4 shows the place of *Chemical Abstracts* in the Chemical Abstracts Service Information System. Brief attention is directed later to most of these entries. It should be noted that some are issued in printed form, some in microform, and some as computer-readable files.[10]

---

[10]The importance and utility of the information system of CAS is receiving international recognition. Agreements have been made to expand international participation in the production (limited to the United Kingdom) and marketing of the CAS data base. Important organizations involved are the Royal Society of Chemistry in the United Kingdom, the Internationale Dokumentations-Gesellschaft für Chemie in West Germany, the Centre National de l'Information Chimique in France, and the Association for International Chemical Information in Japan. See R. J. Rowlett, Jr., *Chem. Br.* **16**, 425 (1980).

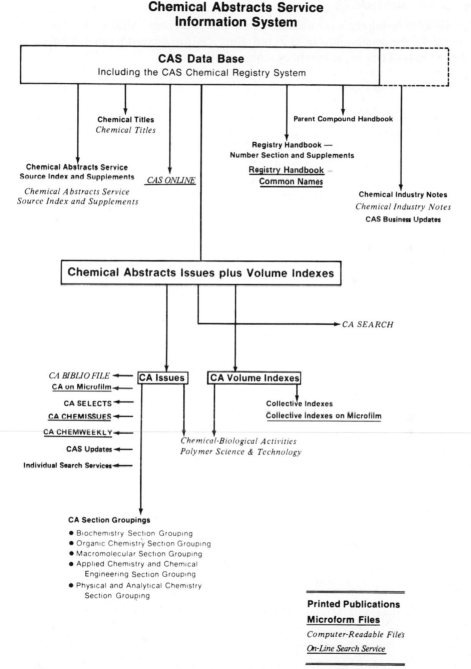

Chemical Abstracts Service Information System

Figure 4
Chart showing parts of the Chemical Abstracts Service Information Service (used with permission).

SELECTED ABSTRACTS

Before considering the various kinds of volume and other indexes, brief mention is made of the availability of selected abstracts for a considerable number of special areas. They include bibliographic citations.

CA SELECTS[11]

Starting in 1976, *CA SELECTS* has been issued biweekly to provide a current-awareness service for those having special interests. Abstracts for individual *CA SELECTS* topics are selected automatically by computer from the information base of Chemical Abstracts Service by means of a special, general-interest search profile, designed by experts in the field. Each profile is tailored to achieve the fullest retrieval of information on a given subject, drawing upon *Chemical Abstracts* and its associated indexes.

Table V shows examples of the 123 topics for which abstracts are available, as of June, 1981. Frequent additions are announced. The complete list is obtainable from the Marketing Department of Chemical Abstracts Service, together with subscription rates for these topic abstracts.

In advertising the utility of this service, Chemical Abstracts Service grouped the topics as follows: Agrochemicals, Analytical Chemistry, Applied Chemistry, Biochemistry, Chemical Reactions, Classes of Chemical Compounds, Energy Resources and Storage, Environmental Pollution and Waste Treatment, General Topics, Pathology, Safety and Toxicity, and Spectroscopy.

Chemical Abstracts Service (CAS) and Bioscience Information Service (BIOSIS) is now producing a series of current awareness publications under the general title *BIOSIS/CAS SELECTS*.

Tailored Services

CAS Updates, CAS Business Updates, and CAS Individual Search Services are current-awareness, custom-tailored services which are available to users with specific, selective needs. Information about them is available from the CAS Marketing Division.

Chemistry Industry Notes

Another very selective publication, designed to transmit certain information rapidly, is *Chemistry Industry Notes*. This weekly compilation of brief, informa-

---

[11]See J. E. Blake, V. J. Mathias, and J. Patten, *J. Chem. Inf. Comput. Sci.*, **18**, 187 (1978).

TABLE V
EXAMPLES OF *CA SELECTS* TOPICS

Amino Acids	Ion Exchange
Cosmochemistry	Liquid Crystals
Distillation Technology	Organotin Chemistry
Food Toxicity	Subatomic Particles
Herbicides	Synfuels

TABLE VI
EXAMPLES OF PERIODICALS COVERED IN
CHEMISTRY INDUSTRY NOTES

Business Week	*Materials Engineering*
Chemical Age (London)	*Papeterie*
Chemische Rundschau	*Plastics World*
Europa Chemie	*World Mining*
Fortune	*World Oil*

tive summaries of articles covers some 80 major trade and industrial periodicals. Examples are listed in Table VI. The articles selected relate to the management, investment, marketing, and production aspects of the chemical industry.

The extracts are arranged in eight sections: Production, Pricing, Sales, Facilities, Products and Processes, Corporate Activities, Government Activities, and People. Each issue has two indexes: the Keyword Index, which contains keyword phrases highlighted in the extracts, and the Corporate Name Index, which lists the names of all organizations cited in the extracts. There is a volume index.

Each issue is printed on Friday and mailed to reach most subscribers the following Monday.

"Index Guide"

An understanding of the nature and efficient use of the various indexes is so important that Chemical Abstracts Service issues the "Index Guide" to aid searchers in using *Chemical Abstracts* and derived services. The first one appeared in 1968, for volume 69. It now appears at the beginning of each 5-year collective index period, the latest one being the "1977 Index Guide." (When this book went to press, the "1982 Index Guide" was not yet available for inspection.) There are 1,450 pages of small type. Annual supplements will update the material.

Users of *Chemical Abstracts* are advised to study this guide and its supplements before trying to use the *Chemical Substance Index* and the *General Subject Index*. The general nature of the contents of the "Guide" follows:

Introduction A brief section describes cross references, synonyms, and index policy notes covers some 1,200 pages. It is the main part of the work. The listing leads the user from names of substances and subject terms found in the literature to the systematic names and headings adopted for the *Chemical Substance Index* and the *General Subject Index*.

The following simple examples of cross references illustrate the general kind of usage employed:

Aspirin
    See *Benzoic acid, 2-(acetyloxy)* = [50-78-2][12]

---

[12]The Registry Numbers are discussed later.

HMG 50
    See *Carbon fibers,* graphite
Monel 506
    See *Nickel alloy,* base Ni, C, Cu, Fe, Mn, Si, (Monel 506) [11105-31-0]
WOLGA
    See *Computer program*

Appendix I This listing of 74 pages consists of 66 divisions of headings for the *General Subject Index,* arranged in hierarchical order to aid in generic searching. There is an index to hierarchical families of headings.

Appendix II The organization and use of the main indexes are discussed, namely, the *Chemical Substance Index,* the *General Subject Index,* the *Formula Index,* and the *Index of Ring Systems.*

Appendix III A short section covers the content of the *General Subject Index.*

Appendix IV Many details are summarzied for the rules used in deriving systematic CA Index Names for the chemical substances. This section ends with a list of many references on chemical nomenclature.

Modern chemical nomenclature has become so involved that the subject is a study in itself. Deriving appropriate systematic names is probably the most difficult assignment in the entire process of editing *Chemical Abstracts.* An important paper discusses some of the problems involved.[13]

INDIVIDUAL AND MULTIVOLUME INDEXES

Chemical Substance Index

This index relates the systematic Index Names of chemical substances used in *Chemical Abstracts* to the abstract numbers for documents in which the substances are mentioned. The assigned Registry Numbers also appear. The nomenclature used provides Index Names which are as descriptive as possible of the precise molecular structure of the substances. It tends to bring the names of structurally related substances into juxtaposition in the alphabetical listing.

As an example, the familiar compound carbon tetrachloride, CCl_4, is indexed under the heading *Methane, tetrachloro-.* Other related compounds, such as bromoform, chloroform, and iodoform, are listed under methane, as related substituted derivatives. Included with the index name is additional information derived from the content of the document itself.

Prior to 1972, entries now segregated in the *Chemical Substance Index* appeared in the single alphabetical subject index of *Chemical Abstracts.*

---

[13]N. Donaldson, W. H. Powell, R. J. Rowlett, Jr., R. W. White, and K. V. Yorka, *J. Chem. Doc.,* **14,** 3 (1974).

General Subject Index

This index contains, in alphabetical order, all index entries which do not refer to specific chemical substances, e.g., apparatus, applications, concepts, general classes of chemical substances, processes, properties, reactions, uses, and biological and biochemical subjects. The context of each document and the corresponding abstract number are supplied.

In contrast to the keyword indexes for weekly issues, the rigidly controlled vocabulary evolved by Chemical Abstracts Service over the years is employed in this index. Effective use implies, therefore, understanding the practice followed in deciding on the entries adopted. Index users will be successful in their search if they select the terms used by the index builder; to that end they are urged to use the "Index Guide."

The following example, from volume 85, shows the nature of the entries:

Lotions

formulation of, R 25,298w
for hands, R 83,116w
   quaternary ammonium polymers in, P 10,313y
   skin, acylamino acid prepns. as, P 148,984z
   superoxide dismutase in, P 25,285q
   thickening agents for, cross-linked maleic anhydride-vinyl
    acetate polymers as, P 78, 947p

R is for review, P for patent, and B for book.[14]

Formula Index

The molecular formulas of chemical compounds indexed in the *Chemical Substance Index* are listed in alphabetical order by element symbol in the *Formula Index,* together with the index name and registry number for each substance and the corresponding abstract number. As the Hill system is used (see Chapter 4), with compounds containing carbon, C comes first, followed by H if hydrogen is present. An example of an inorganic and an organic compound follow, both taken from volume 85:

$AlH_3O_8V_2$

Vanadoaluminic acid ($H_3AlV_2O_8$
   tricesium salt [60039-53-4], 71,469a

---

[14]See R. D. Nelson, W. E. Hensel, D. E. Baron, and A. J. Beach, *J. Chem. Inf. Comput. Sci.,* **15**, 85 (1975), for a discussion of computer editing of general subject headings for a volume index of *Chemical Abstracts.*

C₃H₇ClO₂S

1-Propanesulfonyl chloride [10147-36-1] , 5,139s, 77,014v, P 77,648e

The abstracts can then be consulted directly; or by means of the index name contained in the *Formula Index,* the user can also enter the *Chemical Substance Index* and find a description of the document's context, as well as references for chemical derivatives of the compound and related compounds which appear nearby in the alphabetical listing.

A special 27-year index includes references to molecular formulas reported from 1920 through 1946.

For the period 1967–1971 there was included the HAIC *(Hetero-Atom-in-Context)* Index.

Author Index

Principal authors, patentees, and patent assignees are listed in alphabetical order with titles of their articles or patents, along with abstract numbers. Names of co-authors are cross-referenced to the names of the principal authors, where full entries appear.

A number of details should be kept in mind. Specific items, with examples, are described in the Introduction to the author index of each volume, as well as the *Ninth Collective Index.* Especially to be noted are the following details: (1) several entries having the same surname but different given names (this includes corporate names) are ordered according to the initials of the first and middle names; (2) surnames with the prefix "Mc" and "M'" are alphabetized as though they began with "Mac"; and (3) "ae," "oe," "ue," and "oe," respectively, are used instead of the German umlauts ä, ö, ü, and the Danish and Norwegian ø (all other diacritical marks are eliminated).

There may be an occasional uncertainty with names, such as H. M. N. H. Irving, the English chemist. In his papers in the primary literature all four initials may be used, or one, two, or three of the four initials, or simply Harry Irving, may appear. In *Chemical Abstracts* all of these papers are brought together under one heading.

Index of Ring Systems

This index provides a means of determining the systematic names used in *Chemical Abstracts* for specific ring systems, as well as the nonsystematic index names for cyclic natural products containing these ring systems. The arrangement of this index is based on a hierarchy of ring data: number of rings, sizes of rings, and elemental analysis of rings. The index provides entry into the *Chemical Substance Index* to obtain references for substances whose structures include rings. An example of a simply ring entry follows:

6, 6, 6   C₄NSe, C₆, C₆
1*H*-Phenoselenazine
Phenoselenazin-5-ium

This entry designates a complex ring structure of three components, each of six members. The first is heterocyclic and contains four carbon atoms, one nitrogen atom, and one selenium atom. The other two are carbocyclic rings, each containing six carbon atoms. Parent compounds of this configuration will be found in the *Chemical Substance Index* under the names given.

Parent Compound Handbook

This compilation is a kind of index covering some 50,000 ring parents (unsubstituted ring systems) presently listed in the Chemical Registry System of Chemical Abstracts Service. Also included are details about some 2,000 cyclic and acyclic stereoparents, and 100 boron and iron cage parents.[15]

These parent compounds are specific molecular skeletons used as building blocks for index nomenclature in *Chemical Abstracts*. As such, they serve as index parent headings in the *Chemical Substance Index*. There are two working sections, as issued in 1976. Supplements update the file every 4 months.

First is the Parent Compound File which lists the ring systems. The arrangement is by "Identifiers," according to the following letter system:

BBBBB—BPZZY	Cage parents
BQBBR—BZZZP	Acyclic stereoparents
CBBBC—DZZZR	Cyclic stereoparents
FBBBF—ZZZZK	Ring parents

These identifiers enable a user to reach the entry in the file. The following example entry, having the identifiers KTXFD, illustrates the information elements included:

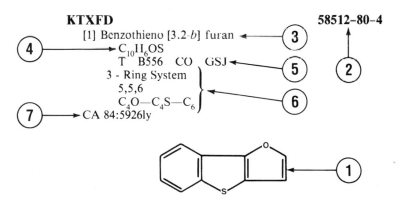

1 Chemical structural diagram illustrating the nomenclature locant numbering system

2 CAS Registry Number

---

[15]The Handbook replaces the "Ring Index" by A. M. Patterson, L. T. Capell, and D. F. Walker, American Chemical Society, 1960. See J. E. Blake et al., *J. Chem. Inf. Comput. Sci.*, **20**, 162 (1980), for development and usefulness of the Handbook.

3 Current CA Index Name used in the Chemical Substance Index to CA

4 Molecular formula

5 Wiswesser Line Notation (for ring parents, cyclic stereoparents, and some acyclic stereoparents)

6 Ring analysis data: number of rings, size of rings, and elemental analysis of rings (for ring parents and cyclic stereoparents)

7 CA reference (if the parent compound was referenced in a CA abstract after volume 78–1973)

If one knows only the molecular formula, $C_{10}H_6OS$, one must know the specific compound sought, as there are 20 other compounds listed under this formula in the *Formula Index.*

The second section of the Handbook consists of a set of seven indexes. There is the *Ring Analysis Index, Ring Substructure Index, Parent Name Index, Parent Formula Index, Wiswesser Notation Index, Parent Registry Number Index,* and *Stereoparent Index.* Thus, these indexes enable the user to find a parent compound in the *Parent Compound File* via the name, or portions of the name, ring analysis, ring structure (i.e., the sequence of carbon and the heteroatoms in the compound rings of the total ring system), Wiswesser line notation, molecular formula, and Registry Number.

The cumulate Index of Parent Compounds includes some 50,000 compounds through December, 1980.

Numerical Patent Index

The Numerical Patent Index, started in 1935, was used through 1980. This index now links specification number of each patent abstracted to its corresponding abstract number in *Chemical Abstracts.* In the index the primary sequence is the country of issue; then under each country, alphabetically arranged, are listed the patent numbers, in increasing numerical order, together with the corresponding abstract number for each.

A typical entry in Volume 87 is

Danish

Number	CA reference
133,369	151,848p
584	198,170b

As shown in the second patent number, only the last three digits are given when the earlier digits do not differ from those for the preceding number. The small letter at the end of the abstract number is a means for computer checking of the accuracy of the citation.

In most countries patents are numbered continuously from year to year. Thus, the United States patents have passed the 4 million mark. Patents in Japan, the Netherlands, and South Africa begin anew each year with number 1. The year is then included as part of the number. For these three countries the patent number, as it appears in *Chemical Abstracts,* is either seven or eight digits in length. The first two digits designate the year, with the last five or six digits forming the sequential portion of the specification number.

Patent Concordance

Equivalent patents, granted in different countries, are linked to one another and to the abstract number in *Chemical Abstracts* of the first patent received and abstracted. The Concordance, started in 1963, continued through 1980.

The following example shows that the British patent was the one abstracted:

Belgian

Patent number	Corresponding patent		CA Reference Number
794,757	Brit	1,408,127	80:98,977y
	Fr	2,215,489	
	Jap K	74,106,444	
	Neth	7,400,779	

Combined Patent Index

Beginning with volume 94 (1981), the new Patent Index replaces the Numerical Patent Index and the Patent Condordance. It provides extended identification of all chemical patents processed by Chemical Abstracts Service and indicates the priority relationships between patent documents in a patent family and the stages of patent examination represented by the documents. Entries for regional and international documents indicate the countries in which the documents are applicable.

This expanded index appears both in the weekly issues of *Chemical Abstracts* (but not in the Section Groupings) and as part of the semiannual volume indexes. It contains patent number entries for both (1) new patent documents abstracted in *Chemical Abstracts* during the period covered by the index and (2) patent documents processed during the period that are equivalent or related to patents abstracted earlier. Each time a new equivalent or related patent is referenced, a complete listing of all patent documents in the "family" of documents on the invention appears at the entry for the abstracted patent.

In the Introduction to the issue indexes the Illustrative Key shows the arrangement of details and the kind of information included in the combined index.

Collective General Indexes

These comprehensive indexes combine into single, organized listings, the content of the individual volume indexes for specified periods. Any corrections or additions which have been noted in the preceding volume indexes are incorporated into these cumulative indexes. Using collective compilations reduces the repetitive work that otherwise would be needed to make a volume-by-volume literature search.

Ten-year (decennial) collective indexes were published from 1907 to 1956. Since 1956, there have been 5-year (quinquennial) cumulations. Table VII lists the kinds of indexes included in the different collectives.

Registry Numbers

CAS Registry Numbers are unique identifying numbers for each distinctly definable chemical substance. The CAS Registry Numbers are used for linking the CA Index Name, the molecular formula, the associated names under which this chemical substance has appeared in the original literature, and the CA references in the data base. For example, the number [75-07-0] was assigned to

TABLE VII
COLLECTIVE INDEXES OF *CHEMICAL ABSTRACTS*

Collective index	1st	2d	3d	4th
Years	1907–1916	1917–1926	1927–1936	1937–1946
CA volumes	1–10	11–20	21–30	31–40
Author Index	X	X	X	X
Subject Index	X	X	X	X
Index Guide				
Chemical Substance Index				
General Subject Index				
Formula Index		*	*	*
Index of Ring Systems	‡	‡	‡	‡
Numerical Patent Index				*
Patent Concordance				
Patent Index				

*Two special Collective Indexes are available for searches of patents and formulas: The *Ten-Year Numerical Patent Index to Chemical Abstracts* (1937-1946) and the *Twenty-Seven-Year Collective Formula Index to Chemical Abstracts* (1920-1946).

‡For the 1st through the 6th Collective indexes, Ring System Information was included in the introduction to the Subject Index; for the 7th through th 11th Collective Indexes, the Index of Ring Systems was bound with the Formula Index.

the substance acetaldehyde. This CAS Registry System began with volume 62 (1965) and is complete to date.[16]

Increasing recognition of the utility of this numerical registration system is shown by its adoption by the *Journal of Organic Chemistry, Inorganic Chemistry, Angewandte Chemie,* the "Merck Index," the "United States Pharmacopeia," the "National Formulary," the TOXLINE and CHEMLINE computer services of the National Library of Medicine, and several Federal agencies, such as the Environmental Protection Agency and the Food and Drug Administration. An example of the use of such numbers by industry is the 1981 "Eastman Organic Chemicals Catalog."

Not included thus far are many substances indexed in *Chemical Abstracts* prior to 1965. Under way is a study of the feasibility of registering the content of the pre-1965 indexes.

Beginning in November, 1980, it became possible to search the CAS Chemical Registry System structure file on-line to identify substances that contain particular structural characteristics. A description of CAS On-Line and its features, as well as how to make arrangements for use of the service, are given by N. A. Farmer and M. P. O'Hara in *Database,* **3** (4), 10 (1980).

The "Registry Handbook—Number Section" (22 volumes) lists over 5 million substances registered from 1965 through 1980. The numbers run from [35-66-5]

[16]For the general design of the chemical registry system, see P. G. Dittmar, R. E. Stobaugh, and C. E. Watson, *J. Chem. Inf. Comput. Sci.,* **16,** 111 (1976).

5th	6th	7th	8th	9th	10th	11th
1947–1956	1957–1961	1962–1966	1967–1971	1972–1976	1977–1981	1982–1986
41–50	51–55	56–65	66–75	76–85	86–95	96–105
X	X	X	X	X	X	X
X	X	X	X			
			X	X	X	X
				X	X	X
				X	X	X
X	X	X	X	X	X	X
‡	‡	X	X	X	X	X
X	X	X	X	X		
		X	X	X		
					X	X

through [76081-79-3]. Listed in ascending order of registry numbers are the index name used in *Chemical Abstracts* and the corresponding molecular formula, if known. Thus, efficient access can be gained to the *Chemical Substance Index* and the *Formula Index*.

Annual supplements add some 350,000 chemical substances per year. Included is an update for all changes in numbers to date. The Introduction explains details of the system and contains advice on using the listings.

The most recent "Registry Handbook—Registry Number Update" should always be consulted before using the "Registry Handbook—Number Section" or supplements. The occurrence of a number in the handbook, or any supplements, should not be interpreted to mean that the number is currently valid. More recent information about the substance may have required a cross reference from the published Registry Number to a replacing Registry Number.

In August, 1980, announcement was made of the 5-millionth substance in the system. The assigned number is [74631-22-4], and the systematic name of the compound is 10-Hydroxy-4,5,6-trimethoxy-9H-azuleno[1,2,3-ij]isoquinolin-9-one. Compared to use of the systematic name, the advantage of using the Registry Number is obvious. The common name is grandirubrine. The original reference is *Heterocycles* **14**, 943 (1980).

The "Registry Handbook—Common Names" is another useful handbook.[17] A new edition, available only on microfilm or microfiche, contains an alphabetical listing of 965,749 common, trade, and semisystematic names for 612,988 unique chemical substances and provides the CAS Registry Number corresponding to each name. Also it lists the 391,457 substances in Registry Number order and provides the CA index name for each substance.

Chemical Abstracts Service *Source Index*

The CAS *Source Index* (CASSI) 1907–1979 contains information on bibliographic holdings for the scientific and technical primary literature sources relevant to the chemical sciences. The work is designed to aid a searcher in locating copies of the primary sources, i.e., journals, patent documents, technical reports, and edited conference proceedings cited by *Chemical Abstracts* since 1907. Included in the list are the journals cited in *Chemisches Zentralblatt* from 1830 to 1940, and by "Beilstein's Handbuch der Organischen Chemie" through 1965.

The information on holdings was compiled by 369 major resource libraries, 300 being in the United States and 69 in 28 other countries. Each entry includes the full title of the publication, the standard title abbreviation of the International Organization for Standardization, the CODEN, the ISSN or ISBN code, the title catalogued according to standard library practice, as well as the history, frequency of appearance, and language used. Entries for current publications contain the name and address of the publisher or other source from which the publication may be ordered. The arrangement is alphabetical by *abbreviated*

---

[17]See R. J. Rowlett, Jr., and D. W. Weisgerber, *J. Chem. Doc.*, **14**, 92 (1974).

titles. Many changes in titles are included. Supplements cover ongoing changes in the data.

The Introduction contains details: (1) Information on Entries, (2) Abbreviations, and (3) Availability of Listed Publications.

The cumulation for 1907–1974 included a list of the 1,000 primary journals cited most frequently in *Chemical Abstracts*.

The following two excerpts from CASSI illustrate the kind of information provided for each periodical covered. In this case, the journal is not widely held, and the name has been changed.

Informacion de Quimica Analitica. IPQAAZ ISSN 0367-777X [Suppl. to ION (Madrid)]. In Span., Eng. Span. sum. v. 1, 1947, v. 27 n 6 N/D 1973.
    INFORMACION DE QUIMICA ANALITICA, MADRID. changed to *Quim. Anal,* which see

DLC	1957 1959	Library of Congress, Washington, DC
InLP	1948–	Purdue University Library, West Lafayette, IN
MaKL	1966–	Linda Hall Library, Kansas City, MO
NN	1963–	New York Public Library, New York, NY
NNCC	1947–1959	Chemists' Club Library, New York, NY
Uk	1962–	British Library Science Reference Library, London, Eng.
GuGIC	1957–	Instituto de Nutricion de Centre America y Panama, Guatemala City
JTJ	1962–	Japan Information Center of Science and Technology, Tokyo, Japan
ItRC	1951–1959	Consiglio Nazionale Ricercha-Biblioteca, Rome, Italy
RuMG	1960–	Gosudarstvennaya Publichnaya Nauchno-Tekhnicheskaya Biblioteka SSSR, Moscow, USSR
JTNDL[18]		National Diet Library, Tokyo, Japan
SpBaU-SQ		Universidad de Barcelona, Barcelona, Spain[19]

Quimica Analitica Puna y Aplicada, QMANAT ISSN 0210-4334 (Formerly *Inf. Quim. Anal.*) In Span., Eng. Span. sum., v 28 n 1 Ja/F 1974+, bm 31 1977. Quimica Analitica, San Bernardo, 62, Madrid 8, Spain
    QUIMICA ANALITICA, MADRID
    Doc. Supplier: UMI
    InLP, MaKL, NN, Uk, GuGIC, JTJ, RuMG, SpBaU-SQ[20]

Publication Interrelationships

The chart in Fig. 5 shows the connections and interrelationships involved among *Chemical Abstracts* and the various kinds of indexes issued by Chemical Abstracts Service. The letters (A to F) represent six kinds of general areas involved in many searches in *Chemical Abstracts*. Each type of access term is

[18]No date indicates that the library holds a complete set.
[19]CASSI, p. 892.
[20]CASSI, p. 1816.

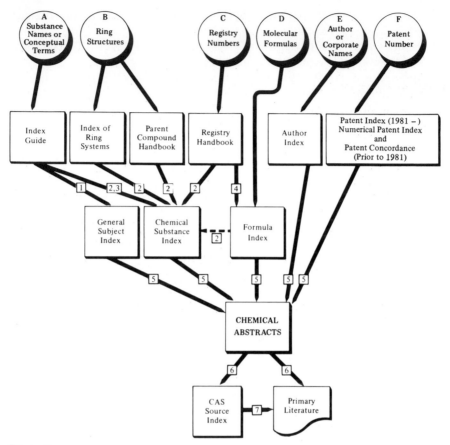

Figure 5
Chart showing the interrelationships of *Chemical Abstracts* and its indexes (used with permission).

connected to an appropriate index, which, in turn, leads to *Chemical Abstracts* itself. From there, if necessary, the searcher can go to the *Source Index* for information which leads to the primary source of the abstract.

The numbers (1 to 7) refer to the kind of information retrieved: (1) *General Subject Index* headings, (2) CA Index Names, (3) CAS Registry Number, (4) Molecular Formula, (5) CA Abstract Number, (6) bibliographic citation, and (7) library holdings.

CAS Printed Access Tools 1980

This work is a 250-page problem/solution workbook which shows various step-by-step approaches to searching the printed services of Chemical Abstracts Service (especially the volume indexes) to solve typical problems encountered by those who search the chemical literatures. Samples of actual pages are included from the publications consulted in solving the problems.

The workbook also contains background information about the printed services and other useful supplementary material which provides insight into editorial policies of Chemical Abstracts Service.

MICROFORM FILES OF *CHEMICAL ABSTRACTS*

Two microform files of *Chemical Abstracts* are available, one as microfiche and one as microfilm. Perhaps the most important characteristic of these two forms is the relatively small space required for storage. The disadvantage is the necessity of using a magnifying reader, and inconvenience in handling cards and film rolls is also possible. For example, an abstract in the middle of a film roll might refer to another possible reference in another roll. The researcher might have to refer to the second roll and then return to the first one.

Only brief mention is made here of these two files. Further information, including arrangements for a license to copy them, is available from Chemical Abstracts Service.

CA CHEMISSUES

This file of weekly issues of *Chemical Abstracts* is distributed on microfiche, 105 by 148 mm. Each microfiche has a human-readable header to indicate its contents. A microfiche reader-printer enables one to read, and copy if desired, individual abstracts.

The first ten collective indexes are available as microfiche and microfilm.

CA on Microfilm

A file of all of the abstracts (without indexes) published in *Chemical Abstracts* from 1907 to date is available on 16mm film. It is packaged in cassettes, each one containing some 30,000 abstracts. Updating is done by issuing an additional cassette whenever 30,000 abstracts or 3,000 abstract pages are published. Through volume 87 there are some 26 cassettes.

CA CHEMWEEKLY

This file of weekly issues of *Chemical Abstracts* is available on either microfiche or mircofilm and includes printed Volume Indexes.

COMPUTER-READABLE FILES

In addition to the printed and microform products of Chemical Abstracts Service, there is available a variety of information services in computer-readable form. The files are designed to transfer information to computer searching systems which search and manipulate large quantities of information for rapid retrieval. Chapter 15 is devoted to data bases and their use.

Individuals and organizations may use their own computers to search the files or may request one of the Information Centers licensed by Chemical Abstracts Service to perform the search. Further information regarding such services is available from Chemical Abstracts Service. The Marketing Department will provide, on request, a current list of more than 40 Information Centers. A license fee is charged for use of any of these files (see "CAS Information Tools 1981" for details).

Only brief attention is directed here to the nature of several of these services. Their use is considered in Chapter 15.

CA SEARCH

This biweekly service provides bibliographic information: CA section/subsection number, section cross references, keyword index entries, volume index entries, and CA abstract number for all documents abstracted in *Chemical Abstracts*.

Users can rapidly scan all new chemical and chemical engineering information to retrieve the portion pertaining to their fields of interest. If desired, the printed abstract or source document may be consulted. A yearly subscription includes 26 biweekly issues on magnetic tape.

As part of its user education program, Chemical Abstracts Service has made available a manual, "CA Search for Beginners." The information focuses on the content, structure, and format of the computer-readable file as it appears on-line. Also included is a discussion of CAS indexing policies used for both the controlled vocabulary and the natural language terms used to access the file.

Three kinds of information about documents abstracted in *Chemical Abstracts* are the basis of the file's flexibility: controlled vocabulary entries to Volume Indexes, natural language entries to Keyword Indexes, and bibliographic citations.

Further information on uses of the file, and on arrangements for licensing, are available from the Marketing Division of Chemical Abstracts Service.

CA Biblio File

This information service makes the bibliographic citations for documents abstracted in *Chemical Abstracts* available in a separate file.

REG CAN

This file links the unique CAS Registry Numbers for chemical substances with the corresponding abstract numbers for documents which contain information on the substances.

CAS On-Line

This new service for substructure searching makes available to the public direct access from remote terminals to chemical substances in the CAS Registry System.

Other Computer-Readable Files

A number of other printed access tools are available in computer-readable files. Included are *Chemical Titles, Index Guide, CAS Source Index,* and *Chemical Industry Notes.*

Document Delivery Service

CAS now provides copies of papers and patent documents cited in its publications and computer files. Copies of most documents cited in *Chemical Abstracts, Chemical Industry Notes,* and other CAS publications and computer files since 1972 are available, as well as copies of most Russian documents since 1960. CAS pays publishers' copy fees or, in cases where publishers' permissions to copy have not been given or where copyright status is not clear, CAS loans the original document from its library collection.

Additional information and order forms are available from CAS Customer Service, Box 3012, Columbus, OH 43210.

Educational Programs

CAS has an expanding education program. It is designed to educate users with the wealth of information contained in *Chemical Abstracts,* and with the means available to use the various services based on the CAS Information Base. All of this information is summarized in a helpful new 56-page booklet, "Information Tools 1981."

Advice on Searching

A new booklet of 23 pages, "SEARCHING CA," is designed to show quickly and simply how to search *Chemical Abstracts* for different kinds of questions. Copies are available for $1.00 each from SEARCHING CA Booklet, P.O. Box 3000, Hilliard, OH, 43026.

SECONDARY SOURCES—
BIBLIOGRAPHIES

Every investigation must begin with a bibliography and end with a better bibliography.

George Sarton

Inspection of any current issue of *Chemical Abstracts* will show the great number and variety of subjects for which reports of research are presented. It is likely that few, if any, of the subjects are new enough to refer to no background information, i.e., prior publications. Even Bartlett's brief report of the first compound of a noble gas listed three references.[1]

Making a thorough résumé of the current state of knowledge of the subject to be investigated is the first task of any serious research worker. The researcher should find or compile a partial or complete list of references relating to the subject. Such a list is known as a bibliography.

A carefully prepared bibliography enables one to determine what has already been reported and thus to build upon the knowledge accumulated by others. Without it, one may spend valuable time and effort rediscovering facts previously published.

On the importance of bibliographies in consulting practice, the director of one well-known laboratory wrote:

> The compilation of bibliographies forms a very important part of our work. When a new subject comes up for investigation, the library is requested to compile a list of references to the important literature on that subject. . . . Experience teaches that the

[1]N. Bartlett, *Proc. Chem. Soc.* (London), **1962,** 218.

chemist prefers personally to examine and digest the list on a subject under investigation, rather than to have someone else do it for him.

Nature and Kinds of Bibliographies

Although a bibliography is essentially a list of references, such lists differ enough in their characteristics to warrant considering some of the details involved.

Scope

Bibliographies may be either partial or complete (or as nearly so as it is possible to make them). In the latter case, everything found relating to the subject is included, while in the former case some limitation is set in the compilation, such as language, country, journals, books, patents, authors, time covered, uses of material, preparation of substances, or other similar schemes.

Information Included

In the usual bibliography, part or all of the following information is given for each entry: author's name; title of publication; some statement, in addition to the title, indicating the exact location in the publication of the material cited (the last includes for a book, the volume, page, and year, or number of the edition; for a patent, the name of the country issuing the patent, the number, and the date; for a public document, the name of the division issuing the document, together with proper designation, number, and date; and for a journal article, the title of the article, series, volume, page, and year); and an annotation indicating the nature of the material to be found in the source to which reference is made.

If the bibliography is a separate publication, preferably all the above information should be included. If it is part of a book, often the author's name, together with some statement regarding the contents of the cited publication, occurs in the body of the discussion. This is the general practice in the annual reviews published in *Analytical Chemistry*. An annotation for each entry is the exception outside separately published bibliographies.

Regardless of the part of the above data included for a given entry, whatever is included is usually designated as a reference. In any case, the data should indicate unmistakably the essential points of those mentioned, such as volumes and pages; but, at the same time, they should be as brief as possible in the interest of efficiency in reading, proofreading, writing, and preventing error. Unfortunately, the literature of chemistry abounds in annoying examples of digression from this general principle. As instances of this kind, one need mention only such things as the use of Roman numerals, the inclusion of the issue number for a continuously paged journal, the omission of the year, or more serious, the omission of the series number. It is a matter of confusion in the older literature and a lack of uniformity in present practice.

As an example of what happens in scientific publications, two citations are quoted from an article[2] calling attention to some desirable reforms. These two citations refer to the same article but appeared in different periodicals: (1) *Ann. Appl. Biol.,* (**24**(1923). No. 2. pp. 151–193, pls. 3, figs. 31); and (2) *Ann. Appl. Biol.* **24**:151–193, 3 pl. 31 fig. 1923. The second contains 16 fewer characters without sacrificing anything in essential information.

The following examples are representative forms, respectively, for (1) periodicals (see Chapter 2 for variations found), (2) technical reports, (3) patent documents, and (4) books (including doctoral dissertations and manufacturers' technical publications).

1 T. W. Carr, *Anal. Chem.,* **49,** 828 (1977); J. M. Hunt, *Adv. Org. Geochem.,* **1973,** 593; M. Palazzi *Ann. Chim. Paris* [15], **3,** 47 (1978).
2 W. J. Campbell and D. L. Neylan, *U.S. Bur. Mines Rep. Invest.,* **R. I. 7773** (1973); Rohm and Haas Company, "Amberlite XAD-4," Techn. Bull. (1974).
3 R. J. Perry, U.S. Patent 3,736,307, May 29, 1973.
4 J. A. N. Friend, "Textbook of Inorganic Chemistry," **9,** 381–389, Charles Griffin Co., Ltd., London, 1920; N. A. Lange, "Handbook of Chemistry," McGraw-Hill Book Company, New York, 1961, p. 33; American Society for Testing and Materials, "1977 Annual Book of Standards," Part 24, p. 770, Method D 2887-73.

Citations to some multivolume treatises are awkward, such as $VIII_{1a\alpha}$ for a particular section of the work by W. Fresenius and G. Jander. It involves a Roman numeral, with a triple subscript consisting of the Arabic numeral 1, and the lower case letters, Roman a and Greek α.

Published bibliographies exhibit some editorial variation. A bibliography prepared for a specific publication should conform to the practice of the editor.

Arrangement

Defining a bibliography, essentially only a list of references, implies no specification regarding the arrangement of the separate entries. Although a list not systematically arranged is preferable to none at all, the advantages of a definite order are well worth the extra time and effort required to put the references in this form.[3] An examination of the bibliographies published each year reveals variation in their usefulness owing to the kind of data included and/or the arrangement used. An outstanding example of a poor publication is a bibliography of over 400 pages, including many hundreds of entries, published by a national society. Because of its alphabetic arrangement by author, without subject index, very much of its possible usefulness is practically lost.

For most bibliographies, usually one of the following schemes is followed in listing references:

1 *Sequential order.* The simplest arrangement is to list the references in the sequence in which they are mentioned or found. Although this order is

[2]E. D. Merrill, *Science,* **62,** 419 (1925).
[3]J. L. St. John, *Science,* **70,** 217 (1930).

rarely, if ever, used in separately published bibliographies, it is fairly common in some important treatises[4] and in many scientific papers and dissertations. One may have to read the accompanying discussion in order to determine the significance of a given citation. In a large bibliography considerable time might be needed to find a given reference in the list.

2 *Chronological order.* This scheme may be desirable, especially if time of publication of the contributions is an important point, as in patent searches. It serves also to indicate the historical development of a subject. References published in the same year may be given a serial number,[5] or they may be arranged alphabetically by authors.[6] In any chronologically arranged bibliography the subdivision by years should be prominent on the page, as by a center heading. The entries may also be made in reverse chronological order.

A good example of a bibliography of this kind is a manufacturers' technical bulletin issued as "L and N Bibliography of Polarographic Literature." The 2,208 references, for 1903 to 1950, are first arranged in chronological order from 1903 and then alphabetically by authors in each year. Each entry has a number. The author and subject indexes refer to these numbers.[7]

3 *Alphabetical order.* There are several possibilities for making alphabetical arrangements:

  a One may alphabetize by name or title of the publication containing the contribution. There are few bibliographies of this kind.

  b Items may be alphabetized by the names of the countries issuing the publications listed. Again, such arrangements are rare.[8]

  c The most common and useful alphabetical arrangement is by authors of the publications listed. Preferably the order should be that followed in the author indexes of *Chemical Abstracts.* For example, names beginning with "Mc" and "Mac" are handled in a particular way in this publication.

A simple unnumbered list is satisfactory for small numbers of references. However, when the number of entries is large, too much time is required to ascertain whether a reference to some specific point is included. For example, one has to read Branner's book[9] to determine whether china-marking ink is mentioned.

To make such an arrangement readily useful, the references should first be assigned successive numbers after alphabetization. Then a comprehensive subject index, based upon the material in each entry, should

[4]J. W. Mellor, "Treatise on Inorganic and Theoretical Chemistry," Longmans, Green & Co., Inc., New York, 1922–1937; R. Abegg et al., "Handbuch der anorganischen Chemie," S. Hirzel Verlag, Leipzig, 1905–1939.

[5]J. L. Howe, Bibliography of the Metals of the Platinum Group, *Smithsonian Inst. Publs.,* No. 1084 (1897).

[6]C. J. West, "Reading List on Vitreous Enameling of Iron and Steel," A. D. Little, Cambridge, MA, 1921.

[7]"Bibliography E-90," Leeds and Northrup, Philadelphia, 1950.

[8]The patent index of *Chemical Abstracts,* although not a bibliography, is arranged alphabetically by countries.

[9]J. C. Branner, "A Bibliography on Clays and the Ceramic Arts," American Ceramic Society, Columbus, OH, 1906.

be prepared. The numbers of all references pertaining to the subject should follow any index heading.[10] The subject indexes of *Chemical Abstracts* are superb examples illustrating the best practice of selecting index headings. It should be emphasized that these indexes are of subjects rather than of words or titles.

4 *Functional order.* References on many subjects may be arranged in some logical or functional order, which represents a kind of classification of the references in terms of the nature of the material they contain.[11] Thus, the literature for lithium might be arranged as follows: discovery (history), preparation, physical properties, chemical properties, and uses.[12]

During bibliographic work it may be convenient to arrange references according to the nature of the publications in which they occur, e.g., periodicals, public documents, patents (subdivided by countries), and books. Collections in large libraries are usually arranged in this way in separate quarters.

Some arrangement, such as alphabetical order by authors, is desirable within each subdivision. Each reference may be numbered and a subject index added.

In this kind of arrangement any card containing material belonging to more than one subdivision necessitates either as many cards as there are divisions referred to or cross references to accomplish the same end.

The following classification was used for a bibliography of references on water glass:[13]

Bibliography
Patent literature
History
General
Properties
Manufacture
Applications
  General and miscellaneous
  Agglutinants
    General and miscellaneous
    Abrasives
    Cements

[10]N. Van Patten, "Bibliography on Corrosion," N. Van Patten, Marblehead, MA, 1923; C. J. West and H. Gilman, "Organomagnesium Compounds in Synthetic Chemistry," National Research Council, Washington, DC, 1922.

[11]P. Borgstrom et al., "Bibliography of Organic Sulfur Compounds," American Petroleum Institute, Baltimore, MD, 1930; D. E. Cable, "1936 Bibliography of Rubber Literature," *Rubber Age,* New York, 1937.

[12]For example, the author's card bibliography on colorimetric methods of analysis is first divided into chemistry and instruments. Then each of these subjects is subdivided into some 100 sections. The division tabs serve as a kind of index which enables one to remove very quickly all references on a particular item.

[13]M. Schrero, "Water Glass," Carnegie Library, Pittsburgh, PA, 1922.

Detergents
  Analysis
Egg preservation
Glass and ceramics
Medicine and surgery
Paper
Structural materials
  General and miscellaneous
  Artifical stone
  Concrete
  Fireproofing-preservation of wood
  Paints and preservative coatings
  Textiles
Analysis

An arrangement devised by the author[14] for a bibliography covering methods used in determining the composition of amalgams illustrates the application of a different scheme to a special case.

For general, and more or less extensive, bibliographies, the most useful arrangements seem to be alphabetical order by authors and functional order, particularly if there are accompanying annotations—that is, critical statements of the content and value of the material contained in each publication listed. In locating, for example, the references dealing with the corrosion of copper alloys in a general bibliography on corrosion, one would turn in the former case directly to the index and look for the words "alloy" and "copper." In the latter case it would be necessary to locate the division dealing with copper alloys. Probably in most cases an individual using such a bibliography wants information only for special purposes and consequently is not interested in reading through even 50 articles to find that the valuable material is contained in some 3 or 4 which might have been immediately evident in a well-arranged bibliography. If the list has no such arrangement, researchers must examine all references given in order to select the desired ones or to assure themselves that none of value are included.

This simple example emphasizes the importance of having and using terms defined as precisely as possible. Careless usage and/or disregard of accepted definitions may result in a very defective computer-generated bibliography.

For example, the *analysis* of calcium means ascertainment of the minor components in the metal. In contrast, the *determination* of calcium implies the analysis of something else, such as milk, water, or rocks, for calcium. Because of rather general disregard of this distinction, a search of a computer file for reference on the analysis of calcium will likely yield a considerable number of

[14]M. G. Mellon, *Proc. Indiana Acad. Sci.*, **34**:157 (1925).

references, only a small fraction of which relate to the composition of calcium metal.

A different kind of bibliography is contained in N. Gaudenzi's "Guida Bibliographica Internazionale per il Chimico," Sansoni Edizioni Scientifiche, Milan, 1952. Although now dated, this work provides an interesting international perspective on books (Libri) and journals (Revista) dealing with chemistry and chemical industry. The works listed are in Italian, German, French, and English.

The classification consists of 33 subdivisions, numbers 17 to 31 following closely the divisional numbers of *Chemical Abstracts* around 1950. The introduction, including the classification outline, contains an English and a German version.

The alphabetical indexes include journal titles, authors, subjects, and publishers.

NONPRINT BIBLIOGRAPHIES

The previous discussion relates to bibliographies in printed form. They are numerous, widely dispersed, and found in most kinds of publications. Individual bibliographies range from a few to thousands of entries.

Most writers and researchers prepare bibliographies in nonprint form. These are open-end collections of references accumulated prior to starting, and usually continuing through, a writing or research project. Completion of the project may end the bibliography, as in many doctoral dissertations of graduate students.

In contrast, such a "working" bibliography may cover a general research program and be built up over a period of several decades. As an example, the author's primary research interest for some 35 years was the application of molecular spectrometric methods to the determination of various substances. Nearly all the spectrophotometric methods studied involved colored solutions. Unless the desired constituent, or something chemically equivalent to it, was itself colored, a chromogenic reagent was used to develop a colored system, the absorptivity of which was then measured.

As the work progressed, a bibliographic classification system was adopted for the references accumulated. Part I, Chemistry, comprised separate sections, one for each constituent of interest, with an appropriate index card. There were subheadings for the different reagents used to develop colors. Thus, under iron, there were references for at least a score of different reagents.

Part II, Physics, comprised sections for all of the different kinds of measuring instruments of interest. They included both filter photometers and spectrophotometers. The work began with visual instruments and continued through three decades of introduction of photoelectric instruments. There were separate index cards for each kind of instrument, including their basic components, such as light sources, filters, monochromators, detectors, and recorders.

With the bibliography thus classified and indexed for chemistry and physics, any individual reference found could be inserted in its appropriate place. The bibliography was readily checked to determine whether a given reference was already included. In case of an actual new one, an additional index card was added for it.

The biennial additions to this bibliography formed the basis for the reviews on light absorption spectrometry published in *Analytical Chemistry* for more than two decades. An additional section was included in them for references dealing with the measurement and specification of color.

BIBLIOGRAPHIC PRACTICE[15]

Mention has already been made of the items of information included in a citation to the literature. Preparation of a bibliography on a given subject involves searching for the references (see Chap. 14) and then arranging the information obtained in some usable form. Although the bibliography problem in Chapter 16 involves the selection of some systematic way of recording each reference, this aspect of the problem is discussed here, along with other aspects of completed bibliographies.

Occasionally someone insists that all details of a reference found should be copied into a bound notebook. For a bibliography of any size this practice is too cumbersome to follow. The alternative is to use a separate card or slip of paper for each reference. The compiler, of course, must guard against loss of individual cards. Two variations of the card technique are of interest.

Plain Cards

Many people use plain cards, convenient sizes being 3 by 5, 4 by 6, and 5 by 7 inches. Preferably the larger sizes at least should be ruled. Industrial libraries often use cards with some printing to indicate the space for entering certain parts of the bibliographical details. One arrangement is illustrated in Fig. 6. Research-

[15]See M. Schrero, *Spec. Lib.*, **30**:302 (1939), for advice on bibliographic technique.

Figure 6
One arrangement for a bibliography card.

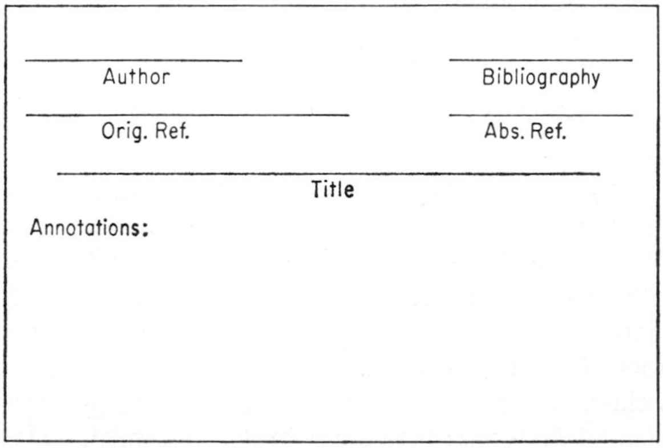

ers may prefer to put the different items in some other arrangement. Some searchers use different-colored cards for the different kinds of sources, such as periodicals, public documents, patents, and books. For a working bibliography, all the cards are then arranged in one of the ways described. Stiff separator cards with projecting tabs serve to indicate the different sections. This system is very adaptable in that the separator cards may be changed or added to at any time.

Punched Cards

A newer development is the use of punched cards, one for each reference cited. There are books describing the preparation, coding, and use of such cards.[16]

An example of a simple marginally punchable card is shown in Fig. 7. There may be one or two rows of the holes. Various sizes of cards are available. The printed numerals and letters may be designed for a variety of specific requirements.[17] The space within the border of holes serves for recording the bibliographic data.[18]

When the bibliography has progressed to the point that no further indexing subdivisions or facets seem likely to be found, a coding system is adopted. This necessitates laying out a classification or code, with each aspect or facet of the subject assigned to a given position or hole on the card. For example, in the bibliography on methods of chemical analysis, hole 1 of the top row of the card shown might be assigned to gravimetric methods of measurement. Then a special hand punch, similar to those used by train conductors, is used to notch away the edge of the card to the inside of this hole for every card involving a gravimetric method. Machine punches are also available. Other holes would be assigned to other aspects of the subject.

In order to select the cards on gravimetric methods from such a set, one first aligns them on edge, all in the same position, so that a long sorting needle resembling an ice pick can be pushed through hole 1 of the cards. Then by holding the needle and shaking the cards apart somewhat, all the cards clipped at hole 1 will drop out. If only gravimetric methods for nickel are wanted, these first fallout cards must be shaken again, this time with the needle in a hole which must be clipped out for nickel. If it is to be gravimetric methods for nickel in steel, a third shake is necessary, this time a shake of the nickel cards with the needle pushed through still another hole clipped out for steel. Thus, three clipped-out holes are necessary to sort for (1) method, (2) element, and (3)

---

[16]R. S. Casey, J. W. Perry, M. M. Berry, and A. Kent, "Punched Cards: Their Applications to Science and Industry," Reinhold Publishing Corporation, New York, 1958. See also section in R. E. Kirk and D. F. Othmer, "Encyclopedia of Chemical Technology," vol. 12, pp. 511–528, Interscience Publishers, Inc., New York, 1967.

[17]For a description of one system, see "Keysort Punching and Sorting Manual," Royal McBee Corporation, New York.

[18]See M. L. Soeder and R. W. Soeder, *Planta Med.*, **30**, 189 (1976), for a punched card system on flavonoid compounds.

Figure 7
Keysort card. (Reproduced with permission of the Royal McBee Corporation.)

material. The total number of holes determines the number of possible codable facets or aspects of the subject.

This simple example indicates that the coding system must be elaborately worked out before the holes can be clipped. The coding facets necessary are comparable to the division and subdivision tabs necessary for a classified card file.[19]

There are two chief advantages of punched cards over plain cards: (1) the cards never need to be in any particular order and (2) no duplicate or cross-reference cards are needed.

For a very large and/or many-faceted file such punched cards are not adequate. Too many cards become inconvenient for hand sorting, and only a limited number of holes are available for coding. Thus, it would be impossible to accommodate a general bibliography on chemical analysis covering all kinds of methods of separation and measurement for all elements, radicals, and compounds in all kinds of materials.

Cards of this type provided for International Business Machines (IBM) have had extensive use. They are designed for machine sorting. However, if enough space is reserved for punching the necessary holes, there is a problem of providing enough space for the desired bibliographic data.

Several extensive descriptions of punched card systems have been published.[20]

A coordinate indexing system employs Termatrex optical coincidence cards.[21] As an example, in the set for infrared absorption spectra,[22] each card represents a single characteristic of infrared curves, such as peak present or absent, element present or absent, and the usual functional groups present or absent. Holes drilled in a given card represent the reference compounds in the index having the characteristic represented by this card.

In use one selects the cards which represent the characteristics of the unknown curve and superimposes them over a light source. The points at which light shines through all of the superimposed cards represent the reference compounds which have the characteristics of the unknown.

Another example of this kind of card is the Keydex information retrieval system.[23]

SOURCES OF BIBLIOGRAPHIES

Unless one's problem is very specialized or of very recent interest, the chances are that somewhere there are more or less complete lists bearing directly or

---

[19]See A. G. Newcombe, *Science,* **140:**1312 (1963), for suggestions on increasing the efficiency of coding notched cards.

[20]Casey et al., op. cit.: M. Scheele, "Punch Card Methods in Research and Documentation; with Special Reference to Biology," Interscience Publishers, Inc., New York, 1961; R. R. Shaw (ed.), "The State of the Library Art," vol. 4, Rutgers University Press, New Brunswick, NJ, 1961.

[21]Jonkers Business Machines, Inc., Gaithersburg, MD.

[22]F. H. Dyke, *Develop. Appl. Spec.,* **2:**6 (1963).

[23]Royal McBee Corporation, New York.

indirectly upon the subject. Difficulty may be encountered, however, in finding them. They may appear either as separate publications or in connection with some other publication, such as treatises, encyclopedias, monographs, journal articles, or bulletins. Those appearing separately are usually either more or less general in nature or contain many entries, while the shorter ones appear in connection with a chapter in a book or at the end of an article.

Although bibliographies are widely scattered and in many cases can be found only after long searches, there are several publications available which are valuable aids in such work.

H. C. Bolton's "A Select Bibliography of Chemistry"[24] covers the titles of the principal books on chemistry published in America and Europe from the first appearance of chemical literature to the close of the year 1892. The term "chemistry" is taken in its fullest significance, and the bibliography contains books in every department of this early period. The bibliography is confined to independent works and their translations.

To facilitate reference, the work is divided into eight sections: bibliography; dictionaries; history; biography; chemistry, pure and applied; alchemy; periodicals; and academic dissertations. The first volume covers the field from 1492 to 1892, while the first supplement takes the literature on to 1897 and includes items omitted from the first volume. Out of about 18,000 titles appearing in these volumes, approximately 375 relate to bibliographies. A third volume comprises a list of academic dissertations printed independently between 1492 and 1897. The second supplement continues the work of the other volumes through 1902.

C. J. West and D. D. Berolzheimer's "Bibliography of Bibliographies on Chemistry and Chemical Technology"[25] is the most important bibliographic collection for the chemist and chemical engineer for the period covered. The main work covers 1900 to 1924, and two supplements go on through 1931. In its compilation about 100 periodicals and treatises were searched for lists of references on special topics. The material was assembled alphabetically by subject and classified as follows: bibliographies, abstract journals and yearbooks, general indexes of serials, bibliographies on special subjects, and personal bibliographies.

Bibliography of Reviews in Chemistry was issued for 1958–1963 by Chemical Abstracts Service. It lists all the abstracts of reviews, and others containing bibliographies, as published in *Chemical Abstracts*.

Two other works of general bibliographic interest are "A World Bibliography of Bibliographies"[26] and the "Bibliographic Index."[27]

A number of specific bibliographies have already been cited as examples of

---

[24]*Smithsonian Inst. Publs. Misc. Collections*, No. 1253 (1893–1904).

[25]National Research Council, Washington, 1925; two supplements, 1924–1931.

[26]T. Besterman, "A World Bibliography of Bibliographies and of Bibliographical Catalogues, Calendars, Abstracts, Digests, Indexes, and the Like," Societas Bibliographica, Geneva, 1955.

[27]H. W. Wilson Company, New York, cumulative vols., 1937–

particular kinds of bibliographic practices. The following additional examples may be of interest:

Beale, H. P. (ed.): "Bibliography of Plant Viruses," Columbia University Press, New York, 1976. Included are more than 29,000 references, dating from 1892 to 1970. There is a subject index.

Duveen, D. I., and H. S. Klickstein: "A Bibliography of the Works of A. L. Lavoisier," William Davenport and Sons, Ltd., London, 1954.

Gould, R. F. (ed.): *Adv. Chem. Ser.,* **30,** 1961 and *Adv. Chem. Ser.,* **78,** 1968. Many bibliographies are included.

Hawkins, D. T., L. S. Bernstein, W. E. Falconer, and W. Plemperer: "Binary Fluorides: Free Molecular Structure and Force Fields," 1959–1975, NSRDS Bibliographic Series, IFI/Plenum Data Company, New York, 1976.

———: "Physical and Chemical Properties of Water, "1956–1974, IFI/Plenum Data Company, New York, 1976. More than 3,600 references are included.

Hochstein, A. R.: "Bibliography of Chemical Kinetics and Collision Processes," IFI/Plenum Data Company, New York, 1969.

Lutz, G. J., R. J. Boreni, R. S. Maddock, and J. Wing: "Activation Analysis: A Bibliography through 1971," National Bureau of Standards, Washington, DC, NBS Technical Note 467.

Mavrodineanu, R.: "Bibliography on Flame Spectroscopy: Analytical Applications," 1800–1966, National Bureau of Standards, Washington, DC, NBS Miscellaneous Publication 281, 1967.

McLafferty, F. W., and John Pinzelik: "Index and Bibliography of Mass Spectrometry," 1963–1967, Interscience Publishers, New York, 1967.

Smith, M. F.: Two bibliographies with abstracts, "Gas Dynamic Lasers," Report No. NTIS/PS-78/0829/8GA; and "Neodymium YAG Lasers," Report No. NTIS/PS-78-0808/2GA.

Swann, S., Jr., and R. Alkire: "Bibliography of Electro-Organic Syntheses," 1801–1975, The Electrochemical Society, Inc., Princeton, NJ, 1980.

Touloukian, Y. S., J. K. Gerritsen, and N. Y. Moore (eds.): "Thermophysical Properties Research Literature Retrieval Guide," 3 v., IFI/Plenum Data Company, New York, 1967. Two supplements, with some change in editorship and each of six volumes, have been issued, one in 1973 and the other in 1979.

———: "Publication List (NSRDS) 1964–1977," National Bureau of Standards, Washington, DC. Included are more than 40 critical bibliographies, the majority being NBS Special Publications.

———: "Radiation Chemistry," Defense Documentation Center, Alexandria, VA, DDC/BIB-77/13, 1977.

———: "Coal Bibliography and Index," 2 v., Gulf Publishing Company, Houston, TX, 1980.

Computer-generated bibliographies are available on request from a number of sources, such as the IFI/Plenum Data Company and the American Society for Metals. The latter service is prepared to cover 66 subject areas.

Reference was made in Chapter 3 to bibliographical products of the National Technical Information Service. SRIM Category 99 (Selected Research in Microfiche) covers chemistry. The NTISearch bibliographies cover some 500 categories.

The U.S. Government Printing Office, Washington, DC, is a source of some 300 subject bibliographies. A complete list of subjects is available. Recent examples, with the subject bibliography number, follow:

SB 046   Air Pollution, 1978.
SB 050   Water Pollution and Water Resources, 1978.
SB 095   Solid Waste Management, 1977.
SB 151   Minerals and Mining, 1979.
SB 203   Industrial Water and Mine Drainage, 1978.
SB 227   Pesticides, Insecticides, Fungicides, and Rodenticides, 1979.

Each publication listed in these bibliographies gives the title, year, number of pages, price, and stock number of ordering. Many of the entries include annotations.

SECONDARY SOURCES— REFERENCE WORKS I. TABULAR COMPILATIONS, DICTIONARIES, ENCYCLOPEDIAS, AND FORMULARIES

Books are the masters who instruct us without rods and ferules, without hard words and anger, without clothes or money. If you approach them, they are not asleep; if investigating you interrogate them, they conceal nothing; if you mistake them, they never grumble; if you are ignorant, they cannot laugh at you. The library of wisdom, therefore, is more precious than all riches, and nothing that can be wished for is worthy to be compared with it. Whosoever therefore acknowledges himself to be a zealous follower of truth, of happiness, of wisdom, of science, or even of the faith, must of necessity make himself a lover of books.[1]

Richard de Bury

In a general sense, any published source to which one turns for information is a work of reference. Many primary and secondary publications contain citations to one or more prior publications. Frequently these cited works are designated as "References" or "Literature Cited." Such cited publications may include practically any kind of primary or secondary source. Most often cited are primary periodicals, technical reports, dissertations, reviews, patents, bibliographies, monographs, and textbooks.

Chapters 9 and 10 employ more restricted usage. First, only secondary sources are included, and second, these works are compilations of data selected and arranged according to some general plan.

Chemical facts, and their interpretations, usually appear first in primary sources. The first step in bringing the material from an unorganized to an

[1]From "Philobiblon," 1344, the first English book on the joys of reading.

organized state is taken in indexing and abstracting journals. The contribution of reviewing journals follows and is closely related to the other two. Likewise, bibliographies are more or less related to indexing journals.

Important as their work is, these organizing agencies do not bring the material into a readily usable form. They are the reapers, which collect the individual stalks of grain and tie them in bundles; but a thresher or separator is needed to bring out the kernels in a form suitable for ordinary consumption. The separators of chemical literature constitute the large majority of the so-called "books" on chemistry (not bound periodicals and bulletins).

These publications may be specialized and limited presentations related to some narrow phases of work; or the treatment may be sufficiently comprehensive and exhaustive to involve a whole field of chemistry. They are the works to which one turns when in need either of specific facts or detailed discussions encompassing relationships and general significance of facts.

According to J. K. Gates,[2] the term "reference book" has come to mean a specific kind of publication, to be consulted for items of information, rather than read throughout. It contains facts which have been brought together from various sources and organized for quick and easy use.

In the opinion of J. W. Mellor,[3] a work of reference should not only give the authorities for the statement of facts, but it should also indicate what knowledge has been gleaned on the particular subject in question. To do this in a practicable manner, attention must be directed to the original publications on the subject. This naturally makes the compilation extremely laborious; in some cases scores of independent references are involved in the statement of one particular fact.

These works touch practically every phase of chemistry and chemical technology. They vary widely in quality, dependability, usefulness, arrangement, comprehensiveness, and kind of information included. Their quality and dependability depend largely upon the perspective and discernment of the authors. As such works are not published frequently, the information in many cases is not up to date. The arrangement, comprehensiveness, and kind of information included are decided upon by the author; but the objective for which the book is produced is the guiding principle.

Works of reference, depending upon the general arrangement and manner of presenting the material, may be roughly divided into five groups: tabular compilations, dictionaries, encyclopedias, formularies, and treatises. Chapter 10 is devoted to treatises.

In considering each of these groups, the general plan of this book is followed, namely, describing the nature or characteristics of the group and then giving typical examples, or listing the more important or representative contributions. The latter point is carried a little further in this chapter in order to mention some of the works in special fields of chemistry.

---

[2]"Guide to the Use of Books and Libraries," McGraw-Hill Book Company, New York, 1979.

[3]"Treatise on Inorganic and Theoretical Chemistry," vol. 1, p. viii, Longmans, Green and Company, Inc., New York, 1922.

A. TABULAR COMPILATIONS

There are enormous numbers of physical constants which, when collected and properly arranged simply as statements of facts, are of frequent and wide use. This material is usually arranged in tabular or graphical form and includes such items as atomic weights, molecular weights, boiling and melting points, solubilities, absorption spectra, and many other constants.

There seem to be no distinctive points of difference among the various works of this type, except in the amount and kind of material included, and there is not much tendency to classify the material on the basis of fields of chemistry. The reason, of course, is obvious. The solubility of calcium citrate, for example, may be of just as much importance to the biochemist as to the physical chemist.

Some of the well-known works are described below. Statements are included to indicate the general nature of the contents and to direct attention to any special features.

Handbooks[4]

Various small works of this type contain the information thought to be of most general interest, such as physical constants for a limited number of organic and inorganic substances and other similar data in considerable variety. Some contain much more than physical data. Unfortunately, in those more definitely chemical in nature, different authors have selected different items, so that one may need several such works where a considerable range of facts is necessary. Thus, one book has an extensive classified list of books, another has much on physics and mathematics, another emphasizes metallurgy, and still another includes outlines of analytical methods. In most cases original references are not included. Some of the more common general works are listed below[5]:

Condon, E. U., and H. Odishaw: "Handbook of Physics," McGraw-Hill Book Company, New York, 1967.

Dean, J. A. (ed.): "Lange's Handbook of Chemistry," McGraw-Hill Book Company, New York, 1979.

Gordon, A. J., and R. A. Ford: "The Chemist's Companion," John Wiley & Sons, Inc., New York, 1972.

Gray, D. E. (ed.): "American Institute of Physics Handbook," McGraw-Hill Book Company, New York, 1972.

Helbing, W., and A. Burkart: "Chemical Tables for Laboratory and Industry," Halsted Press, New York, 1980.

Roses, A. J.: "The Practicing Scientist's Handbook," Van Nostrand Reinhold, Princeton, NJ, 1978.

Rossini, F. D.: "Fundamental Measures and Constants," CRC Press, Inc., Boca Raton, FL, 1974.

---

[4]For works in English the word "handbook" usually is used for the smaller collections of physical data. Frequently they are bound, single volumes. In contrast, the German "Handbuch" is applied to broad-range, comprehensive treatises, such as those of Beilstein and Gmelin (see Chapter 10).

[5]For an annotated bibliography of over 2,000 scientific and technical handbooks, see R. H. Powell's "Handbooks and Tables in Science and Technology," Oryx Press, Phoenix, AZ, 1979.

Weast, R. C. (ed.): "Handbook of Chemistry and Physics," CRC Press, Inc., Boca Raton, FL, annual.

As more and more data accumulate, and as the need arises, compilations are devoted to more limited areas. The following works are examples:

Albert, A., and E. P. Serjeant (eds.): "Ionization Constants of Acids and Bases," John Wiley & Sons, Inc., New York, 1962.

Barrell, B. G., and B. F. C. Clark: "Handbook of Nucleic Acid Sequences," Joynson Bruvvers, Oxford, England, 1975.

Bretherick, L.: "Handbook of Reactive Chemical Hazards," Butterworths, Woburn, MA, 1979.

Dayhoff, M. O. (ed.): "Atlas of Protein Sequence and Structure," National Biomedical Research Foundation, Silver Springs, MD, 1972.

Denisov, E. T.: "Liquid-Phase Reaction Rate Constants," IFI/Plenum Data Company, New York, 1974.

Dobos, D.: "Electrochemical Data," Elsevier Publishing Company, Amsterdam, 1975.

Horvath, A. L.: "Physical Properties of Inorganic Compounds," Edward Arnold, London, 1975.

Kertes, A. S.: "Solubility Data Series," Pergamon Press, New York, 1979– .

King, R. W., and J. Magid: "Industrial Hazard and Safety Handbook," Butterworth Publishers, Inc., Woburn, MA, 1979.

Klyne, W., and J. Buckingham: "Atlas of Stereochemistry," Oxford University Press, New York, 1978.

Lederer, C. M., and V. S. Shirley: "Table of Isotopes," Wiley-Interscience, New York, 1978.

Linke, W. F.: "Solubilities of Inorganic and Organic Compounds," Van Nostrand Reinhold, Princeton, NJ, 1958– .

Meites, L. (ed.): "Handbook of Analytical Chemistry," McGraw Hill Book Company, New York, 1963.

———, and P. Zuman (eds.): "Handbook of Electrochemistry," 4 v., CRC Press, Inc., Boca Raton, FL, 1977–1980.

Smith, R. M., and A. E. Martell (eds.): "Critical Stability Constants," vol. 1, Amino Acids; vol. 2, Amines; vol. 3, Other Organic Ligands; vol. 4, Inorganic Complexes, Plenum Publishing Company, New York, 1974–1977.

Steere, N. V.: "Handbook of Laboratory Safety," CRC Press, Inc., Boca Raton, FL, 1971.

Touloukian, Y. S., et al. (eds.): "Thermophysical Properties Research Data Series," IFI Plenum Press, New York, 1970– .

Spectral data are a striking example of extensive compilations limited to spectrometric techniques. The current diversity grew out of instrumental developments.

For many decades following Sir Issac Newton's demonstration of the solar spectrum, the word "spectrum" designated the range of wavelengths (frequencies) which give rise to the sensation of color in the normal human eye. As measuring techniques developed, the meaning of the term broadened to cover the whole range of wavelengths in the electromagnetic spectrum from cosmic rays to radio waves.

Further extension of coverage was the application to electrically charged masses (mass spectrometry). Other areas followed. Indexes to volume 88 of *Chemical Abstracts* lists, among others, the following entries under spectrometry for kinds of spectra[6]: absorption, atomic, Auger, electronic, emission, EQR, ESCA, ESR, line, mass, molecular, Mössbauer, rotational, vibrational, and x-ray fluorescence.

In addition, another kind of entry refers to the region covered. Thus, for the electromagnetic spectrum, one finds, in order of increasing wavelength, gamma, x-ray, ultraviolet, visible, infrared, microwave, and radio.

In all these cases spectra represent numerical data (often curves), or meter readings, of some kind.[7]

The following works are examples of some of the collections of spectral data. They range from single and multibound volumes to open-end, loose-leaf collections of many volumes:

Beynon, J. H., and A. E. Williams: "Mass and Abundance Tables for Use in Mass Spectrometry," Elsevier Publishing Company, New York, 1963.
———, R. A. Saunders, and A. E. Williams: "Table of Meta-Stable Transitions for Use in Mass Spectrometry," Elsevier Publishing Company, New York, 1965.
Briggs, D.: "Handbook of X-Ray and Ultraviolet Photoelectron Spectroscopy," Heyden and Son, Inc., Bellmawr, NJ, 1977.
Brügel, W.: "Handbook of NMR Spectral Parameters," Heyden and Son, Inc., Bellmawr, NJ, 1979.
Cole, A. R. H. (ed.): "Tables of Wavenumbers for the Calibration of Infrared Spectrometers," Pergamon Press, New York, 1976.
Cornu, A., and R. Massot: "Compilation of Mass Spectral Data," Heyden and Son, Inc., Bellmawr, NJ, 1975.
Craver, C. D. (ed.): "Evaluated Infrared Reference Spectra," 8 v. (14 parts), 8,000 spectra. Compiled by the Coblentz Society and published by the Sadtler Research Laboratories, Philadelphia, PA, 1969–
Dolphin, D. (ed.): "Tabulation of Infrared Spectral Data," John Wiley & Sons, Inc., New York, 1977.
Grasselli, J. G., and W. M. Ritchey: "Atlas of Spectral Data and Physical Constants for Organic Compounds," 6 v., CRC Press, Inc., Boca Raton, FL, 1975.
Greenwood, N. N., and E. J. F. Ross: "Index of Vibrational Spectra of Inorganic and Organometallic Compounds," Butterworths, London, 1972.
Harrison, G. R. (ed.): "Wavelength Tables," MIT Press, Cambridge, MA, 1969.
Kirschenbaum, D. M. (ed.): "Atlas of Protein Spectra in the Ultraviolet and Visible Regions," 2 v., IFI/Plenum Publishing Company, New York, 1972–1974.
Kuba, J., L. Kucera, F. Pizak, M. Doorak, and J. Mraz: "Coincidence Tables for Atomic Spectroscopy," CRC Press, Inc., Boca Raton, FL, 1965.
Lang, L. (ed.): "Absorption Spectra in the Ultraviolet and Visible Region," 19 v., Akadémiai Kiado, Budapest, 1959–   .
Miller, R.: "Infrared Structural Correlation Tables," 11 v., Heyden, London, 1964–1973.

[6]There is no heading for "spectroscopy" or "spectrum," but "spectra" appears.

[7]Unfortunately, misapplication of the word "spectrum" can cause confusion. For example, one advertisement advises broad-spectrum vitamin tablets, while another shows cuts for a spectrum of analytical instruments. Presumably in these cases what is meant is a range or variety of vitamins or instruments.

Nakamoto, K.: "Infrared and Raman Spectra of Inorganic and Coordination Compounds," John Wiley & Sons, Inc., New York, 1978.

Parsons, M. L., B. W. Smith, and G. E. Bentley: "Handbook of Flame Spectroscopy," Plenum Press, New York, 1975.

Phillips, J. P., et al. (eds.): "Organic Electronic Spectral Data," John Wiley & Sons, Inc., New York, 1960–   .

Pouchert, C. J.: "The Aldrich Library of Infrared Spectra," 1981, with J. M. Campbell, "The Aldrich Library of NMR Spectra," 1974, Aldrich Chemical Company, Milwaukee, WI.

Robinson, J. W. (ed.): "Handbook of Spectroscopy," 3 v., CRC Press, Inc., Boca Raton, FL, 1974–1980.

Safe, S., and O. Hutzinger: "Mass Spectrometry of Pesticides and Pollutants," CRC Press, Inc., Boca Raton, FL, 1978.

Stenhagen, E., S. Abrahamson, and F. W. McLafferty: "Atlas of Mass Spectral Data," 3 v., 1969, "Registry of Mass Spectral Data," 1974, John Wiley & Sons, Inc., New York.

Suchard, S. N. (ed.): "Spectroscopic Data," 2 v. in 3, IFI/Plenum Data Company, New York, 1975–1976.

————: "Sadtler Standard Spectra," Sadtler Research Laboratories, Philadelphia, PA. The Sadtler collections comprise extensive compilations for the infrared (prism and grating), ultraviolet, NMR, and carbon-13. The dates vary and the collecting continues. There are several different indexes.

The reliability of spectral data may vary considerably with the preparational, instrumental, and operational conditions maintained in their determination. The user should know that the conditions would yield results adequate for the particular purpose.

The chemical technologist and engineer may need information in fields closely related to chemistry, such as physics and engineering of several kinds. Thus, a chemical engineer may be the only individual with any general engineering knowledge in a chemically oriented plant.

Likely to be of use are the types of handbooks suggested below. They vary considerably in nature. Some are almost entirely tabular. Others contain much additional material. For example, the Perry and Chilton work, "Chemical Engineers' Handbook," has 26 sections covering mathematics, physical and chemical data, materials, processes, and operations. It duplicates a work such as Lange's only to a limited extent.

Chemical Engineering

Cremer, H. W., and T. Davies (eds.): "Chemical Engineering Practice," 17 v., Butterworths, Woburn, MA, 1956–1965.

Cheremisinoff, N. P., and P. N. Cheremisinoff: "Fiberglass-Reinforced Plastics Deskbook," Ann Arbor Science Publishers, Ann Arbor, MI, 1978.

Kent, J. A. (ed.): "Riegel's Handbook of Industrial Chemistry," Van Nostrand Reinhold Publishing Corporation, New York, 1974.

Pecsok, R. L., K. Chapman, and W. H. Ponder: "Chemical Technology Handbook," American Chemical Society, Washington, DC, 1977.

Perry, R. H.: "Engineering Manual: A Practical Reference of Design," McGraw-Hill Book Company, New York, 1976.

—— and C. H. Chilton (eds.): "Chemical Engineers' Handbook," McGraw-Hill Book Company, New York, 1980.

Schweitzer, P. A. (ed.): "Handbook of Separation Techniques for Chemical Engineers," McGraw-Hill Book Company, New York, 1980.

Stobaugh, R. B.: "Petrochemical Manufacturing and Marketing Guide," Gulf Publishing Company, Houston, TX, 1968.

Civil Engineering

Blake, L. S. (ed.): "Civil Engineer's Reference Book," Transatlantic Arts, Inc., Levittown, NY, 1975.

Jackson, N. (ed.): "Civil Engineer's Materials," Scholium International, Inc., New York, 1977.

Merritt, F.: "Standard Handbook for Civil Engineers," McGraw-Hill Book Company, New York, 1976.

Urquhart, L. C.: "Civil Engineering Handbook," McGraw-Hill Book Company, New York, 1959.

Electrical Engineering

Croft, T., C. Carr, and J. Watt: "American Electrician's Handbook," McGraw-Hill Book Company, New York, 1970.

Fink, D. G., and H. W. Beaty (eds.): "Standard Handbook for Electrical Engineers," McGraw-Hill Book Company, New York, 1978.

Hughes, L. E. C., and F. W. Holland: "Handbook of Electronic Engineering," CRC Press, Inc., Boca Raton, FL, 1977.

McPartland, J. F. (ed.): "National Electrical Code Handbook," McGraw-Hill Book Company, New York, 1978.

Say, M. G.: "Electrical Engineers Reference Book," Transatlantic Arts, Levittown, New York, 1975.

General Engineering

Boltz, R. E., and G. L. Tuve: "Handbook of Tables for Applied Engineering Science," CRC Press, Inc., Boca Raton, FL, 1973.

Brady, G. S., and H. R. Clauser: "Materials Handbook," McGraw-Hill Book Company, New York, 1977.

Braker, W., and A. L. Mossman: "Matheson Gas Data Book," Matheson Gas Products, East Rutherford, NJ, 1980.

Considine, D. M. (ed.): "Energy Handbook," McGraw-Hill Book Company, New York, 1977.

Souders, M., and O. W. Esbach: "Handbook of Engineering Fundamentals," John Wiley & Sons, Inc., New York, 1975.

Mechanical Engineering

Baumeister, T. (ed.): "Mark's Standard Handbook for Mechanical Engineers," McGraw-Hill Book Company, New York, 1978.

Boustead, I., and G. F. Hancock: "Handbook of Industrial Energy Analysis," John Wiley & Sons, Inc., New York, 1978.

Parrish, A. (ed.): "Mechanical Engineer's Reference Book," CRC Press, Inc., Boca Raton, FL, 1973.

Rohsenow, W., and J. P. Hartnett: "Handbook of Heat Transfer," McGraw-Hill Book Company, New York, 1971.

Metallurgical Engineering

Hampel, C. A.: "Rare Metals Handbook," Reinhold Publishing Company, New York, 1961.

Smithells, C. J.: "Metals Reference Book," Plenum Press, New York, 1975.

————: "Metals Handbook," American Society for Metals, Metals Park, OH, 1979.

————: "Unified Numbering System for Metals and Alloys," American Society for Testing and Materials, Philadelphia, PA, 1977.

Some of the manufacturers' technical publications mentioned in Chapter 5 contain, in addition to general technical information, engineering data on the selection, installation, and operation for the equipment described. Examples are air conditioning, pollution, refrigeration, vacuum drying, and heat transfer.

Comprehensive Works

The large, comprehensive works include, as far as possible, all the reliable data of this kind. They usually cite references to the original literature so that one may consult experimental details, if desired.

H. Landolt and R. Börnstein (eds.), "Zahlenwerke und Funktionen aus Physik, Chemie, Astronomie, Geophysik, und Technik," Springer-Verlag, Berlin (New York). This famous compilation of numerical data began in 1883, with the first edition in one volume. It was compiled by H. Landolt and R. Börnstein. For nearly 100 years, through six editions, it has been *the* collection of such data.

Begun in 1950, the sixth edition was completed in 1981, in four volumes of 28 parts. Throughout this period editors followed the original arrangement to fit the entire body of data into a rigid overall plan. The language has been German, but English titles are given in the sixth edition.

The general outline of the sixth edition follows:

I Atomic and Molecular Physics, 5 parts, 1950–1952
II Properties of Matter in its Aggregated State, 10 parts in 13
III Astronomy and Geophysics, 1 part, 1952
IV Technology, 4 parts in 10

In 1961 the "New Series" was started. The original plan, no longer considered feasible, was changed to a flexible scheme in which each volume is complete in itself. Supplements can be issued as needed. This plan now envisages 36 parts. The general outline follows, as of 1977[8]:

Group I   Nuclear and Particle Physics, 8 Volumes, in 10 Parts   1961–1974
Group II   Atomic and Molecular Physics, 10 Volumes, in 12 Parts   1965–
Group III   Crystal and Solid State Physics, 12 Volumes, in 22 Parts   1966–
Group IV   Macroscopic and Technical Properties of Matter, 3 Volumes, in 4 Parts   1974–
Group V   Geophysics and Space Research, 2 Volumes
Group VI   Astronomy, Astrophysics, and Space Research, 1 Volume   1965

In an accompanying rapid-reference chart, the volumes relevant to a particular combination of system-class and properties are entered in the boxes at the intersections of the rows (properties) and the columns (system and material classes).

E. W. Washburn, "International Critical Tables," McGraw-Hill Book Company, New York, 1926–1930. This set was the first comprehensive compilation of tabular data in the English language. The data selected came largely from the first seven volumes of the French annual tables (discussed below).

The program covered all available information of value for (1) pure substances, (2) mixtures of definite composition, (3) the important classes of industrial materials, (4) many natural materials and products, and (5) selected bodies or systems, such as the earth and its main physical subdivisions, the solar and stellar systems, and certain biological organisms, including people. No supplements have been issued. Lists of errata were published with each volume. An index volume enables one to locate easily the various types of data. Certain peculiarities of the index are explained in the introduction.

Specific data for a given system may or may not be easily found. In some cases a general formula has to be used to calculate the desired data: in other cases the difficulty comes in interpreting the tabular system used. Often it centers in employing the "key-number" formula for compounds. This "standard arrangement" of data in "A," "A–B," "B," and "C" tables, explained in volume I, page 96, and volume III, page viii, should be mastered by users of the set.

The principal explanatory text is in English, French, German, and Italian. Citations to the literature appear at the end of each section, where reference is then made to the list of publications at the end of each volume. An introductory paragraph explains the method of handling references. Volume VII gives a complete list of publications cited. It should be kept in mind that the references included are not likely to be later than 2 to 3 years preceding the date of publication of the individual volume consulted.

This compilation cannot be current later than 1930. It was intended to be a

---

[8]See "Outline 1977" for the titles of the separate parts.

continuing project, but no supplementary volumes were ever issued. A closed set of critical tables of the best data, at a given time, seemed no longer viable. Instead, isolated data compilation efforts were started and several notable collections published.[9]

In 1963 the National Standard Reference Data System (NSRDS) was established, with the National Bureau of Standards (NBS) to be responsible for overall planning and coordination, through the Office of Standard Reference Data (OSRD).

About a dozen centers have been established as sources of data. In general, each one is rather specialized. The "Publication List"[10] covers the period 1964–1977. Included are the reprints from the *Journal of Physical Chemical Reference Data,* a periodical started in 1972.

A monthly publication, *Reference Data Report,* contains news briefs on data activities (NBS). A special publication, NBS Technical Note 947, is entitled "Critical Evaluation of Data in the Physical Sciences." It is an NSRDS report, as of January, 1977.

International developments were reported by P. S. Glaeser (ed.), "The Proceedings of the Seventh Biennial International CODATA Conference," Pergamon Press, New York, 1981.

"IUPAC Chemical Data Series." Sponsored by the International Union of Pure and Applied Chemistry, this series complements the compilations of the National Standard Reference Data System. It is published currently by the Pergamon Press, New York.

From its beginning in 1974, some 25 volumes have appeared to 1980. In general, the data are very specialized, with applications to specific systems. Two examples of the series follow: "Ion Exchange Equilibrium Constants" and "Critical Survey of Stability Constants of EDTA Complexes."

The Center for Information and Numerical Data Analysis and Synthesis (CINDAS), sponsored by industry and located at Purdue University, is a specialized data center of particular interest to many engineers. It maintains a national data base on thermophysical, electronic, electrical, magnetic, and optical properties of materials. In addition to compiling such data, CINDAS disseminates the information through publication and direct technical and bibliographic inquiry services. Publication is by the Plenum Publishing Corporation.

The following three extensive compilations of data are available:

1 "Thermophysical Properties of Matter," 13 v. and "Master Index of Materials and Properties," 1970–1979. For the literature involved mention was made in Chapter 8 of the "Thermophysical Properties Research Literature Retrieval Guide," 12 v., including supplements.

---

[9]See D. R. Lide, Jr., and S. A. Rossmassler, *Ann. Rev. Phys. Chem.,* **24,** 135 (1973), for a brief account of the intervening developments between 1930 and 1963, and for a report of the status of critical compilations at that time.

[10]Available from NSRDS of NBS.

2 "Electronic Properties of Materials, a Guide to the Literature," 3 v., 1965–1971, with the accompanying "Electronic Properties Research Literature Retrieval Guide," 4 v., 1979.

3 "Handbook of Electronic Materials," 9 v., 1971–1972.

C. Marie (ed.), "Tables annuelles de constants et donnée numérique," Gauthier-Villars & Cie, Paris. The "Annual Tables" were started to bring together each year all the numerical data published in chemistry, physics, biology, and technology. Although the summaries have not appeared annually or as promptly as desirable, the results achieved by the international commission are probably all that could be done under the circumstances. The indexes and text for each table are given in both French and English. Two cumulative indexes (in French) cover the first 10 volumes. Since 1926 this set supplements "International Critical Tables."

The years covered by each volume follow: I (1910); II (1911); III (1912); IV (1913–1916); V (1917–1922); VI (1923–1924); VII (1925–1926); VIII (1927–1928); IX (1929); X (1930); XI (1931–1934); XII (1935–1936).

In 1947 publication was resumed, under different auspices, and with the title "Tables de constantes et données numériques. Constantes sélectionnées." The following single volumes have appeared in the new series:

1 Wavelengths of Emission and Discontinuities in Absorption of X-Rays (1947)

2 Nuclear Physics (1948)

3 Magnetic Rotatory Power, and Magneto-Optic Effect of Kerr (1951)

4 Spectroscopic Data for Diatomic Molecules (1951)

5 Atlas of Characteristic Wavelengths for Emission and Absorption Bands of Diatomic Molecules (1952)

6 Optical Rotatory Power I. Steroids (1956)

7 Diamagnetism and Paramagnetism; Paramagnetic Relaxation (1957)

8 Oxidation-Reduction Potentials (1958)

9 Optical Rotatory Power II. Tri-Terpenoids (1958)

10 Optical Rotatory Power III. Amino Acids (1959)

11 Optical Rotatory Power IV. Alkaloids (1959)

12 Semiconductors (1961)

13 Radiolytic Yields (1963)

14 Optical Rotatory Power: Steroids (1965)

15 Sesquiterpenoids (1966)

16 Metals: Thermal and Mechanical Data (1969)

17 Spectroscopic Data: Diatomic Molecules (1970)

There are other more specialized compilations, such as the following:

Braker, W., and A. L. Mossman (eds.): "The Matheson Unabridged Gas Data Book," 4 v., Matheson Gas Products, East Rutherford, NJ, 1974.

Hultgren, R., et al.: "Selected Values of Thermodynamic Properties of Metals and Alloys," John Wiley & Sons, Inc., New York, 1963.

Lonsdale, K., et al. (eds.): "International Tables for X-Ray Crystallography," 5 v., Kynoch Press, Birmingham, England, 1952–   .

Wyckoff, R. W. G.: "Crystal Structures," 6 v. in 7, Interscience Publishers, New York, 1963–1971.

B. DICTIONARIES AND ENCYCLOPEDIAS

Another class of reference works includes dictionaries and encyclopedias. These publications are grouped together since the words "dictionary" and "encyclopedia" have been used more or less synonymously in chemical publications.

In current dictionaries of the English language a dictionary is defined as a work containing the *words* belonging to some province of knowledge. It is concerned with the word itself. Thus, a chemical *dictionary* is one which includes a list of chemical terms with their definitions and usage, rather than one dealing with subjects arranged in alphabetical order.

An encyclopedia is a comprehensive summary of knowledge or a branch of knowledge. It deals primarily with subjects and is concerned with the thing the word represents. Then a chemical *encyclopedia* is composed of separate discussions of topics for the whole field of chemistry or a large, representative portion. The usual arrangement is alphabetical by subjects. The general scheme is to summarize for each topic the information that is considered to be of the most general value. The articles are historical, descriptive, explanatory, and statistical. References may or may not be given to sources where more extensive information is to be found.

With this distinction between dictionaries and encyclopedias in mind, an attempt has been made to separate them into two lists, as shown below. Some mistakes may have been made, since not all the works mentioned were available for examination.

Dictionaries[11]

Ballentyne, D. W. G., and D. R. Lovett: "A Dictionary of Named Effects and Laws in Chemistry, Physics, and Mathematics," Chapman and Hall, London, 1980.

Bennett, H. (ed.): "Concise Chemical and Technical Dictionary," Chemical Publishing Company, Inc., New York, 1974.

Challinor, J.: "A Dictionary of Geology," University of Wales Press, Cardiff, Wales, 1978.

Cooke, E. I., and W. I. Cooke: "Gardner's Chemical Synonyms and Trade Names," CRC Press, Inc., Boca Raton, FL, 1978.

Corrigan, E. L., J. D. Shoff, and M. T. Rasmussen (eds.): "APhA Drug Names," American Pharmaceutical Association, Washington, DC, 1979.

Crowley, E. T., and R. C. Thomas (eds.): "Acronyms and Initialisms Directory," Gale Research Company, Detroit, MI, 1973.

---

[11]The earliest chemical dictionary, "Lexicon Chymicum," by William Johnson, was published in two volumes in London in 1652.

Davis, C. W., and A. M. James: "A Dictionary of Electrochemistry," Macmillan Company, London, 1976.

Denny, R. C.: "A Dictionary of Spectroscopy," John Wiley & Sons, Inc., New York, 1973.

De Sola, R.: "Abbreviations Directory," American Elsevier Company, New York, 1973.

Ernst, R.: "Dictionary of Engineering and Technology," vol. I, English-German; vol. II, German-English, Oxford University Press, New York, 1978.

Gennaro, A. R., et al. (eds.): "Gould's Medical Dictionary," McGraw-Hill Book Company, New York, 1979.

Grant, J. (ed.): "Hackh's Chemical Dictionary," McGraw-Hill Book Company, New York, 1969.

Gray, H. J., and A. Isaacs (eds.): "New Dictionary of Physics," Longman, Inc., New York, 1975.

Hampel, C. A., and G. G. Hawley: "Glossary of Chemical Terms," Van Nostrand Reinhold Company, New York, 1976.

Hawley, G. G.: "The Condensed Chemical Dictioanry," Van Nostrand Reinhold Company, New York, 1981.

James, A. M.: "A Dictionary of Thermodynamics," Macmillan Company, London, 1976.

Kelly, K. L., and D. B. Judd: "The ISCC-NBS Method of Designating Colors and a Dictionary of Color Names," Circular 553, National Bureau of Standards, Washington, DC, 1955.

Lapedes, D. N. (ed.): "Dictionary of Scientific and Technical Terms," 1974; "Dictionary of Physics and Mathematics," 1978, McGraw-Hill Book Company, New York.

Maerz, A., and M. R. Paul: "A Dictionary of Color," McGraw-Hill Book Company, New York, 1950.

Marler, E. E. J.: "Pharmacological and Chemical Synonyms," Excerpta Medica, Amsterdam, 1978.

Neumüller, O. A.: "Römpp's Chemisches Wörterbuch," 6 v., Frank'sche Verlag, Stuttgart, Germany, 1971–1974.

Richter, F.: "Organic Name Reactions," John Wiley & Sons, Inc., New York, 1964.

Spencer, D. D.: "Acronyms Handbook," Prentice-Hall, Inc., Englewood Cliffs, NJ, 1974.

Stenesh, J.: "Dictionary of Biochemistry," John Wiley & Sons, Inc., New York, 1975.

Titus, J. B.: "Trade Designations of Plastics and Related Materials," National Technical Information Service, Springfield, VA, 1978.

————: "EJC Thesaurus of Engineering Terms," Engineers Joint Council, New York, 1967.

————: "Paint/Coatings Dictionary," Federation of Societies for Coatings Technology, Philadelphia, 1978.

Encyclopedias

The compilations listed here are examples of such works.

Bard, A. J., and H. Lund (eds.): "Encyclopedia of Electrochemistry of the Elements," vols. I–X, Inorganic; vol. XI, Organic, Marcel Dekker, New York, 1973–1979.

Brandrup, J., and E. H. Immergat (eds.): "Polymer Handbook," John Wiley & Sons, Inc., New York, 1975.

Clason, W. E.: "Dictionary of Chemical Engineering," 2 v., Elsevier Publishing Company, New York, 1969.

Clauser, H. R.: "Encyclopedia of Engineering Materials, Parts, and Finishes," Reinhold Publishing Company, New York, 1976.

Collocott, T. C., and A. B. Dobson: "Chamber's Dictionary of Science and Technology," 2 v., W. and R. Chambers, Edinburgh, Scotland, 1978.

Considine, D. M.: "Chemical Process and Technology Encyclopedia," McGraw-Hill Book Company, New York, 1974.

—— (ed.): "Van Nostrand's Scientific Encyclopedia," Van Nostrand Reinhold, Princeton, NJ, 1976.

Glasby, J. S.: "Encyclopedia of the Alkaloids," 3 v., Plenum Press, New York, 1976–1977.

Gray, P. (ed.): "The Encyclopedia of the Biological Sciences," Reinhold Publishing Company, New York, 1970.

Hampel, C. A., and G. G. Hawley: "The Encyclopedia of Chemistry," Van Nostrand Reinhold, Princeton, NJ, 1973.

Hey, D. H. (ed.): "Kingzett's Chemical Encyclopedia," Van Nostrand Reinhold, Princeton, NJ, 1966.

Lapedes, D. N. (ed.): "Encyclopedia of Environmental Science," 1974; "Encyclopedia of Energy," 1981; "Encyclopedia of Science and Technology," 15 v., 1977; "Encyclopedia of the Geological Sciences," 1978, McGraw-Hill Book Company, New York.

Lerner, R. G., and G. L. Trigg (eds.): "Encyclopedia of Physics," Addison Wesley, Reading, MA, 1980.

Leung, A. Y.: "Encyclopedia of Common Natural Ingredients Used in Food, Drugs, and Cosmetics," John Wiley & Sons, Inc., New York, 1980.

Mark, H. F., N. G. Gaylord, and N. M. Bikales (eds.): 15+ v., "Encyclopedia of Polymer Science and Technology," Interscience Publishers, New York, 1964–1972.

Morvan, R. G. (ed.): "Encyclopédie Internationale des Sciences et des Techniques," Les Presses de la Cité, Paris, 1969–1975.

Pecsok, R. L., K. Chapman, and W. H. Ponder (eds.): "Chemical Technicians' Handbook," American Chemical Society, Washington, DC, 1975.

Robinson, J. W. (ed.): "Handbook of Spectroscopy," CRC Press, Inc., Boca Raton, FL, 1974.

Sax, N. I.: "Dangerous Properties of Industrial Materials," Reinhold Publishing Company, New York, 1979.

——: "Cancer Causing Chemicals," Van Nostrand Reinhold Company, NY, 1981.

Shugar, G. J., R. A. Shugar, L. Bauman, and R. S. Bauman: "Chemical Technicians' Ready Reference Handbook," McGraw-Hill Book Company, New York, 1981.

Sittig, M.: "Organic Chemical Process Encyclopedia," Noyes Development Corporation, Park Ridge, NJ, 1969.

Thewlis, J. et al. (eds.): "Encyclopedia Dictionary of Physics," 9 v. and suppls., Pergamon Press, New York, 1961–   .

Uhlein, E., and O. A. Neumüller: "Römpp's Chemie-Lexikon," 6 v., Frank'sche Verlag, Stuttgart, Germany, 1972–1977; there is a two-volume condensation by O. A. Neumüller, 1977.

Williams, R. J., and E. M. Lansford, Jr.: "The Encyclopedia of Biochemistry," Reinhold Publishing Company, New York, 1977.

Windholz, M. (ed.): "The Merck Index," Merck and Company, Rahway, NJ 1976; a companion work is "Table of Molecular Weights," 1979.

Zweig, G.: "Modern Plastics Encyclopedia," McGraw-Hill Book Company, New York, 1941–   .

———: "Guide for Safety in the Chemical Laboratory," (Manufacturing Chemists Association) Van Nostrand Reinhold Company, NY, 1972.

———: "Safety in Academic Chemical Laboratories," American Chemical Society, Washington, DC, 1974.

———: "Fire Protection Guide on Hazardous Materials," National Fire Protection Association, Boston, 1975.

Zweig, G., and J. Sherma: "Handbook of Chromatography," 2 v., CRC Press, Inc., Boca Raton, FL, 1972.

Because of their technological importance and comprehensiveness, three multivolume encyclopedias are listed in Chapter 10 under Industrial Chemistry and Chemical Engineering. They are the Kirk-Othmer "Encyclopedia of Chemical Technology," McKetta and Cunningham's "Encyclopedia of Chemical Processing and Design," and Ullmann's "Encyklopädie der technischen Chemie."

A number of older encyclopedias, important in their day, are listed in the third edition of this book, page 146. They are chiefly of historical interest now.

In addition to these encyclopedias dealing with chemistry, general encyclopedias, such as the "Encyclopaedia Britannica," frequently contain valuable discussions of chemical subjects.

The works known as "almanacs" provide another source of information of possible chemical value. These publications contain a wide variety of facts in tabular and graphic form. Since they often appear annually, they serve as supplements to the large encyclopedias. The following selection is representative. The year appearing at the left indicates the date of the first appearance of the work.

1868	"World Almanac," *New York World Telegram and Sun*, New York.
1869	"Whitaker's Almanack," Whitaker, London.
1908	"New International Yearbook," Dodd, Mead & Company, Inc., New York.
1910	"American Yearbook," Appleton-Century-Crofts, Inc., New York.
1923	"American Annual," Encyclopedia Americana, New York.
1947	"Information Please Almanac," Doubleday & Company, Inc., New York.

C. FORMULARIES

As a result of the work of practical experimenters and manufacturing chemists, there has been accumulated an extensive collection of formulas and recipes consisting of directions for performing given reactions or making certain products. In general, such information is intended to be distinctly practical in nature, with emphasis upon commercial methods. In the following representative publications may be found a large variety of recipes:

Ash, M., and I. Ash: "Formulary of Paints and other Coatings," Chemical Publishing Company, Inc., New York, 1978.

Bennett, H. (ed.): "The Chemical Formulary," 1933–   ; "New Cosmetic Formulary," 1970; "Chemical Specialities," 1978, Chemical Publishing Company, Inc., New York.

Chalmers, L., and P. Bathe: "Household and Industrial Specialties," Chemical Publishing Company, Inc., New York, 1978.

Freeman, M.: "Practical and Industrial Formulary," Chemical Publishing Company, Inc., New York, 1962.

Goldschmidt, H.: "Practical Formulas for Hobby and Profit," Chemical Publishing Company, Inc., New York, 1973.

————: "National Formulary," American Pharmaceutical Association, Washington, DC, 1975 and suppl.

————: "American Hospital Formulary Service," American Society of Hospital Pharmacists, Washington, DC, loose-leaf compilation.

SECONDARY SOURCES— REFERENCE WORKS II. TREATISES AND MONOGRAPHS

If I have seen further . . . it is by standing upon the shoulders of Giants

Sir Isaac Newton

If one wants only some physical constant for a given element, compound, or system, generally the required data may be found in a tabular compilation, such as those described in Chapter 9. In many cases, however, it may be necessary to know about the occurrence of an element or compound, methods of preparation and purification, chemical properties of systems, or other facts.

Information of this kind might be found in some textbook, if it is some simple fact about a common element, in a chemical dictionary, or in an encyclopedia. But, in general, the best place to look for such facts is in the works designated here as treatises.[1] In them one finds facts, together with discussions of the facts. If the discussion is critical, in the sense that an opinion is given regarding the merit of the material presented, one has the highest type of treatise. Whether critical or not, there is generally an extensive presentation of the material of a given field, based upon some general outline, such as groups of elements or classes of compounds. The more important original references are included.

In these publications the usual practice is to limit the material to a broad field of chemistry, such as biological, inorganic, or organic.

H. Kayser's treatise, "Handbuch der Spektroskopie," 6 v., S. Hirzel, Leipzig,

---

[1]Some authors use the word "compendium" or "handbook" ("Handbuch" in German).

1900–1913, is an example of limiting the material to the theory and application of a single instrument in chemistry and physics.

Anyone unfamiliar with a treatise should examine the introductory section for any peculiarities of the set, such as the system of arrangement, abbreviations of titles of periodicals and words, details of the table of contents, and any limitations regarding the information included.

The great treatises of Beilstein and of Gmelin, along with lesser organic and inorganic works, respectively, comprise the fundamental chemistry of the chemical elements and their compounds. That is, they represent our knowledge of what is basic to the periodic table.

What chemistry, then, makes up the various subdivisions, such as analytical, biological, ceramic, environmental, metallurgical, pharmaceutical, physical, and others? It must be organic and/or inorganic (including organometallic), if these areas include the chemistry of all the elements and all their compounds. Strictly, therefore, there is no ceramic or pharmaceutical chemistry, for example, if thereby one means a different kind of chemistry. In such cases we isolate from the organic and/or inorganic areas what is relevant to the area of interest.

Early projections of coverage in multivolume treatises may not be followed because of unforeseen developments.

Time lag is most disconcerting in these large works of reference.[2] Publication of such sets in up-to-date form is probably the most difficult task in the realm of chemical literature. Examination of the dates of publication of the volumes of the inorganic treatises described in this chapter reveal the present situation for any element that may be selected. When such sources are used, their dates of publication should be noted. Better yet is to know the literature closing date, as given for each part of the Gmelin set described later.

Some of the important treatises will now be considered. First they are divided according to well recognized fields of chemistry. Then in several of these divisions a distinction is drawn between the multivolume sets and those having only a few volumes.

ORGANIC CHEMISTRY

Attention is directed first to the works which deal with the chemistry of the compounds of carbon (excepting a few, such as limestone). There are several extensive treatises.

F. K. Beilstein (ed.), "Handbuch der Organischen Chemie." The outstanding treatise on organic chemistry[3] was originated by Friedrich Konrad Beilstein, a Russian chemist.[4] The first edition of two volumes, in five parts, was completed in 1882, after 20 years of work. The third edition, consisting of four volumes, four supplements, and an index, was published between 1892 and 1906. It was

[2] M. G. Mellon, *J. Chem. Educ.,* **10,** 284 (1933).
[3] See E. H. Huntress, *J. Chem. Educ.,* **15,** 303 (1938).
[4] For a biography, see *Chem. Unserer Zeit.,* **4,** 115 (1970).

the last one by Beilstein, and is the one to which reference is made so often in the Richter "Lexikon." Actually, Paul Jacobson edited the four supplements. There was a second edition of three volumes, 1886–1890.

Because of the difficulty of knowing the names of compounds used in the subject indexes, M. M. Richter designed one based on empirical formulas. His "Lexikon der Kohlenstoff-Verbindungen," published between 1910 and 1912, covered all organic compounds known to December 31, 1909. Some 144,150 are listed. Supplementing it for 1910–1921 is H. Stelzner's "Literatur-Register der Organischen Chemie."

This Richter formula index was important here, but more so in its later use for the collective indexes of *Chemisches Zentralblatt* from 1922 through 1954 (see Chapter 6).

In the Richter system[5] the compounds listed are arranged according to their molecular formula, water of crystallization being neglected. The details of arrangement, as described in the work, are as follows:

1 Formulas are divided into groups according to the number of carbon atoms present. All compounds containing one carbon atom are listed first—the "C_1-Gruppe."

2 Formulas of each group are divided into classes on the basis of the number of elements, in addition to carbon, contained in the compound. In the "C_1-Gruppe" are listed first all compounds containing one element other than carbon, the "C_1-Gruppe mit einem Element." Then follows the "C_1-Gruppe mit zwei Elementen," and so on through all compounds containing only one carbon atom.

3 Formulas are arranged in each group in alphabetical order, according to the elements present. The alphabet of the system, or the succession of the elements combined with carbon as determined by the frequency of occurrence, is not the ordinary alphabet, but one worked out by Richter. Its order is *C, H, O, N, Cl, Br, I, F, S, P,* and then the others in ordinary alphabetical order from *A* to *Z,* according to symbols.

4 Formulas are arranged finally according to the number of atoms of each element which, in addition to carbon, are contained in the compound. All compounds containing two carbons and two hydrogens are considered before those containing two carbons and three hydrogens.

Suppose one is looking for information on ethyl cyanide, C_2H_5CN. The formula is arranged in the form C_3H_5N. At the top of the pages are Arabic and Roman numerals, which serve as a guide. The former indicate the number of carbon atoms in the formulas listed on the pages, and the latter indicate the number of other elements combined with the carbon. In this case, one would turn to the page with "3 II" upon it and look for the formulas with five hydrogens and one nitrogen.

If there are isomers, they are all listed under the appropriate formula. Thus, for the entry $C_4H_9O_2N$ there are 27 isomers listed in the Richter index (to 1910)

---

[5]The Richter system does not provide for inorganic compounds.

and 44 in the Beilstein index (to 1920). The first five in the latter work are (German names used) 1-Nitro-butan, 2-Nitro-butan, 1-Nitro-2-methyl-propan, 2-Nitro-2-methyl-propan, and Saltpetrigsäure-butylester.

The fourth edition, edited initially by F. Richter, was issued by the Deutsche Chemische Gesellschaft until after World War II, when it was taken over by the Beilstein Institute.

Exclusive of the naturally occurring substances of unknown composition (Division IV), 27 volumes were required to cover the literature to 1910. They constitute the Hauptwerk. Fifteen supplementary volumes (Ergänzungswerk I) bring the literature for the first three divisions to 1920. The second supplement (Ergänzungswerk II), that is, to 1930, comprises 26 more volumes.

The third and fourth supplements (Ergänzungswerk III and IV), planned to cover the literature to 1950 and 1960, respectively, are still incomplete. In 1981, completion is projected in 17 volumes.

The outline following shows the total parts issued for each of the 27 volumes, together with Ergänzungswerk (E_3, E_4) not yet completed in 1981. A new General Register is planned with completion of Ergänzungswerk IV.

A fifth supplement (Ergänzungswerk V), to appear in the English language, will cover the period 1960–1979.

The general arrangement is based on kinds of compounds, such as hydrocarbons and other types. For the fourth edition the classification was modified so that 4,877 sections (Systemnummer) was intended to meet all requirements. An accompanying outline shows the distribution of the material in the different volumes, together with a general indication of the scheme of classification used.

Indexes, abbreviations, a classified table of contents, and corrections for previous volumes are given in each volume. References to the literature are in the body of the text. At the top of each page of a supplement is the page number of the main volume in which the same compounds are discussed (if known then).

Divisions

The compounds are arranged first in four main divisions on the basis of whether their structure is known, and, if so, how the carbon atoms are bound together.

Division I.   Acyclic Compounds (Nos. 1–449)
(Stem Nuclei[6] with Carbon Alone Bound in Chain Form)

  I Hydrocarbons and   —OH,   —C≡, and   =C(OH)—C=derivatives 12
  II Carboxylic acids   8
  III Carboxylic acid derivatives   6
  IV Sulfonic acids, amines, other N compounds, organometallic compounds   8

[6]"Stem nuclei" refer to the ultimate groups obtained by substituting hydrogen for all other elements attached to carbon, without disrupting any rings.

Division II.   Isocyclic Compounds (Nos. 450–2,358)
(Compounds with Carbon-to-Carbon Rings)

V Hydrocarbons   11

VI Hydroxy compounds, including alcohols, phenols, phenol alcohols
20

VII Carbonyl compounds, including aldehydes, ketones, quinones   7;
E_4

VIII Hydroxycarbonyl compounds   9; E_4

IX Carboxylic acids   8; E_4

X Hydroxycarboxylic acids   10; E_4

XI Other acids, including S, Se, Te derivatives   3; E_4

XII Monoamines   8; E_4

XIII Polyamines   5; E_4

XIV Carbonyl amines, amino acids   8; E_4

XV Hydroxylamines, hydrazines   3; E_4

XVI Other nitrogen compounds, organometallic compounds   5; E_4

Division III.   Heterocyclic Compounds (Nos. 2,359–4,720)
(Compounds with Other Elements Besides Carbon in the Ring)

XVII One cyclic oxygen (nuclei, —OH, —C= compounds, S, Se, Te
derivatives)   9

XVIII One cyclic oxygen (—C=, acidic, nitrogen, organometallic
compounds)   13

XIX Two (or more) cyclic oxygens   12

XX One cyclic nitrogen (nuclei)   8

XXI One cyclic nitrogen (—OH, —C= compounds)   9

XXII One cyclic nitrogen (other derivatives)   13

XXIII Two cyclic nitrogens (nuclei, —OH compounds)   7

XXIV Two cyclic nitrogens (—C= compounds)   2; E_3; E_4

XXV Two cyclic nitrogens (other compounds)   3; E_3; E_4

XXVI Three to eight cyclic nitrogens   2; E_3; E_4

XXVII Compounds with both cyclic oxygen and nitrogen atoms   3; E_3; E_4

XXVIII Subject index ⎱ Divisions I–III, including first and second supple-
XXIX Formula index ⎰ ments

Division IV.   Compounds of Uncertain Structure (Nos. 4,721–4,877)

XXX Caoutchouc, gutta-percha, balata, carotinoids (literature to 1934
E_3; E_4

XXXI Monosaccharides, oligosaccharides (literature to 1919) E_2; E_3; E_4

XXXII Polysaccharides (   )

XXXIII Alkaloids (   )

The two index volumes (XXVIII and XXIX) cover the main work and the
first supplement (to 1920). The formula index to this point uses the Richter

system.[7] Volume XXVIII of the second supplement is a collective index of the main work and the first two supplements. Likewise, volume XXIX of the second supplement is a collective formula index.[8] In contrast to the previous formula index, this one uses the American Hill system (see Chapters 4 and 7).

These index volumes list all organic compounds reported up to 1930. From 1930 on each volume of the supplements has its own subject and formula index.

Generally an inspection of the structural formula will indicate in which of the three divisions a compound of known structure belongs. In the more complicated and uncertain cases the formula may be broken down into the simpler stem nuclei by substituting hydrogen for all the other elements bound to carbon, without disrupting any ring. According to the principle of latest position, the compound will be found in the last division to which any of its nuclei belong.

Subdivisions

Only the heterocyclic division (III) has subdivisions. They are arranged in arbitrary sequence according to the kind and number of hetero atoms, such as oxygen and nitrogen, in the ring. Thus, compounds with oxygen in the ring come before those with nitrogen, and both of these before compounds containing oxygen and nitrogen in the ring.

Classes

Divisions I and II and the heterocyclic subdivisions are then arranged in classes on the basis of functioning groups.[9] Class 1 includes only stem-nuclei compounds, such as ethane, benzene, and pyridine. In each of the others the compound contains a group having at least one hydrogen atom replaceable by another substituent. The classes for volume IV, arranged in the order of discussion, are as follows:

1 Stem nuclei

2 Hydroxy compounds   —OH

3 Carbonyl compounds $=O$ or $\left(\begin{smallmatrix}\diagup OH\\ \diagdown OH\end{smallmatrix}\right)$

4 Carboxylic acids $=O(OH)$ or $\left(\begin{smallmatrix}\diagup OH\\ -OH\\ \diagdown OH\end{smallmatrix}\right)$

5 Sulfinic acids   —SO(OH)

---

[7]See F. L. Taylor, *Ind. Eng. Chem.*, **40**:470 (1948), for a proposed numerical index key for the Beilstein treatise.

[8]Formula indexes complete the fourth supplement of each volume.

[9]In a given division there may be fewer or more than 38 numbered classes. In some cases compounds containing all the functioning groups are not known. Then again a separate class number may be assigned each metallic element, which may make more than 38 actual classes. Since those included are numbered consecutively, a given type of compound cannot be assigned a class number unless the set is consulted. Volume IV has 38 classes, but vol. XVI, supplement 2, has 49.

6 Sulfonic acids	—$SO_2(OH)$
7 Seleninic and selenonic acids	—$SeO(OH)$ and —$SeO_2(OH)$
8 Amines	—NH_2
9 Hydroxylamines	—NHOH
10 Hydrazines	—$NH·NH_2$
11 Azo compounds	—N:NH
12–18 Other nitrogen compounds	
19–38 Organometallic compounds	

Under each class is included certain general information on nomenclature, properties, and derivatives.

Subclasses

In arranging compounds in subclasses the sequence is (1) decreasing saturation in stem nuclei, as C_nH_{2n+2}, C_nH_{2n}, . . . compounds; (2) the number of single characteristic functioning groups, as mono-, di-, tri- . . . compounds; and (3) the increasing number of different characteristic groups, as carbonyl, hydroxycarbonyl, amino-hydroxycarbonyl, . . . compounds.

Rubrics

Decreasing saturation is also the basis of arrangement within the subclasses, rubric 1 being the most highly saturated type. Each rubric may then be expressed in terms of a general formula, as $C_nH_{2n+2}O$ for monohydroxy alcohols.

Series

In each rubric group the compounds of the same degree of saturation are arranged in the order of increasing number of carbon atoms. The series designation, as C_3, shows the number of carbon atoms in the compound. The individual members of the series are the index or parent compounds.

For each index compound the available material is arranged as follows: (1) structure, configuration, historical; (2) occurrence, formation, preparation; (3) properties (color, crystallography, physical constants); (4) chemical properties; (5) physiological action; (6) uses; (7) analytical data (detection, examination, estimation); (8) addition compounds and salts; and (9) derivatives.

Derivatives

Derivatives of index compounds require special consideration. There are three types.

1 *Functional Derivatives.* These are produced, in effect, by replacing the characteristic hydrogen of the functioning group by an organic or inorganic group (atom). The process involves the reaction of the index compound with some coupling compound, either organic or inorganic, with the elimination of water. Organic coupling compounds contain a hydroxyl group attached to carbon. Inorganic coupling compounds function through a hydroxyl group as in hydrogen peroxide and oxygenated acids, or a hydrogen directly attached to a halogen, nitrogen, or certain other elements in the periodic system. Such hydrides couple only with hydroxyl in the index compound.

2 *Nonfunctional Derivatives.* These are produced by substituting hydrogen of the stem nucleus by one or more of the nonfunctioning groups, —F, —Cl, —Br, —I, —NO, —NO_2, —N_3.

3 *Replacement Derivatives.* These are produced by replacing the oxygen of a functioning group by sulfur, selenium, or tellurium.

All these relationships for two index compounds, methyl and ethyl alcohol, are illustrated below.

Division I.   Acyclic Compounds

Subdivision	
Class 2	Hydroxy compounds
Subclass *A*	Monohydroxy compounds
Rubric 1	Monohydroxy alcohols ($C_nH_{2n+2}O$)
Series *A*	CH_4O—methyl alcohol
	a Structure, historical
	b Occurrence, formation, preparation
	c Physical properties
	d Chemical properties
	e Physiological behavior
	f Uses
	g Analytical methods
	h Addition compounds and salts
	i Derivatives
	a′ Functional
	a.″ Organic coupling
	b.″ Inorganic coupling
	b′ Substitution
	c′ Replacement
Series *B*	C_2H_6O—ethyl alcohol
Rubric 2	Monohydroxy alcohols ($C_nH_{2n}O$)
Subclass *B*	Dihydroxy compounds
Class 3	Carbonyl compounds

Finding a Compound

Ability to locate facts for a given compound is the test of one's knowledge of the set. Finding some compounds is a real puzzle, but familiarity with the set will yield results worth the effort required to master the scheme.

If the compound is listed in the Formula Index (volume XXIX), the problem is easy. However, this volume covers the literature only to 1930. Compounds not listed must be located otherwise until another index appears.

One approach is to find the index compound. As an example of a triple derivative (functional, nonfunctional, and replacement), 4-nitrothioanisole will serve. The formula is

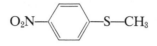

Sulfur in the position shown indicates a replacement derivative, and the nitro group shows a nonfunctional derivative. Replacing the S with O, and the NO_2 group with H, gives

Reversing the dehydration process producing such an ether

$$\langle\ \rangle\text{·O·CH}_3 + \text{H}_2\text{O} \rightarrow \langle\ \rangle\text{·OH} + \text{CH}_3\text{OH}$$

yields phenol, C_6H_5OH, and methanol, CH_3OH. As phenol is discussed in the treatise after methanol, phenol, a class 2, or hydroxy, compound, is the index compound under which to look.

Probably the best advice on the use of this great treatise is the brochure, "How to Use Beilstein," published by the Beilstein-Institut.[10] There are 34 pages of detailed instructions, including Table VIII, which shows the volumes containing the main kinds of compounds. Supplementing this brochure is the German-English "Beilstein Dictionary," which contains an alphabetical listing of the most frequently used German words, with their English equivalents. There is a helpful wall chart (3 by 4 feet), "Contents of Beilstein, Handbook of Organic Chemistry." It shows graphically the kinds of compounds in each of the 27 volumes.

More recent is an extensive explanation of the treatise by Reiner Luckenbach, Director of Beilstein-Institut. It is published in *CHEMTECH*, **9**, 612 (1979).

The following paragraph, reproduced with permission, is from Band I, Part 3, page 1244, of the fourth Ergänsungswerk. It shows the format used, with original references within the text.

Tetradeuteriomethanol CD₄O = CD₃OD (E III 1187)

B. Aus Kohlenmonoxid und Deuterium in Gegenwart eines Kupfer-Zink–Chromoxid-Katalysators (*Corval, Viallard*, Bl. **1954** 484; vgl. E III 1187). Aus C,C,C,-

---

[10]Springer-Verlag, New York.

Trideuteriomethanol und Deuteriumoxid bei 200° (*Venkateswarlu et al.*, J. Chem. Physics **23** (1955) 1195; Org. Synth. Isotopes **1958** 1339).

Potentialschwelle der inneren Rotation: *Swalen*, J. Chem. Physics **23** (1955) 1730. Trägheitsmomente des Moleküls: *Venkateswarlu, Gordy*, J. Chem. Physics **23** (1955) 1200.

Kp: 65.3-65.6 (*Co, Vi*, l.c. S. 485), n_D^{15}: 1.3280; n_D^{20}: 1.3260 (*Co, Vi*, l.c. S. 486). Mikrowellenabsorption bei 36674 MHz und 36799 MHz: *Ve.* et al.

Where this brief outline is inadequate, see the following references: (1) Introduction to the set in volume I; (2) B. Prager, D. Stern, and K. Ilberg, "System der Organischen Verbindungen," Springer-Verlag OGH, Berlin, 1929, a 246-page explanation listing the 4,877 types of compounds, common names, and type index; (3) O. Runquist, "A Programmed Guide to Beilstein's Handbuch," Burgess Publishing Co., Minneapolis, MN, 1966; (4) O. Weissbach, "The Beilstein Guide," Springer-Verlag, New York, 1976; (5) R. F. Brown, "Supplement to Brown's Organic Chemistry," pp. 210–224, Wadsworth Publishing Company, Belmont, CA, 1975; (6) R. Luckenbach, *J. Chem. Inf. Comput. Sci.*, **21**, 82 (1981).

The basic series and supplements I and II are available in two microform editions: (1) microfiche or microopaque cards; and (2) 16mm or 35mm microfilm.

E. Rodd (ed.), "Chemistry of Carbon Compounds," Elsevier Scientific Publishing Company, New York. In comprehensiveness this English work lies between the Beilstein treatise and reasonably comprehensive textbooks. For many purposes the compact coverage is adequate. The second edition is S. Coffey's revision of the original set. The supplements, edited by M. F. Ansell, are indicated by the second date of publication. The projected set follows:

Volume I—Aliphatic Compounds

IA General Introduction; Hydrocarbons; Halogen Derivatives (1964, 1975)

B Monohydric Alcohols, Their Ethers and Esters; Sulfur Analogues; Nitrogen Derivatives; Organometallic Compounds (1965, 1975)

C Monocarbonyl Derivatives of Aliphatic Compounds; Their Analogues and Derivatives (1965, 1973)

D Dihydric Alcohols; Their Oxidation Products and Derivatives (1965, 1973)

E Trihydric Alcohols; Their Oxidation Products and Derivatives (1976)

F Penta- and Higher Polyhydric Alcohols; Their Oxidation Products and Derivatives; Saccharides (1967)

G Tetrahydric Alcohols (1976); cumulative index.

Volume II—Alicyclic Compounds

A Monocarbocyclic Compounds to and Including Five-Ring Atoms (1967, 1974)

B Six and Higher Membered Monocarbocyclic Compounds (1968, 1974)

C Polycyclic Compounds, Excluding Steroids (1969, 1974)

TABLE VIII
CONTENTS OF THE 27 VOLUMES OF THE BEILSTEIN HANDBOOK

	Registry compound	Functional group	Beilstein volume number for main division							
			A (Acyclics)	**B (Isocyclics)**	**C (Heterocyclics)** Type and number of ring heteroatoms					
					10*)	20*), 30*),....	1N	2N	3N, 4N,....	1N, 10*; 1N, 20*; 2N, 10*; 2N, 20*; further heteroatoms†
①	Compounds without functional groups	–	1	5	17	19	20	23	26	27
②	Hydroxy-compounds	–OH		6						
③	Oxo-compounds	=O; =O + –OH		7						
④	Carboxylic acids	$\overset{=O}{-OH}$; $\overset{=O}{-OH}$ + –OH; $\overset{=O}{-OH}$ + =O; $\overset{=O}{-OH}$ + =O + –OH	2, 3	8, 9, 10	18		21	24		
⑤	Sulfinic acids	–SO$_2$H	4	11			22	25		
⑥	Sulfonic acids	–SO$_3$H								
⑦	Seleninic acids, selenonic acids, tellurinic acids	–SeO$_2$H and –SeO$_3$H, –TeO$_2$H								

⑧ Amines — $-NH_2$, $[-NH_2]_n$; $-NH_2 + -OH$; $-NH_2 + =C$; $-NH_2 + \overset{O}{\underset{OH}{\diagdown}}$; $-NH_2 + \cdots$ (12, 13, 14)

⑨ Hydroxylamines and dihydroxyamines — $-NH-OH$, $-N\overset{OH}{\underset{OH}{\diagdown}}$

⑩ Hydrazines — $-NH-NH_2$ (15)

⑪ Azo-compounds — $-N=NH$

⑫ Diazonium compounds — $-N\equiv N]^{\oplus}$

⑬ Compounds with groups of 3 or more N-atoms — $-NH-NH-NH_2$, $-N(NH_2)_2$, $-N=N-NH_2$, etc.   (16)

⑭ Compounds containing carbon directly bonded to P, As, Sb, and Bi — e.g., $-PH_2$, $PH-OH$, $-P(OH)_2$, $-PH_4$, $\cdots -PO(OH)_2$

⑮ Compounds containing carbon directly bonded to Si, Ge, and Sn — e.g., $-SiH_3$, $-SiH_2(OH)$, $\cdots$

⑯ Compounds containing carbon directly bonded to elements of the 3rd–1st A-groups of the periodic table — e.g., $-BH_2$, $-BH(OH)$, $\cdots$, $-Mg^{\oplus}$

⑰ Compounds containing carbon directly bonded to elements of the 1st–8th B-groups of the periodic table — e.g., $-HgH$, $-Hg^{\oplus}$, $\cdots$

*Instead of O also S, Se, Te (cf. pp. 25 f.).
†E.g., B, Si, P, but not S, Se, Te (cf. pp. 25 f.).

D. Barton and W. D. Ollis, "Comprehensive Organic Chemistry," Pergamon Press, New York, 1979. This medium-sized work lies between a large textbook and the multivolume treatises. It is designed and written by specialists, to reflect current development of modern organic chemistry. Each volume has a separate editor. Over ninety topics are covered in the following volumes:

1 Stereochemistry; Hydrocarbons; Halo Compounds; Oxygen Compounds
2 Nitrogen Compounds; Carboxylic Acids; Phosphorus Compounds
3 Sulfur, Selenium, Silicon, Boron, and Organometallic Compounds
4 Heterocyclic Compounds
5 Biological Compounds
6 Indexes (author, formula, subject, reagent, reaction)

M. Dub (ed.), "Organometallic Compounds," Springer-Verlag, New York. This treatise consists of a compilation of methods of synthesis, with a consideration of physical constants and chemical reactions. Each of the three volumes has a supplement, as indicated by the second year of publication:

I Compounds of Transition Metals (1966, 1975)
II Compounds of Ge, Sn, Pb (1967, 1973)
III Compounds of As, Sb, Bi (1968, 1972)

There is a formula index for volumes I to III, 1969.
F. Korte (ed.), "Methodicum Chemicum," Academic Press, New York. In the translation this work is stated to be a critical survey of proven methods and their application to chemistry, natural science, and medicine. It is primarily a treatise on organic syntheses, with the treatment divided as shown.

Part I. General Material

1 Parts A and B. Analytical Methods (1974)
2 Planning of Syntheses (   )
3 Types of Reactions (   )

Part II. Systematic Syntheses

4 Synthesis of Skeletons (   )
5 Formation of C-O Bonds (1974)
6 Formation of C-N Bonds (1975)
7A Elements and Their Compounds: Main Groups O to IV (1976)
7B Elements and Their Compounds: Main Groups V to VIII (1978)
8 Preparation of Transition Metal Compounds (1976)

Part III. Special Syntheses

9 Nonsynthetic Fibers (   )
10 Synthetic Compounds (   )

11/1 Nucleic Acids, Proteins, and Carbohydrates (1976)
11/2 Antibiotics, Vitamins, and Hormones (1977)
11/3 Steroids, Terpenes, and Alkaloids (1978)

E. Müller (ed.), "Houben-Weyl's Methoden der Organischen Chemie," Georg Thieme Verlag, Stuttgart. The fourth edition, projected for 16 volumes, has appeared in some 60 parts thus far. It is a monumental set covering the working methods of organic chemistry. The volumes and parts, issued or projected, follow:

I Allgemeine Laboratoriumsmethoden
   I_1 Methoden zur Stofftrennung (1958)
   I_2 Methoden zur Stoffzerkleinerung (1959)
II Analytischen Methoden (1953)
III Physikalische Forschungsmethoden
   III_1 Mechanische, Thermische, . . . Isotopen-methoden (1955)
   III_2 Elektrische, Optische, . . . Methoden (1955)
IV Allgemeine Chemische Methoden
   IV_{1a} Nichtmetallische Oxidationsmittle (   )
   IV_{1b} Metallische und Organische Oxidationsmittle (1975)
   IV_{1c} Reduktion (1980)
   IV_2 Chemische Methoden (1955, 1967)
   IV_3 Carbocyclische Dreiring-Verbindungen (1971)
   IV_4 Isocyclische Vierring-Verbindungen (1971)
   IV_{5a} Photochemie (Teil I) (1975)
   IV_{5b} Photochemie (Teil II) (1975)
V Chemie der Stoffklassen
   V_{1a} Alkane—Cycloalkane (1970)
   V_{1b} Alkene, Cycloalkene, Arylalkene (1972)
   V_{1c} Konjugierte Diene (1970)
   V_{1d} Offenkettige und Cyclische Polyene (1972)
   V_{2a} Alkine, Di- und Polyine, Allene (1977)
   V_{2b} Aromaten. Carbocyclische π-Elektronen Systeme (   )
   V_3 Fluor- und Chlor-Verbindungen (1962)
   V_4 Brom- und Jod-Verbindungen (1960)
VI Sauerstoff-Verbindungen I
   VI_{1a} Alkohole I (1979)
   VI_{1b} Alkohole II (1980)
   VI_{1c} Phenole (1976)
   VI_{1d} Enole und deren Derivate (1978)
   VI_2 Metallverbindungen organischer Hydroxy-Verbindungen (1963)
   VI_3 Äther, Acetale, Orthoester, Oxoniumsalze (1965)
   VI_4 Cyclische Äther und Acetale (1966)
VII Sauerstoff-Verbindungen II
   VII_1 Aldehyde (1954; 1975)
   VII_{2a} Ketone I (1973)

Other Works on Organic Chemistry

Many other works have been published which seem to belong to the reference category. The following list includes a number of current interest:

Adams, R., et al. (eds.): "Organic Reactions," John Wiley & Sons, Inc., New York,

1942–   . Widely used organic reactions, such as those of Grignard, are critically discussed.

ApSimon, J.: "The Total Synthesis of Natural Products," John Wiley & Sons, Inc., New York, 1973–   .

Buehler, C. A., and D. E. Pearson: "Survey of Organic Syntheses," 2 v., John Wiley & Sons, Inc., New York, 1970–1977.

Dolphin, D.: "The Porphrins," 7 v., Academic Press, New York, 1978.

Elderfeld, R. C. (ed.): "Heterocyclic Compounds," 9 v., John Wiley & Sons, Inc., New York, 1950–1967. The subject is treated in terms of modern concepts for the correlation and explanation of observed facts.

Fieser, L. F., and Mary F. Fieser: "Reagents for Organic Syntheses," John Wiley & Sons, Inc., New York, 1967–   .

Glasby, J. S. (ed.): "Encyclopedia of the Alkaloids," 3 v., Plenum Press, New York, 1975–1977.

Harrison, I. T., and S. Harrison: "Compendium of Organic Synthetic Methods," John Wiley & Sons, Inc., New York, 1971–   .

Kazanskii, B. A., et al. (eds.): "Organic Compounds, Reactions, and Methods," 3 v., Plenum Press, New York, 1973–1975.

Olah, G. A. (ed.): "Friedel-Crafts and Related Reactions," Interscience Publishers, Inc., New York. This work is a critical evalution evaluation and review of the useful information on the Friedel-Crafts reaction, as summarized in the following volumes:

I General Aspects (1963)
II Alkylation and Related Reactions (1964)
III Alkylation and Related Reactions (cont.) (1964)
IV Miscellaneous Reactions. Index (1965)

Pollack, J. R. A., and R. Stephens (eds.): "Dictionary of Organic Compounds," 5 v., Oxford University Press, Fairlawn, NJ, 1965. Included are some 40,000 compounds, selected as being of most interest. Supplement 15 appeared in 1979.

Reid, E. E.: "Organic Chemistry of Bivalent Sulfur," 6 v., Chemical Publishing Company, Inc., New York, 1958–1965.

Theilheimer, W.: "Synthetic Methods of Organic Chemistry," Interscience Publishers, Inc., New York, 1948–   . This set consists of yearbooks, each of which lists abstracts of methods considered important. Chemical reactions involve breaking one bond, followed by making a new bond. There are cumulative indexes each 5 years. Through 1981 35 volumes had appeared.

Weissberger, A. (ed.): "The Chemistry of Heterocyclic Compounds," Interscience Publishers, Inc., New York, 1950–1978. Numbering 40 volumes in 1981, this set illustrates further the wealth of information available on heterocyclic chemistry.

Weissberger, A. (ed.): "Techniques of Chemistry," Wiley-Interscience, New York, 1971–   . This set succeeds and incorporates the material of "Physical Methods of Organic Chemistry," and a later set on inorganic chemistry. It includes physical methods important in the study of all varieties of chemical systems. The general content of the first eight volumes follows:

I Physical Methods of Chemistry (5 parts)
II Organic Solvents

III Photochromism

IV Elucidation of Organic Structures by Physical and Chemical Methods (3 parts)

V Technique of Electroorganic Synthesis

VI Investigation of Rates and Mechanisms of Reactions (2 parts)

VII Membranes in Separations

VIII Solutions and Solubilities (2 parts)

Wender, I., and P. Pino: "Organic Syntheses via Metal Carbonyls," 2 v., John Wiley & Sons, Inc., New York, 1968–1977.

Whistler, R. L. (ed.): "Methods in Carbohydrate Chemistry," 7 v., Academic Press, New York, 1962–1976. Tested preparative and analytical methods are covered.

Wiltbecker, E. L.: "Organic Syntheses," John Wiley & Sons, Inc., New York. Since 1921 organic chemists have issued an annual compilation of tested laboratory methods for preparing organic compounds. Usually the editorship changes every year. Five cumulative volumes cover the first 49 volumes, 1921–1969, with a cumulative index for these volumes.

———: "Macromolecular Syntheses," John Wiley & Sons, Inc., New York. Beginning in 1963, this work on macromolecules is the counterpart of "Organic Syntheses." Seven volumes had been published by 1979.

———: "The Colour Index," Society of Dyers and Colourists, Bradford, England. A vast amount of data deal with homogeneous dyes and pigments in current use. The American Association of Textile Chemists and Colorists cooperated for the third edition (1974).

INORGANIC CHEMISTRY

The general arrangement of treatises on organic chemistry is according to kinds of classes of compounds. In contrast, treatises on inorganic chemistry follow more or less closely the order of the elements of the family groups in the Mendeleff form of the periodic table. Smaller works show a trend toward a long form of the table, such as that of Bohr, or one of its modifications.

L. Gmelin (ed.), "Handbuch der Anorganischen Chemie." As Beilstein's treatise is the preeminent set for organic chemistry, that of Gmelin holds a comparable position for inorganic chemistry.

Leopold Gmelin, professor of chemistry at Heidelberg University, published the first edition of his "Handbuch der Theoretischen Chemie," in three volumes, in 1817–1819. Two more editions continued both inorganic and organic chemistry. The title was changed to "Handbuch der Chemie" for the fourth edition of 10 volumes, 1843–1870.

Starting with the fifth edition, organic chemistry was omitted, to be taken over soon in the Beilstein treatise.

In the sixth edition, 1871–1886, the title became "Handbuch der Anorganischen Chemie," with Karl Kraut as editor. He continued with the seventh edition, 1905–1915, which was often referred to as the "Gmelin-Kraut Handbuch."

With the start of the present eighth edition in 1922 the title became "Gmelin's Handbuch der Anorganischen Chemie" to honor the founder. Publication has continued, under several editors, at the Gmelin Institut for Inorganic Chemistry of the Max Planck Society for the Advancement of Science, in conjunction with the German Chemical Society.[11]

As a kind of index or supplement to the seventh edition, M. K. Hoffmann published his "Lexikon der Anorganischen Verbindungen" in three volumes (four parts) in 1910–1919. It is really a formula index listing all inorganic compounds known to April, 1909. From this date until 1920, when the formula index to *Chemical Abstracts* begins, there is no published systematic compilation of inorganic compounds by formula. They are included in the files of the Hill formula index in the U.S. Patent and Trademark Office (see Chapter 4).

Hoffmann's table of system numbers is now of use only for the seventh edition. A new system of 71 numbers, instead of Hoffmann's 81, was adopted for the eighth edition.

Instead of trying to update and extend the information contained in the seventh edition, the plan of the eighth edition was to reexamine the entire field of inorganic chemistry beginning about the middle of the eighteenth century.[12] The intention was to make the treatise an archival compilation of the entire literature of inorganic chemistry, with the inclusion only of facts critically evaluated in terms of current knowledge.

The material selected is arranged mainly on the basis of the substances involved, that is, in terms of the chemical elements and their compounds. Within the substance classification, the arrangement is then in terms of inorganic and physical chemistry, subdivided thereafter into the following sequence of subjects, with numerous tables and graphs summarizing individual reports (arranged alphabetically in German): analytical chemistry; atomic physics; ore dressing; chemical technology; iron and steel; electrochemistry; geochemistry; history of chemistry; colloid chemistry; coordination chemistry; corrosion and passivity; crystallography; geology; metallography; metallurgy; mineralogy; nonferrous metals; physical properties of elements, compounds, and alloys (crystallographic, mechanical, optical, magnetic, electrical); toxicity and hazards; economic and statistical data.

In a special publication[13] the systematic "Subject Matter Index" is explained in detail. A table shows the order in which all elements of the periodic table (including ammonium, NH_4) are considered. Then follows the general outline used in presenting the various aspects of each subject discussed.

The extensive subdivision of topics under chemical analysis shows the importance attached to this area. Those who maintain that qualitative chemical analysis (detection in Gmelin) is passé should study this topic for the element iron.

---

[11]See W. Lippert, *Chem. Ztg.*, **94**, 47 (1970); *J. Chem. Doc.*, **10**, 174 (1970); *J. Chem. Inf. Comput. Sci.*, **15**, 249 (1975); also H. Becke-Göhring, *J. Chem. Educ.*, **50**, 406 (1973).

[12]J. L. Howe's famous "Bibliography of the Metals of the Platinum Group" begins with 1748.

[13]"Systematik der Sachverhalte," Verlag Chemie, GMBH, Weinheim, Germany, 1957.

The substances are arranged according to the Gmelin principle of the "last position." The elements are assigned characteristic sequential numbers (Systemnummern in German) in such a way that the anion-forming elements have smaller numbers than those forming cations. The material under each element contains all information on the element itself as well as compounds with other elements which precede this element in the Gmelin system. Thus, the volume for the element having system number n contains all the compounds of this element, with all other elements having system numbers between 1 and $n-1$.

Table IX lists the symbols and English names of the elements, with their corresponding Gmelin Systemnummern. Using only the 71 numbers shown, some grouping was necessary. Number 1 includes the noble gases; 39 the rare earths (including scandium and yttrium); and 71 the transuranium elements. There is a number for ammonium, NH_4.

In each case the element itself is discussed first. Then follow the binary compounds with all the elements preceding it in the table. More complicated compounds follow in the same manner.

Referring to Table IX, the compound with the formula HCl is discussed under chlorine, as the number for chlorine, 6, follows that for hydrogen, 2. Zinc chloride, $ZnCl_2$, is discussed under zinc, number 32. Zinc chromate, $ZnCrO_4$, comes under chromium, as the descending element numbers are 52, 32, 3. Any compound or combination of elements thus appears in the volume having the highest system number.

The Gmelin Systemnummern have no relationship to the atomic numbers of the elements (see Bohr form of periodic table, Fig. 8).

References to the original literature, included in parentheses in the text, are complete to the listed closing date of the given volume. The "Price List, Gmelin Handbook of Inorganic Chemistry" gives the literature closing date and the date of publication for each volume, section, supplement, or member of the New Supplement Series. These dates are generally close together, but occasionally several years may elapse.

The following excerpt, reproduced with permission, is from the volume on tungsten (Wolfram in German), Systemnummern 54, page 212. It shows the format for this treatise.

Normales Natriumwolframat Na_2WO_4

Bildung und Darstellung

Wasserfrei Na_2WO_4 kann durch Zusammenschmelzen von 1 Mol Na_2CO_3 mit 1 Mol WO_3 oder durch völliges Entwässern von wasserhaltigen Na_2WO_4 bei 100° darstellt werden, F. Hoermann (*Z. anorg. Ch.* **177** (1929) 150); vgl. dazu V. Forcher (*Ber. Wien. Akad.* **44** II (1861) 173; *J. pr. Ch.* **86** (1862) 239), B. Pawlewski (*Ber.* **33** (1900) 1223), R. Funk (*Ber.* **33** (1900) 3700). Die Umwandlung von $Na_2WO_4.2H_2O$ in wasserfrei Salz kann nach B. Pawlewski (1.c.) auch durch längeres Stehen im Exsiccator über H_2SO_4 bewirkt werden.

Auf dem üblichen Wege hergestelltes Na_2WO_4 enthält nach C. Friedheim, R. Meyer (*Z. anorg, Ch.* **1** (1892) 78) 0.40 bezw. 0.46% MoO_3.

TABLE IX
GMELIN SYSTEMNUMMERN

System No.	Element
1	Rare gases
2	Hydrogen
3	Oxygen
4	Nitrogen
5	Fluorine
6	**Chlorine**
7	Bromine
8	Iodine
9	Sulfur
10	Selenium
11	Tellurium
12	Polonium and isotopes
13	Boron
14	Carbon
15	Silicon
16	Phosphorus
17	Arsenic
18	Antimony
19	Bismuth and isotopes
20	Lithium
21	Sodium
22	Potassium
23	Ammonium

System No.	Element
24	Rubidium
25	Cesium
26	Beryllium
27	Magnesium
28	Calcium
29	Strontium
30	Barium
31	Radium and isotopes
32	**Zinc**
33	Cadmium
34	Mercury
35	Aluminum
36	Gallium
37	Indium
38	Thallium and isotopes
39	Rare Earths
40	Actinium and isotopes
41	Titanium
42	Zirconium
43	Hafnium
44	Thorium and isotopes
45	Germanium
46	Tin
47	Lead and isotopes

System No.	Element
48	Vanadium
49	Niobium
50	Tantalum
51	Protactinium and isotopes
52	**Chromium**
53	Molybdenum
54	Tungsten
55	Uranium and isotopes
56	Manganese
57	Nickel
58	Cobalt
59	Iron
60	Copper
61	Silver
62	Gold
63	Ruthenium
64	Rhodium
65	Palladium
66	Osmium
67	Iridium
68	Platinum
69	Technetium
70	Rhenium
71	Transuranium elements

HCl

$ZnCl_2$

$CrCl_2$

$ZnCrO_4$

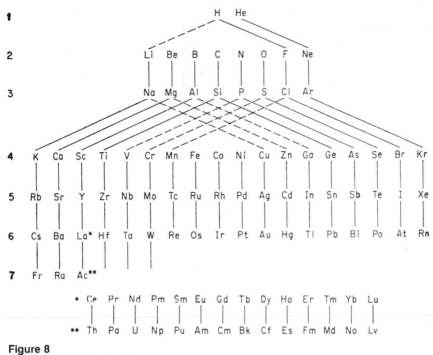

Figure 8
Periodic table (Bohr form).

Bildungswärme, berechnet mit Hilfe der durch Verbrennen von W in Na_2O_2 ermittelten Wärmetönung für die Reaktionsgleichung

$$3\ Na_2O_2 + W = Na_2WO_4 + 2\ Na_2O + 231{,}800\ cal.$$
$$(Na_2O,\ W,\ 3\ O) = 291{,}000\ cal.$$
$$(Na_2O,\ WO_3) = 94{,}700\ cal.$$

W. G. Mixter (*Am. J. Sci.* [4], **24** (1908) 136).

Status of Publication Schedule

Several publications describing the treatise have contained a periodic table of the elements, shown in divided form, to illustrate the status of the publication program. For such a dynamic field, a treatise must be open-ended to provide for changes with passing time.

Most of the main volumes (Hauptwerk) have been published, but a few are still in preparation. A striking example of the latter is volume 59 for iron. Although the first section appeared in 1929, the volume is still incomplete in the 29 sections published. For a number of volumes there are supplements, but the literature closing date may show that a given supplement is far from being up to date. As an example, the main volume for bismuth was published in 1927, and the literature closing date for the supplement is 1960. One would turn, of course, to *Chemical Abstracts* for later references. A given supplement may cover only a specific section for an element.

The New Supplement Series has been under way to take the place, at least partially, of a ninth edition of the treatise. These volumes, while constituting an addition to the eighth edition, are quite distinct from the regular supplement series. Having abandoned a ninth edition covering the entire field, this New Supplement Series is concentrating on specific subjects, particularly those in which advances have been rapid. Examples of such areas are compounds of boron, organometallic compounds, and the transuranium elements. Over 50 volumes of this new series have been issued.

A few volumes, designated "Appended Volumes," do not fit in the regular series. Two examples are "Water Desalting," which cites some 3,000 publications to the end of 1973; and "Oceanic Salt Deposits," in which the literature is cited through 1939, with the Supplement Volume having a literature closing date of 1967.

The "Price List 1980" lists all volumes, parts, sections, reprints, supplements, New Supplement Series, and indexes.

The main series has been issued with the Gmelin number (59 for iron) on the spine. The New Supplement Series does not fit into this scheme. Now a set of "separators" is being provided with the chemical symbol of the element so that any of the new series relating to a specific element can be shelved there.

Indexes

Many of the main volumes have no indexes, but there is a reasonably detailed table of contents showing the subdivisions of the subject. The volumes having indexes are listed in the "Price List 1980." A small explanatory volume (1950) lists abbreviations and symbols.

The General Formula Index has been completed with the issuance in 1980 of volume 12. Each volume shows the elements covered. Thus, volume 1 is Ac–Au, and volume 12 is O–Zr. All elements and defined compounds are listed in alphanumeric order. The entry for zinc peroxydisulfate is

$$O_8S_2Zn \qquad ZnS_2O_8 \qquad 32~(Zn) \qquad Hb—241;~Eb—965$$

This shows that the information is in the main series volume 32, p. 241, and the supplemental volume, p. 965. The introduction to each volume contains instructions for the user.

There is an alphabetical reference wall chart (1.5 by 2 feet), in order of the element symbols.

J. W. Mellor, "Comprehensive Treatise on Inorganic and Theoretical Chemistry," Longmans, Green & Co., Inc., New York. The arrangement of Mellor's treatise follows somewhat the order of the periodic table. Often general principles or theories are discussed in connection with certain elements. The volumes include the material shown.

  I General principles, H, O (1922)
 II F, Cl, Br, I, Li, Na, K, Rb, Cs (1922)
 III Cu, Ag, Au, Ca, Sr, Ba, radioactivity (1923)
 IV Ra family, Be, Mg, Zn, Cd, Hg (1923)
  V B, Al, Ga, In, Tl, Sc, Ce, rare earth metals, C (part 1) (1924)
 VI C (part 2), Si, silicates (1925)
VII Ti, Zr, Hf, Th, Ge, Sn, Pb, noble gases (1927)
VIII N, P (1928)
 IX As, Sb, Bi, V, Nb, Ta (1929)
  X S, Se (1930)
 XI Te, Cr, Mo, W (1931)
 XII U, Mn, Tc, Re, Fe (part 1) (1932)
XIII Fe (part 2) (1934)
XIV Fe (part 3), Co (1935)
 XV Ni, Ru, Rh, Pd, Os, Ir (1936)
XVI Pt, general index (1937)

Six supplements have been issued. They cover the halogens (1956), alkali metals (1961–1965), nitrogen (1964–1967), phosphorus (1971), and boron (Part A) (1980).

The order of presentation under each element is history, preparation, properties, and the hydride, oxide, halide, sulfide, sulfate, carbonate, nitrate, and phosphate compounds. Compounds with elements such as arsenic, carbon, and silicon are described under these elements. Potassium chloride will be found under potassium, but potassium peroxydisulfate is under sulfur. Intermetallic compounds and complex salts discussed under a given element include only compounds with elements already described.

The subject or name of the compound being considered is printed in boldfaced type so that one's eye readily catches it. The original references are grouped at the end of the various sections. The discussion of a given subject usually divides itself into several phases. Suppose it is the first phase being discussed. Mellor has collected all references for this part under "1" at the end of the section, this numeral being mentioned only once—in connection, usually, with the first author named. The numeral is not repeated for the succeeding authors mentioned, but their names will be found by looking through the list under "1."

P. Pascal (ed.), "Nouveau traité de chimie minerale," Masson et Cie, Paris. There are good summaries of selected topics, such as complex compounds, heteropoly acids, and rare earths. The references, collected at the ends of sections, are printed in run-in form in groups of 10, The planned coverage follows:

    I Introduction; air, water; hydrogen and isotopes; noble gases (1956)
  II₁ Li, Na (1966)
  II₂ K (1963)
  III Rb, Cs, Fr, Cu, Ag, Au (1957)
  IV Be, Mg, Ca, Sr, Ba, Ra (1958)
   V Zn, Cd, Hg (1962)
  VI B, Al, Ga, In, Tl (1961)
 VII Sc, Y, rare earths, Ac (1959); 2 parts
VIII₁ Ge, Sn, Pb (1963)
VIII₂ Si (1965)
VIII₃ C (1968)
  IX Ti, Zr, Hf, Th (1963)
   X N, P (1956)
  XI As, Sb, Bi (1958)
 XII V, Nb, Ta, Pa (1958)
XIII O, S, Se, Te, Po (1960); 2 parts
XIV Cr, Mo, W, heteropoly acids (1959)
 XV U, transuranium elements (1960–1962)
      U supplement (1967)
      Transuranium element supplement (1970)
 XVI F, Cl, Br, I, At, Mn, Tc, Re (1960)
XVII₁ Fe (1967)
XVII₂ Co, Ni (1963–   )
XVIII Complexes of Fe, Co, Ni (1959)
 XIX Ru, Rh, Pd, Os, Ir, Pt (1958)
  XX Metallic alloys (1962–1964); 3 parts

A series of supplements (Compléments) is being edited by A. Pacault and G. Pannetier. Thus far, the following volumes have appeared: (1) Rubidium, Cesium, and Francium (1974); (2) Protactinium (1974); (3) Helium (1974); (4) Peroxides and Polyoxides of Hydrogen (1975); (5) Molybdenum (1976); (6) Strontium (1976); (7) Selenium (1977); (8) Tellurium (1977); (9) Mercury (1977); (10) Technetium and Rhenium (1978); and (11) Thallium (1978).

A. F. Trotman-Dickenson (ed.), "Comprehensive Inorganic Chemistry," Pergamon Press, New York, 1973. Quoting the publisher's announcement,

> The physical basis of chemistry is emphasized throughout. . . . Arrangement is by element, in accordance with the periodic table. Each article includes the discovery and history of each element, its occurrence and distribution, production, industrial uses and processes, preparation of pure material, allotropes, nuclear properties, physical properties, chemistry, chemical properties, analytical chemistry, and footnotes. Each volume is indexed separately and a master index to the entire work appears in Volume 5.

The distribution of material in each volume follows:

1 H, noble gases, Li, Na, K, Rb, Cs, Fr, Be, Mg, Ca, Ba, Sr, Ra, B, Al, Ga, In, Tl, C. Si

2 Ge, Sn, Pb, N, P, As, Sb, Bi, O, S, Se, Te, Po, F, Cl, Br, I, At

3 Cu, Ag, Au, Zn, Cd, Hg, Sc, Y, La, Ti, Zr, Hf, V, Nb, Ta, Cr, W, Mo, Tc, Re, Fe, Co, Ni, Pt, Pd, Ir, Rh, Ru, Os

4 Lanthanides; carbonyls, cyanides, isocyanides, nitrosyls; compounds of transition elements involving metal-metal bonds; transition metal hydrogen compounds; nonstoichiometric compounds, tungsten and vanadium bronzes and related compounds; isopolyanions and heteropoly anions; transition metal chemistry; organotransition metal compounds and related aspects of homogeneous catalysis

5 Actinides (elements 89 to 103); General Index

K. A. Gschneidner, Jr., and L. Eyring (eds.), "The Handbook on Physics and Chemistry of Rare Earths," North-Holland Publishing Company, Amsterdam, 1979. This treatise covers the 17 related elements in an integrated manner, each chapter being a comprehensive, up-to-date, critical review of a particular segment of the field. The material in the four volumes is distributed as follows: I, Metals; II, Alloys and Intermetallics; III and IV, Non-Metallic Compounds.

Rotating editors, "Inorganic Syntheses," John Wiley & Sons, Inc., New York. Started in 1938, this set includes tested methods of preparing important inorganic compounds. It is comparable to "Organic Syntheses." The editorship rotates. Twenty volumes had been published by 1980.

PHYSICAL CHEMISTRY[14]

For the present purpose physical chemistry may be considered as the branch which deals with the cause and nature of chemical reactions and with the interpretation of specific properties. As organic and inorganic chemistry include all the chemical elements, and therefore all chemical substances, one would expect to find much discussion belonging to this area in the treatises already considered. Consequently, the tendency has been toward single-volume textbooks and monographs on limited portions of the field, rather than comprehensive treatises. However, several multivolume works are available.

C. H. Bamford and C. F. H. Tipper (eds.), "Comprehensive Chemical Kinetics," Elsevier Scientific Publishing Company, New York. This treatise is designed to provide information on specific reactions, types of reactions, and on more general topics in the field. The following volumes have appeared:

Section 1: The Practice and Theory of Kinetics

1 The Practice of Kinetics (1969)
2 The Theory of Kinetics (1969)
3 The Formation and Decay of Excited Species (1969)

---

[14]As physics and physical chemistry are so closely related, see R. H. Whitford, p. 246.

Section 2: Homogeneous Decomposition and Isomerisation Reactions

    4 Decomposition of Inorganic and Organometallic Compounds (1972)
    5 Decomposition and Isomerisation of Organic Compounds (1972)

Section 3: Inorganic Reactions

    6 Reactions of Nonmetallic Inorganic Compounds (1972)
    7 Reactions of Metallic Salts and Complexes, and Organometallic Compounds (1972)

Section 4: Organic Reactions

    8 Proton Transfer (1977)
    9 Addition and Elimination Reactions of Aliphatic Compounds (1973)
    10 Ester Formation and Hydrolysis and Related Reactions (1972)
    11 Reactions of Carbonyl Compounds (   )
    12 Electrophilic Substitution at a Saturated Carbon Atom (1973)
    13 Reactions of Aromatic Compounds (1972)

Section 5: Polymerisation Reactions

    14 Degradation of Polymers (1975)
14A Free Radical Polymerisation (1976)
    15 Nonradical Polymerisation (1976)

Section 6: Oxidation and Combustion Reactions

    16 Liquid Phase Oxidation (1978)
    17 Gas-Phase Combustion (1977)

Section 7: Selected Elementary Reactions

    18 Selected Elementary Reactions (1976)

Section 8: Heterogeneous Reactions

    19 Simple Processes at the Gas-Solid Phase (   )
    20 Complex Catalytic Processes (1978)
    21 Reactions of Solids with Gases (   )
    21 Reactions in the Solid State (1980)

H. D. Eyring, D. Henderson, and W. Jost (eds.), "Physical Chemistry," Academic Press, New York. This treatise was projected for the following volumes:

1 Thermodynamics (1971)
2 Statistical Mechanics (1967)
3 Electronic Structure of Atoms and Molecules (1969)
4 Molecular Properties (1970)
5 Valency (1970)
6 Kinetics; Gas Reactions (1975)
7 Reactions in Condensed Phases (1975)
8 Liquid State (1971)
9 Electrochemistry (1970)
10 Solid State (1970)
11 Mathematical Methods (1975)

D. Dunitz and J. A. Ibers (eds.), "Perspectives in Structural Chemistry," 4 v., John Wiley & Sons, Inc., New York, 1967–1971.

Many single-volume works have been published on general physical chemistry or on specialized topics. Usually they are textbooks or monographs. Representative titles are listed in Chapter 11.

ANALYTICAL CHEMISTRY

This subdivision of chemistry has an aspect of distinction, at least much of the time, in that it involves both chemistry and physics. Thus, quantitative chemical analysis employs chemistry, whenever necessary, to render measurable the desirable constituent, or something chemically equivalent to it. The final operation of measurement is physics, as it consists of determining the number of times the unit of measurement goes into the unknown. The result is to assign a number to the physical property measured.

Only in recent times has there been available for analytical chemistry a reference set comparable to the comprehensive works mentioned for organic and inorganic chemistry. Shorter sets have been the general practice. The present situation is summarized here.

W. Fresenius and G. Jander, "Handbuch der analytischen Chemie," Springer-Verlag OHG, Berlin. For analytical chemistry this treatise is analogous, in a general way, to the Gmelin set for inorganic chemistry. It covers the elements according to families of the periodic table. In giving separate treatment to methods of detection, separation, and determination, the set seems likely to replace the older, and still incomplete, set of Rüdisüle.

The volumes projected are shown below. They are being issued in smaller parts which may be bound separately. Incomplete parts are shown by italicized symbols. For some of these a second edition is being prepared.

Part I.   General Methods

I Not announced
II Not announced

Part II.   Qualitative Analysis

    Ia H, Li, Na, K, NH$_4$, Rb, Cs (1944)
    Ib Cu, Ag, Au (1955)
    Ibβ Ag, Au (1967)
    II Be, Mg, Ca, Sr, Ba, Zn, Cd, Hg (1955)
    III B, Al, Ga, In, Tl, Sc, Y, rare earths, Ac (1944)
  IVaα C, Si (1963)
  IVaβ Ge, Sn (1956)
  IVb Ti, Zr, Hf, Th (1956)
  Va/b N, P, As, Sb, Bi, V, Nb, Ta, Pa (1956)
    VI O, S, Se, Te, Cr, Mo, W, U (1948)
    VII F, Cl, Br, I, Mn, Tc, Re (1953)
VIIIbβ Pt, Pd, Rh, Ir, Ru, Os (1951)
VIIIbα Fe, Co, Ni (1951)
    IX Vorproben und Trennung der Kationen und Anionen (1956)

Part III.   Quantitative Methods of Separation and Determination

    Ia Li, Na, K, NH$_4$, Rb, Cs (1940)
    Ib Cu, Ag, Au (   )
    IIa Be, Mg, Ca, Sr, Ba, Ra (1940)
    IIb Zn, Cd, Hg (1945)
    III B, Al, Ga, In, Tl, Sc, rare earths, Ac (1942)
IIIaα1 B (1971)
IIIaα2 Al (1972)
 IIIaβ/ Ga, In, Tl, Sc, Y, rare earths, Ac (1956)
  IIIb
  IVaα C, Si (1967)
IVaβδ Ge, Pb (1966)
  IVaγ Sn (1978) (first volume to appear in English)
  IVb Ti, Zr, Hf, Th (1950)
  Vaα N (1957)
  Vaβ P (1953)
  Vaγ As, Sb, Bi (1951)
   Vb V, Nb, Ta (1957)
  VIaα O (1953)
  VIbα Cr (1958)
  VIbβ U (1972)
  VIbγ W (1978)
 VIIaα H, F (1950)
 VIIaβ Cl, Br, I (1967)
  VIIIa Noble gases, Rn (1949)
VIIIbγ Pt, Pd, Rh, I, Ru, Os (1953)

I. M. Kolthoff and P. J. Elving (ed.), "Treatise on Analytical Chemistry," Interscience Publishers, Inc., New York. This American set is designed to be a concise, critical, comprehensive, and systematic treatment of the various aspects of modern analytical chemistry. The outline of the contents follows.

Part I. Theory and Practice

1 Sect. A: Analytical chemistry—its objectives, functions, and limitations Sect. B: Applications of chemical principles (1959)
2 Sect. B: Applications of chemical principles (1959) Sect. C: Separation—principles and technics (1961–1979)
3 Sect. C: Separation—principles and technics (cont.) (1961)
4 Sect. D-1: Magnetic field methods of analysis Sect. D-2: Electrical methods of analysis (1963)
5 Sect. D-3: Optical methods of analysis (1964)
6 Sect. D-3: Optical methods of analysis (cont.) (1965)
7 Sect. D-4: Classical physical methods (1967)
8 Sect. D-4: Classical physical methods (1961) Sect. D-5: Thermal methods (1968)
9 Sect. D-6: Radioactive methods Sect. E: Application of measurement (1971)
10 Sect. E: Application of measurement (cont.) Sect. F: Principles of instrumentation (1972) Sect. G: Preparation for analytical research and utilization (1972)
11 Sect. H: Common techniques and operations Sect. I: Classical methods of analysis. I-1: Quantitative analysis—gravimetric; I-2: Quantitative analysis—titrimetric (1975)
12 Index to volumes 1 to 11 (1976)

Part II. Analytical Chemistry of Inorganic and Organic Compounds

Sect. A. Systematic Analytical Chemistry of the Elements

1 Nomenclature; general concepts; H; H_2O; noble gases; alkali metals (1961)
2 Ga, In, Tl, Si, Ge, Fe, Co, Ni (1962)
3 Cd, Cu, Mg, Hg, Sn, Zn (1961)
4 Al, Ca, Au, Rn, Ra, Ag, Sr, Ba (1966)
5 N, P, Th, Ti, Zr, Hf (1961)
6 Be, Pb, Nb, Ta, Tc, Ac, At, Fr, Po, Pa (1964)
7 S, Se, Te, Cl, F, Br, I, Mn, Re (1961)
8 Rare earths, Bi, V, Cr, platinum metals (1963)
9 U, transuranium actinide elements (1962)
10 Sb, As, B, C, Mo, W (1978)
11 Organic analysis (1965)
12 Organic analysis (cont.) (1965)
13 Organic analysis; functional groups (1966)
14 Organic analysis (1971)

Sect. B-1: Cl, Br, I

Sect. B-2: Unsaturation; Several Groups; Peroxides (1971)

15 Organic analysis (cont.); functional groups (1976)
16 Organic analysis (cont.); functional groups (1980)
17 Index, vol. 1–16 (1980)

Part III. Analytical Chemistry in Industry

1 Sect. A: Analytical chemistry in industry (1967)
2 Sect. B: Industrial toxicology and environmental pollution and its control (1971)
3 Sect. C: Standards and specifications (   )
Sect. D: Physical testing methods for the characterization of materials; thermal and chemical testing (1976)
4 Sect. D: Physical testing methods for the characterization of materials (cont.) (1977)

The following volumes of a second edition have appeared[15]:

Volume 1: A, Analytical Chemistry; B, Analytical Chemistry Methodology; C, General Chemistry and Analytical Chemistry; D, Solution Equilibria and Chemistry (1978)
Volume 2: Solution Equilibria and Chemistry (1979)

C. L. Wilson and D. W. Wilson, "Comprehensive Analytical Chemistry," American Elsevier Publishing Co., New York. The Wilsons edited the first seven volumes, followed by G. Svehla. This work and the treatises edited by Kolthoff and Elving and by Snell and Hilton are together the most generally useful in English.
The following volumes have appeared:

IA Analytical Processes; Gas Analysis; Inorganic Qualitative Analysis; Organic Qualitative Analysis; Inorganic Gravimetric Analysis (1959)
IB Inorganic Titrimetric Analysis; Organic Quantitative Analysis (1960)
IC Analytical Chemistry of the Elements (1962)
IIA Electrochemical Analysis; Electrodeposition; Potentiometric Titrations; Conductometric Titrations; High-Frequency Titrations (1964)
IIB Liquid Chromatography in Columns; Gas Chromatography; Ion Exchangers; Distillation (1968)
IIC Paper and Thin-Layer Chromatography; Radiochemical Methods; Nuclear Magnetic Resonance and Electron Spin Resonance Methods; X-Ray Spectrometry (1971)
IID Coulometric Analysis (1975)
III Elemental Analysis with Minute Samples; Standards and Standardization;

---

[15]The new outline differs somewhat from that in the first edition.

Separations by Liquid Amalgams; Vacuum Fusion Analysis of Gases in Metals; Electroanalysis in Molten Salts (1975)

IV Instrumentation for Spectroscopy; Analytical Atomic Absorption and Fluorescence Spectroscopy; Diffuse Reflectance Spectroscopy (1975)

V Emission Spectroscopy; Analytical Microwave Spectroscopy; Analytical Applications of Electron Microscopy (1975)

VI Analytical Infrared Spectroscopy (1976)

VII Thermal Methods in Analytical Chemistry; Substoichiometric Analytical Methods (1976)

VIII Enzyme Electrodes in Analytical Chemistry; Molecular Fluorescence Spectroscopy; Photometric Titrations; Analytical Applications of Interferometry (1977)

IX Ultraviolet Photoelectron and Photoion Spectroscopy; Auger Electron Spectroscopy; Plasma Excitation in Spectrochemical Analysis (1979)

X Organic Spot Test Analysis; The History of Analytical Chemistry (1980)

XI Application of Mathematical Statistics in Analytical Chemistry; Mass Spectrometry; Ion Selective Electrodes (1980)

F. D. Snell and C. L. Hilton (eds.), "Encyclopedia of Industrial Chemical Analysis," Wiley-Interscience Company, New York, 1966–1974. With volume 8, L. S. Etre replaced C. L. Hilton as coeditor.

Starting with volume 4, this set does resemble the usual encyclopedia in its alphabetical arrangement of topics. However, it seems to belong with other treatises on analytical chemistry.

With the separate sections written by practical industrial analysts, the general treatment is analysis, control, evaluation, and testing. Included are data on commercial specifications, properties, and standards.

The first three volumes cover general techniques, with an index in volume 3. Volume 4 begins, with "Ablative Materials," the alphabetical listing of methods for a wide variety of chemical substances. Volume 19 ends with "zinc" (and Pesticides). Volume 20 is an index for volumes 4 to 19.

A. P. Vinogradov (ed.), "Analiticheskaya Khimiya Elementov." Many volumes of this Russian set have been translated by the Israel Program for Scientific Translations, Keter, Inc., New York. Added to the time lag inherent in the original is the time required to translate and publish the English edition.

For each element the arrangement of material follows: General Information, Chemical and Analytical Properties, Detection, Determination; Separation, Determination in Natural and Industrial Materials, and Determination of Impurities.

Other important analytical reference sets follow.

Duval, C.: "Traité de micro-analyse minérale, qualitative et quantitative," 4 v., Presses scientifiques internationales, Paris, 1954–1956.

Furman, N. H., and F. J. Welcher (eds.): "Standard Methods of Chemical Analysis," 5 v. in 4, D. Van Nostrand Company, Inc. Princeton, NJ, 1962–1966.

Jordan, J.: "Treatise on Titrimetry," vol. 1, Inorganic Titrimetric Analysis, 1971; vol. 2, New Developments in Titrimetry, 1974, Plenum Press, New York.

Li, N. N.: "Recent Developments in Separation Science," 6 v., CRC Press, Inc., Boca Raton, FL, 1972–1978.

McCrone, W. C., and J. G. Delly: "The Particle Atlas," 6 v., Ann Arbor Science, Ann Arbor, MI, 1973–1978.

Richardson, J. H., and R. V. Peterson: "Systematic Materials Analysis," 4 v., Academic Press, New York, 1974–1978.

Sawicki, E., and C. R. Sawicki: "The Analysis of Organic Materials," 5 v., Academic Press, New York, 1975–1977.

Snell, F. D., and C. T. Snell: "Colorimetric Methods of Analysis," 6 v., D. Van Nostrand Company, Inc., Princeton, NJ, 1948–1971.

Strouts, C. R. N., H. N. Wilson, and R. T. Parry-Jones: "Chemical Analysis: The Working Tools," 3 v., Clarendon Press, Oxford, England, 1962.

Zweig, G., and J. Sherma: "Analytical Methods for Pesticides, Plant Growth Regulators, and Food Additives," 9 v., Academic Press, New York, 1963–1978.

There are many equally valuable single-volume reference books in this field. An outstanding example is the edition by H. A. Bright and J. I. Hoffman of "Applied Inorganic Analysis," by W. F. Hillebrand and G. E. F. Lundell. When these books become rather specialized they are hardly distinguishable from monographs (see Chapter 9).

Official methods of testing and analysis for engineering materials, drugs,[16] feeds, fertilizers, foods, and many other commercial products have been issued. They are of great importance in buying, using, and selling products. The following examples are important in the United States:

————: "American Standards," American Standards Association, New York.

————: "ASTM Standards," 48 parts, American Society for Testing and Materials, Philadelphia.

————: "Official Methods of Analysis," Society of Leather Trades' Chemists, Croydon, England.

————: "Standard Methods for Testing Petroleum and Its Products," Institute of Petroleum, London.

Franson, M. A. (ed.): "Standard Methods for the Examination of Water and Waste-water," American Public Health Association, New York.

Horwitz, W. (ed.): "Official Methods of Analysis," Association of Official Analytical Chemists, Washington, DC.

Jolly, S. C. (ed.): "Official, Standardized, and Recommended Methods of Analysis," Heffer and Sons, Cambridge, England.

MacMasters, M. M., et al. (eds.): "Cereal Laboratory Methods," American Association of Cereal Chemists, St. Paul, MN.

Paquot, C.: "Standard Methods for the Analysis of Oils, Fats, and Derivatives," Pergamon Press, New York.

Sallee, E. M., et al. (eds.): "Official and Tentative Methods of the American Oil Chemists' Society," American Oil Chemists' Society, Chicago.

Seligson, D., et al. (eds.): "Standard Methods of Clinical Chemistry," Academic Press Inc., New York.

[16]See the section on pharmaceutical chemistry, p. 231, for pharmacopoeias.

Tuthill, S. M. et al (eds.), "Reagent Chemicals: American Chemical Society Specifications," American Chemical Society, Washington, DC.

West, D. B., et al. (eds.): "Methods of Analysis," American Society of Brewing Chemists, Madison, WI.

To keep these standard methods up to date, they are frequently revised, usually each 5 years. The outstanding example is the "ASTM Standards." The entire work has been reissued each 3 years, and every year a supplement has been issued for each part. Beginning in 1964, the set has been issued annually in 48 parts. Users should note issue dates if they are important for specific purposes.

BIOLOGICAL CHEMISTRY

Over a period of several decades a number of outstanding treatises appeared on biological or physiological chemistry (biochemistry). Their present value as reference sources is largely historical.

Various later sets have appeared, or are appearing. The following are noteworthy in their coverage:

M. Florkin and E. H. Stotz (eds.), "Comprehensive Biochemistry," American Elsevier Publishing Company, New York. This new work is somewhat comparable to the great Abderhalden compilations. It is designed to assemble the historical, experimental, and theoretical material which constitutes the biological and chemical basis of biochemistry, and to project biochemistry into areas of physiology and medicine. It is planned to include the following sections and volumes:

I Physicochemical and Organic Aspects of Biochemistry
    1 Atomic and Molecular Structure (1962)
    2 Organic and Physical Chemistry (1962)
    3 Methods for the Study of Molecules (1962)
    4 Separation Methods[17] (1962)
II Chemistry of Biological Compounds
    5 Carbohydrates (1963)
    6 Lipids; Amino Acids and Related Compounds (1965)
    7 Proteins I (1963)
    8 Proteins II; Nucleic Acids (1963)
    9 Pyrrol Pigments; Isoprenoid Compounds; Phenolic Plant Constituents (1963)
    10 Sterols, Bile Acids, and Steroids (1963)
    11 Water-soluble Vitamins; Hormones; Antibiotics (1963)
III Biochemical Reaction Mechanisms
    12 Enzymes—General Considerations (1964)
    13 Nomenclature of Enzymes (1964)

[17]Limited to countercurrent distribution and chromatography.

14 Biological Oxidations (1966)
15 Group Transfer Reactions (1964)
16 Hydrolytic Reactions; Cobamide and Biotin Coenzymes (1965)
IV Metabolism
    17 Carbohydrate Metabolism (1969)
    18 Lipid Metabolism (1970)
    18S Pyruvate and Fatty Acid Metabolism (1971)
    19 Amino Acid Metabolism and Sulfur Metabolism (1981)
    19B Protein Metabolism (1980)
    20 Metabolism of Cyclic Compounds (1968)
    21 Metabolism of Vitamins and Trace Elements (1971)
V Chemical Biology; General Index
    22 Bioenergetics (1967)
    23 Cytochemistry (1968)
    24 Biological Information Transfer (1977)
    25 Regulatory Functions—Mechanisms of Hormone Action (1975)
    26 Extracellular and Supportive Structures (1968)
    26B Extracellular and Supportive Structures (cont.) (1968)
    26C Extracellular and Supportive Structures (cont.) (1971)
    27 Photobiology; Ionizing Radiations (1967)
    28 Morphogenesis, Differentiation and Development (1967)
    29A Comparative Biochemistry; Molecular Evolution (1974)
    29B Comparative Biochemistry; Molecular Evolution (1975)
    30 A History of Biochemistry I and II (1972)
    31 A History of Biochemistry III (1975)
    32 A History of Biochemistry IV (1977)
    33 A History of Biochemistry V (1979)
    34 General Index (   )

Some of the volumes have several parts, such as A, B, and C.

M. Florkin and H. S. Mason (eds.), "Comparative Biochemistry," Academic Press Inc., New York. This work aims to compare the physicochemical properties of every form of life, both contemporary and ancient. Seven volumes are included.

I Sources of Free Energy (1960)
II Free Energy and Biological Function (1960)
III Constituents of Life—Part A (1962)
IV Constituents of Life—Part B (1962)
V Constituents of Life—Part C (1963)
VI Cells and Organisms (1964)
VII Supplementary Volume (1964)

P. D. Boyer (ed.), "The Enzymes," Academic Press Inc., New York. This is a comprehensive set, but with a more limited area. There are 13 volumes to date, 1970– .

1 Structure and Control
2 Kinetics and Mechanism
3 Hydrolysis; Peptide Bonds
4 Hydrolysis; Other C-N Bonds, Phosphate Esters
5 Hydrolysis (Sulfate Esters, Carboxyl Esters, Glycosides); Hydration
6 Carboxylation and Decarboxylation (Nonoxidative), Isomerization
7 Elimination and Addition, Aldol Cleavage and Condensation, Other C-C Cleavage, Phosphorolysis, Hydrolysis (Fats, Glycosides)
8 Group Transfer, A: Nucleotidyl Transfer, Nucleosidyl Transfer, Acyl Transfer, Phosphoryl Transfer
9 Group Transfer B: Phosphoryl Transfer, One-Carbon Group Transfer, Glycosyl Transfer, Amino Group Transfer, Other Transferases
10 Protein Synthesis, DNA Synthesis and Repair, RNA Synthesis, Energy-Linked ATPases, Synthetases
11 Oxidation-Reduction A: Dehydrogenases (I), Electron Transfer (I)
12 Oxidation-Reduction B: Electron Transfer (II), Oxygenases, Oxidases (I)
13 Oxidation-Reduction C: Dehydrogenases (II), Oxidases (II), Hydrogen Peroxide Cleavage; Author, Subject, and Topical Subject Indexes for volumes 1–13

G. D. Fasman, "Handbook of Biochemistry and Molecular Biology," CRC Press, Inc., Boca Raton, FL, 1976. The following parts have been issued:

Section A—Proteins, 3 v.
Section B—Nucleic Acids, 2 v.
Section C—Lipids, Carbohydrates, and Steroids, 2 v.
Section D—Physical and Chemical Data, 2 v.
"Cumulative Series Index" volume was issued in 1977.

S. Blackburn, "Amino Acid Determination: Methods and Techniques," M. Dekker, New York, 1978.
P. Boulanger and J. Polonovski (eds.), "Traité de Biochimie Générale," 3 v. in 7, Masson et Cie, Paris, 1959–1972.
D. Glick (ed.), "Methods of Biochemical Analysis," John Wiley & Sons, Inc., New York, 1954–   . This set is an annual review of selected methods and techniques.
C. Long, "Biochemists' Handbook," D. Van Nostrand Company, Princeton, NJ, 1961.
H. Neurath (ed.), "The Proteins: Chemistry, Biological Activity, and Methods," Academic Press, Inc., New York, 1975.
————: "Biochemical Preparations," John Wiley & Sons, Inc., New York, 1949–1971.

INDUSTRIAL CHEMISTRY AND CHEMICAL ENGINEERING

In the general area of industrial (applied) chemistry and chemical engineering there is nothing comprehensive comparable to the Beilstein treatise in organic chemistry. It is very difficult, if not impossible, to have a compilation of general nature that is reliable very long after the time of publication. Changes in practice are so frequent and advances in technique so steady that new editions could hardly be arranged with sufficient rapidity to keep up with the pace.

The general practice in works dealing with chemical technology is to limit them to some special field. Many of these, relating to almost every phase of applied chemistry, have been published. Many are monographs. (See works containing lists of books, mentioned in Chapter 12.)

In contrast to previous editions of this book, the two well-known encyclopedias on applied chemistry are listed here as general reference sources rather than with the other encyclopedias. The coverage is very broad and comprehensive. In many cases the subjects are discussed in considerable detail. The Ullmann and the Kirk-Othmer sets are the industrial chemical "Britannicas."

F. Ullman (ed.), "Encyklopädie der technischen Chemie," Verlag Chemie International, Weinheim, Germany. The third edition of 19 volumes, edited by Wilhelm Foerst, appeared in 1951–1969. Revision for the fourth edition began in 1972, with completion projected in 25 volumes. H. Buchholz-Meisenheimer is the general editor.

The first six volumes present an account of general principles and methodology. The following 18 are an alphabetical compilation of articles on key subjects, with volume 25 devoted to an index in German and English. The text is German, but all of the volume titles listed here are in English.

1 General Principles of Process and Reaction Technology (1972)
2 Process Engineering I (Unit Operations) (1972)
3 Process Engineering II and Reaction Plants (1973)
4 Process Development and Design of Chemica Plants—Documentation (with a comprehensive index to volumes 1 to 4) (1974)
5 Physical and Physicochemical Analysis Methods (1980)
6 Measurement Technology, Environmental Control, and Occupational Safety (   )
7 Acaricides to Antihistamines (1973)
8 Antimony to Bread (1974)
9 Butadiene to Cytostatica (1975)
10 Dental Chemistry to Naphtha Petroleum Processing (1976)
11 Naphtha Petroleum to Formazan Dyes (1976)
12 Fungicides to Hydroxylamine (1976) Cumulative Index, volumes 7 to 12
13 Hormones to Ceramics (1977)
14 Ceramic Colors to Cork (1977)
15 Corrosion to Lacquer (1978)
16 Bearing Materials to Milk (1978)

R. E. Kirk and D. F. Othmer (eds.), "Encyclopedia of Chemical Technology," John Wiley & Sons, Inc., New York. This work is the most comprehensive current encyclopedia in English. The second edition, edited by H. Standen, appeared in 22 volumes in 1963–1970. The third edition, with M. Grayson as editor, began in 1978. The 16 volumes published through 1981 end with the entry "Perfumes." Additional volumes are planned each year, with completion expected in 1983 and a cumulative index in 1984.

Interim indexes for the new edition cover four volumes each. They include the *Chemical Abstracts* Registry Numbers for each chemical substance listed.

An interesting discussion of the preparation of such an encyclopedia was published by M. Grayson, D. Eckroth, and R. Kromer.[18]

J. J. McKetta and W. A. Cunningham (eds.), "Encyclopedia of Chemical Processing and Design," Marcel Dekker, Inc., New York. Like the Kirk-Othmer encyclopedia, this work is an alphabetical arrangement of subjects which cover technologically important materials, processes, methods, practices, standards, and products. Economic surveys and future trends are included.

Through 1981, 14 volumes will have appeared, the last item in volume 14 being "Design Codes, Standards, and Recommended Practices."

A number of more specialized works have been published. Examples follow.

R. A. Young and P. N. Cheremisinoff (eds.), "Pollution Engineering and Technology," Marcel Dekker, Inc., New York. The following volumes have been published:

1 Air Pollution Control and Design Handbook (2 parts) (1977)
2 Wastewater Renovation and Reuse (1977)
3 Water and Wastewater Treatment (1977)
4 Biofouling Control Procedures (1977)
5 Energy and Solid Wastes (1976)

R. G. Bond and C. P. Straub (eds.), "Handbook of Environmental Control," CRC Press, Inc., Boca Raton, FL. This compilation deals especially with data on pollutants, their effects, and control measures. The following volumes are involved:

  I Air Pollution (1973)
 II Solid Waste (1973)
III Water Supply and Treatment (1973)
IV Waste Water Treatment and Disposal (1974)
 V Hospital and Health-Care Facilities (1975)
   Series Index (1978)

C. T. Lynch, "Handbook of Materials Science," 4 v., CRC Press, Inc., Boca Raton, FL. This work deals with physical properties of solid state and structural materials. Three volumes are included.

  I General Properties (1974)
 II Metals, Composites, and Refractory Materials (1975)
III Nonmetallic Materials and Applications (1975)

[18]*J. Chem. Inf. Comput. Sci.,* **19,** 117 (1979).

Bockris, J. O'M., B. E. Conway, and E. Yeager, "Comprehensive Treatise of Electro-chemistry," 4 v., Plenum Publishing Company, New York, 1980–1981.

Coulson, J. M. (ed.): "Fluid Flow, Heat Transfer, and Mass Transfer," vol. 1; "Unit Operations," vol. 2, Pergamon Press, New York, 1978.

Cox, K. E., and Williamson, K. D., Jr.: "Hydrogen: Its Technology and Implications," 5 v., CRC Press, Inc., Boca Raton, FL, 1977–1979.

Cremer, H. W. (ed.): "Chemical Engineering Practice," 12 v., Plenum Press, New York, 1956–1965.

Howard, G. M.: "Poucher's Perfumes, Cosmetics, and Soaps," 3 v., John Wiley & Sons, Inc., New York, 1974.

Miles, D. C., and J. H. Briston: "Polymer Technology," Chemical Publishing Company, New York, 1979.

Pojasek, R. B.: "Toxic and Hazardous Wastes Disposal," Ann Arbor Science, Ann Arbor, MI, 1979.

Smoot, L. D., and D. T. Pratt (eds.): "Pulverized Coal Combustion and Gasification," Plenum Press, New York, 1978.

Weissermel, K.: "Industrial Organic Chemistry," Verlag Chemie International, New York, 1978.

PHYSICS

Chemistry and physics are very interrelated, especially in measuring techniques and in theoretical interpretation of facts. The literature of physics, like that of chemistry, has become enormous. R. N. Whitford[19] has discussed the various sources. Only two treatises are mentioned here.

S. Flügge (ed.), "Handbuch der Physik," Springer-Verlag, Berlin, 1955–   . In its way, this great treatise is roughly comparable to the treatises of Beilstein and Gmelin in chemistry.

G. W. Ewing (ed.), "Laboratory Instrumentation and Techniques," Plenum Press, New York, 1974–   . Detailed accounts of the physical principles underlying instrumentation, the design and engineering parameters involved, and applications in research are covered. Various volumes are projected.

PHARMACEUTICAL CHEMISTRY

As with biochemistry and agricultural chemistry, pharmaceutical chemistry comes ultimately to include the chemistry of certain inorganic and organic materials. In this field these materials are the ones that possess preventive, remedial, or curative physiological properties. The basic chemistry of such substances is covered in treatises already considered. The following treatises are representative of those devoted to the pharmaceutical and pharmacological properties of materials:

Huff, B. B. (ed.): "Physicians' Desk Reference," Medical Economics Company, Oradell, NJ, 1977–1978.

Manske, R. H. F., and H. L. Holmes: "The Alkaloids: Chemistry and Physiology," Academic Press, New York, 1950–   .

---

[19]"Physics Literature," Scarecrow Press, Metuchen, NJ, 1968.

Negwer, M.: "Organic-Chemical Drugs and Their Synonyms," Verlag Chemie International, New York, 1978.

Osol, A., and G. E. Farra (eds.): "The Dispensatory of the United States of America," J. B. Lippincott Company, Philadelphia, 1973.

Wilson, C. O., and T. E. Jones: "The American Drug Index," J. B. Lippincott Company, Philadelphia, ann.

—— and ——: "Remington's Pharmaceutical Sciences," Mack Publishing Company, Easton, PA, 1975.

—— and ——: "Index Nominem 1978," 3 v., Societe Suisse Pharmacie, Zurich, 1977.

A rather special type of book is represented by the pharmacopoeias of the various countries. They deal with drugs, including preparation, action, and official methods of testing and analysis. Related to these works are the formularies. The following selections are representative:

"AMA Drug Evaluations," Publishing Sciences Group, Inc., Little, MA, 1977.

"British Pharmaceutical Codex," Pharmaceutical Press, London, 1975.

"British Pharmacopoeia," Pharmaceutical Press, London, 1973 plus suppls.

"Martindale the Extra Pharmacopoeia," Pharmaceutical Press, London, 1977.

"National Formulary," American Pharmaceutical Association, Washington, DC, 1975 plus suppls.

"Pharmacopoeia of the United States of America," American Pharmaceutical Association, Washington, DC, 1975 plus suppls.

"Specifications for the Quality Control of Pharmaceutical Preparations," World Health Organization, Geneva, 1967 plus suppls.

Symposium Series No. 84 of the American Chemical Society is "Retrieval of Medicinal Chemical Information." It is edited by W. I. Howe, M. M. Milne, and A. F. Pennell, (1978).

AGRICULTURAL CHEMISTRY

Agricultural chemistry is largely applied general biological chemistry, which in turn rests upon the general principles of inorganic, organic, physical, and analytical chemistry. Therefore, a number of the treatises already discussed are of value here, especially those dealing with biological chemistry. The "agricultural chemistries" of earlier years seem largely to have been abandoned.

Broadly considered, agricultural chemistry concerns plants and animals. Production of agricultural products involves chemistry in the soil, the water, and the air. Fertilizers facilitate growth of crops. Various "-cides" help control blights, bugs, rodents, weeds, and worms. Raising and caring for the different kinds of farm animals involve various nutritional and remedial problems. Chemicals aid in the preservation and marketing of many products.

Only an encyclopedic compilation could cover these and many other subjects. A hardbound work would soon be out of date in various aspects. Perhaps a loose-leaf, open-end collection of late bulletins is the best answer. The national research laboratories and the state agricultural experiment stations issue many brochures. The latter are particularly relevant for local situations (see Chapter 3).

MONOGRAPHS

Of making many books, there is no end; and much study is a weariness of the flesh.

Ecclesiastes 12:12

Several different kinds of secondary sources have been considered as works of reference. To these may be added monographs.

The term "monograph" is used here to designate a comprehensive survey of contemporary knowledge relating to a given subject. In many instances, at least, these publications present an intensive, current statement of what is known about a specific subject. In this sense, they represent an elaboration and updating of the kind of treatment the same subject is generally allotted in a comprehensive treatise or encyclopedia. Examples are the two monographs on sodium chloride cited below.

The general nature of monographs is indicated by the following statement of the Board of Editors of the American Chemical Society Series of Scientific and Technologic Monographs.

> The development of knowledge in all branches of science, and especially in chemistry, has been so rapid during the last fifty years and the fields covered by this development have been so varied that it is difficult for any individual to keep in touch with the progress in branches of science outside his own specialty. In spite of the facilities for the examination of the literature given by *Chemical Abstracts* and such compendia as "Beilstein's Handbuch der organischen Chemie" and "Gmelin's Handbuch der anorganischen Chemie" and the encyclopedias of chemistry, it often takes a great deal of time to coordinate the knowledge available upon a single topic. Consequently, when men who have spent years in the study of important subjects are willing to coordinate their knowledge and present it in concise, readable form, they perform a service of the highest value to their fellow chemists.
>
> Two rather distinct purposes are to be served by such monographs. The first purpose, whose fulfillment will probably render to chemists in general the most service, is to present the knowledge available upon the chosen topic in a readable form, intelligible to those whose activities may be along a wholly different line. Many chemists fail to realize how closely their investigations may be connected with other work which, on the surface, appears far afield from their own. Monographs enable such men to form closer contact with the work of chemists in other lines of research. The second purpose is to promote research in the branch of science covered by the monograph, by furnishing a well digested survey of the progress already made in that field and by pointing out directions in which investigation needs to be extended. To facilitate the attainment of this purpose, it is intended to include extended references to the literature, which will enable anyone to follow up the subject in more detail. If the literature is so voluminous that a complete bibliography is impracticable, a critical selection is made of those papers which are most important.

An English series states that monographs serve another purpose.

It is difficult in the case of large treatises to keep abreast of so rapidly a growing science by means of new editions. Monographs may be issued more frequently upon the various divisions of a general subject, each publication independent of and yet dependent upon the others, so that from time to time as new material and the demand therefore necessitates, a new edition of each monograph can be issued without reissuing the whole series. In this way, both the expenses of publication and the expense to the purchasers are diminished, and by a moderate outlay it is possible to obtain a full account of any particular subject as nearly current as possible.

Monographs are frequently issued as members of a series, written by individual authors under the general direction of an editor or board of editors. New series and new volumes for old series keep appearing.

In addition, many books not belonging to any monograph series obviously meet the specifications for this kind of publication. They may or may not be called monographs. Except for textbooks and periodical sets, it seems likely that most of the books one sees on library shelves might well be classified as monographs.

When prepared by one who knows the particular field and who can and will write, monographs are among our most useful and dependable sources. Thus, J. G. Vail's monograph "The Soluble Silicates", published by Reinhold Publishing Corporation in 1953, appeared only after the author had many years of experience with the Philadelphia Quartz Company. Few others could have written out of such a background of practical knowledge.

The narrower the field encompassed by an author, the more intensive the treatment is likely to be. Thus, the monograph by N. E. Dorsey, "Properties of Ordinary Water Substance," devoted 673 pages to perhaps our best known chemical, water.

Coverage by a monograph is illustrated further by separate volumes dealing with another common chemical, sodium chloride.

The first work, by D. W. Kaufmann of the International Salt Company, is entitled, "Sodium Chloride" with the subtitle "Production and Properties of Salt and Brine." Published by the Reinhold Publishing Corporation, New York, in 1960, it has 743 pages. It is one of the monographs sponsored by the American Chemical Society.

The second work, "Neptune's Gift," by R. P. Multhauf, was published by The Johns Hopkins University Press, Baltimore, MD, in 1978, with the subtitle "History of Common Salt," as one of a series of historical monographs. This 325-page scholarly book is very different from the one by Kaufmann.

It is interesting to compare the content of these two monographs with what is included on sodium chloride in any beginning text on chemistry, or in the comprehensive inorganic treatises (see pp. 209–217), such as those of Gmelin, Mellor, Pascal, or Trotman-Dickenson.

The following titles of monographs are representative selections:

Analytical Chemistry

Mitchell, J., Jr.: "Aquametry," John Wiley & Sons, Inc., New York, 1977–1981.
Roberts, T. R.: "Radiochromatography," Elsevier Publishing Company, New York, 1978.

Slavin, M.: "Atomic Absorption Spectroscopy," Wiley-Interscience, New York, 1979.

Yau, W. W., J. J. Kirkland, and D. D. Bly: "Size-Exclusion Chromatography," John Wiley & Sons, New York, 1979.

Yau, W. W., et al., "Gel-Exclusion Chromatography," John Wiley & Sons, New York, 1979.

Biochemistry

Coburn, R. F., et al.: "Carbon Monoxide," National Academy of Science, Washington, DC, 1977.

Dwek, R.: "NMR in Biology," Academic Press, New York, 1977.

Grover, P. L.: "Chemical Carcinogens and DNA," CRC Press, Boca Raton, FL, 1979.

Weinstein, M. J., and G. H. Wagman (eds.): "Antibiotics," Elsevier Publishing Company, New York, 1978.

Industrial Chemistry (Chemical Engineering)

Oglesby, S., Jr., and G. B. Nichols: "Electrostatic Precipitation," Marcel Dekker, New York, 1978.

Stafford, D. A., D. L. Hawkes, and H. R. Horton: "Methane Production from Waste Organic Matter," CRC Press, Inc., Boca Raton, FL, 1980.

Throne, L.: "Plastics Process Engineering," Marcel Dekker, New York, 1979.

Wisniak, J., and A. Tamir: "Mixing and Excess Thermodynamic Properties," Elsevier Publishing Company, New York, 1978.

Inorganic Chemistry

Bau, R. (ed.): "Transition Metal Hydrides," American Chemical Society, Washington, DC, 1978.

Bell, C. F.: "Principles and Applications of Metal Chelation," Oxford University Press, New York, 1978.

Puddephatt, R. J.: "The Chemistry of Gold," Elsevier Publishing Company, New York, 1978.

Ray, N. H. (ed.): "Inorganic Polymers," Academic Press, New York, 1978.

Organic Chemistry

Bender, M. L.: "Mechanism of Homogeneous Catalysis from Protons to Proteins," John Wiley & Sons, Inc., New York, 1971.

Brown, H. C.: "Organic Syntheses via Boranes," John Wiley & Sons, Inc., New York, 1975.

Ellis, G. P.: "Chromenes, Chromanones, and Chromones," John Wiley & Sons, Inc., New York, 1977.

Pizey, J. S.: "Synthetic Reagents: Lithium Aluminum Hydride," Ellis Horwood, Ltd., Chichester, England, 1977.

Walsh, C.: "Enzymatic Reaction Mechanisms," W. H. Freeman and Company, San Francisco, 1979.

Physical Chemistry

Bacon, G. E.: "Neutron Scattering in Chemistry," Butterworths, Woburn, MA, 1977.

Harris, D. C.: "Symmetry and Spectroscopy," Oxford University Press, New York, 1978.

Nielsen, L. E.: "Predicting Properties of Mixtures," Marcel Dekker, New York, 1978.

Rhodin, T. N., and G. Ertl (eds.): "The Nature of the Surface Chemical Bond," North-Holland Publishing Company, Amsterdam, 1979.

SECONDARY SOURCES—
TEXTBOOKS AND
NONPRINT MEDIA

Oh, . . . that mine adversary had written a book.

Job 31:35

The sources considered in this chapter consist of printed and nonprinted types. The former are textbooks, and the latter are tapes, filmstrips, cassettes, and similarly recorded media. Although the two types are related in some respects, they seem to merit separate consideration.

TEXTBOOKS

Textbooks belong in a general discussion of works of reference, although usually they are not very useful as reference sources.

In the more comprehensive examples many specific facts may be included, such as sources of the chemical elements, their preparation, physical and chemical properties, and uses. Thus, J. W. Mellor's "Modern Inorganic Chemistry," published in 1917, contains a wealth of information in its 900 pages. Incidentally, he followed the usage of the Chemical Society (London) for the spelling "radicle." It is uncertain how later the word came to be politicized to the "pink" radical.

The word "textbook" is used here to denote a manual of instruction. Similar meaning attaches to the German "Lehrbuch" and the French "traité," as commonly used. Other meanings may be found. For example, J. A. N. Friend's "Textbook of Inorganic Chemistry" (24 v.) and K. Jellinek's "Lehrbuch der

Physikalischen Chemie" (5 v.) are treatises, as the term is used in Chapter 10.

According to W. T. Lippincott,[1] in 1597 Andreas Libavius wrote what appears to be the first textbook of chemistry. Entitled "Alchemica," it envisioned teaching chemistry rather than alchemy. There followed some 200 years of discoveries, speculations, and attempted transmutations by chemists, alchemists, and phlogistonists. Finally, in 1789, A. L. Lavoisier published what is probably the first modern textbook of chemistry, "Traité Elementaire de Chimie."

Usually the common American textbook is a single-volume work. There are three more or less distinct types. Some consist of directions for performing laboratory experiments, the nature of the contents being indicated by the title, as "Experimental Organic Chemistry." Others are devoted entirely to descriptive and/or theoretical matter, an example of the latter being "Chemical Principles." Some contain descriptive and theoretical material, together with directions for performing exercises relating to or illustrating the other portion of the text. Various books on quantitative chemical analysis represent this type.

All such texts are intended to introduce the subject to students starting their study in the respective fields. Even at this level, there may be considerable variation. Some are very elementary in consisting of a little of "this" and a little of "that," for a one-semester course for students with little or no preparation. Others assume good preparatory background and generally are designed for students majoring in chemistry or chemical engineering. For subsequent courses the texts are more advanced and specialized.

Most advanced are texts designed for graduate courses. They may differ little from a monograph in being largely a survey of the present status of some rather narrow field. This topic may be the specialized area of research of some professor, such as the sequencing of amino acids in proteins.

Because of their nature and the manner in which they are produced, textbooks need special attention in evaluating them as part of the chemical literature. As the last of the secondary sources, they stand apart from the others in most respects. They have some value as reference sources, but mainly they are teaching tools.

Of considerable interest in understanding them is the manner followed by probably most authors in writing their texts. A book designed to introduce students majoring in chemistry may be taken as an example. It seems unlikely that any one author would know all the data to be presented. In essence, these data comprise facts and their interpretations. The latter are usually dignified as principles. For example, sodium chloride is identified as a specific entity, with definite properties, such as color, crystal structure, density, solubility in water, and others. Why are the properties what they are? Interpretations are attempts to correlate properties and to relate specific effects to immediately preceding causes. They try to answer the how and the why of things.

The author, then, searches the field which the book is designed to cover and

[1]*J. Chem. Educ.* **56,** 355 (1979).

selects the facts and the principles which it seems desirable to include. These gleanings, seldom including new material, are arranged in the order and the way which the author deems desirable. To promote interest, pictures, graphs, tables, and other illustrative material may be included. Thus, the author's chief function is to select, arrange, and discuss. Often questions are included for the student, along with numerical problems, as in analytical and physical chemistry.

If this viewpoint is tenable, an author bears special responsibility in several directions. The resulting textbook may well be the means of intellectual conditioning of untold thousands of students. In this respect textbooks are unique publications.

No doubt W. Gerard[2] was overly pessimistic in an article entitled "What Are Textbooks Made Of?" But he did caution users that the information may not all be facts, and that the interpretations may not apply to the stated facts. He might have reminded us of the title of a song in *Porgy and Bess,* "It Ain't Necessarily So." H. H. Sisler and C. A. VanderWerf have discussed difficulties of trying to interpret many oxidation-reduction transformations in terms of transfer of electrons.[3]

In general, writers in chemistry have available the following principal means of communication: words, formulas and equations, tables, drawings and graphs, and pictures (or in engineering, small models constructed to scale). Disregarding fictional productions, what is achieved in using these means varies widely; but the general aims of meticulous editors are accuracy, clarity, brevity, consistency, and interest.[4] How well does an author perform?

First in importance is the sampling of material relevant to the area to be covered. Organic chemistry offers a superb illustration of the problem. Several million compounds have been characterized and described. Which ones of this vast list should be selected to discuss in an elementary text for chemistry majors?

The prestigious text by J. B. Conant and A. H. Blatt[5] of Harvard University mentioned almost exactly 4,000 compounds. How did the authors select these? Next comes the question of how many data and how much discussion are to be allotted to each one. Taking methanol as an example, there is no separate discussion. A few facts are included under simple alcohols as a group. In contrast, the Beilstein treatise devotes the following pages to this one compound: Hauptwerk (to 1910) 19; E I (1911–1920) 15; E II (1921–1930) 29; E III (1931–1950) 76; E IV (1951–1960) 16.

It should be noted, of course, that the Beilstein presentation covers functional, nonfunctional, and replacement derivatives of methanol. However, each page contains at least twice as much print, in smaller type, as one by Conant and Blatt.

Along with the question of adequate sampling is the question of the reliability

[2]*Chem. Ind., (London),* **1969,** 1338.
[3]*J. Chem. Educ.,* **57,** 42 (1980).
[4]M. G. Mellon, *J. Chem. Doc.,* **4,** 1, (1964).
[5]"The Chemistry of Organic Compounds," The Macmillan Company, New York, 1959.

of the information included. Careless proofreading of galleys and page proof is unusual, but there have been noteworthy exceptions. The primary concern here is the author's judgment in statement of facts, clarity of definitions, and overall perspective of the field.

The general reliability of textbooks probably rates above the advertised purity of the famous soap (99.44 percent). However, one must be on the alert to the possibility that the other 0.56 percent may be questionable, if not actually wrong. Several examples illustrate error or lack of clarity.

In a widely used book on quantitative chemical analysis the sample is not mentioned until midway through the book, although every experienced analyst knows that an analysis begins with a sample. Then there occurs the amazing statement that all samples must be dried, ground, and dissolved. Yet this book was written in the shadow of a famous hospital and within sight of a large body of water. Where, in any approved methods of chemical analysis, is one directed to dry, grind, and dissolve samples of urine, blood, or water?

A book on organic chemistry contains the statement that the announcement of all new chemical knowledge appears first in journals of research. Yet some 10,000 chemical patents are reported annually in the *Official Gazette* (Patents) of the United States Patent and Trademark Office. Many of these are for organic compounds and/or processes for making them. Novelty, i.e., newness, is the first requirement of patentability. The statement seems to imply, too, that nothing new is coming out of the 779 Federal research laboratories listed in the Whitten Report, except the few articles appearing in the small number of Federal research journals mentioned in Chapter 3.

The perspective of the authors of a book on quantitative chemical analysis of certain commercial products is revealed in their classification of methods: gravimetric, volumetric, and special.

The usual dictionary definition of gravimetric is determination by mass, with no implication about the method of separation employed to obtain the desired constituent, or something chemically equivalent to it, in weighable form. The book first restricts the separation to precipitation, with the further restriction that the precipitate must be at least a binary compound, such as silver chloride. Thus, separation and weighing of certain elements as such, e.g., gold, platinum, and selenium, would not be gravimetric methods.

The volumetric measurements are limited to titrations by measuring the volume of the titrant, although measuring a titrant by mass eliminates several errors inherent in volumetric measurements. There is no mention of the possibility of measuring gases and solids by volume.

Finally, think what the "special" group includes, with about a dozen more or less common methods of separation, and more than a score of methods of measurement available. Once students are conditioned to a certain viewpoint, it is very difficult to decontaminate them.

Probably errors occur because authors try to write about more than they can know personally.

In Chapter 6 note was made of the uncertainties arising from the use of imprecise titles in indexing periodicals. They may occasion inconvenience and loss of time, and one may miss what is wanted. Probably there is no lasting educational damage.

With students exposed to textbooks and teachers, however, the result may be more serious. Individuals usually have faith in scientific books and professors. The use of uninformative, imprecise, misleading, or even wrong terms is generally unfortunate. Once conditioned to them, one does not easily recover. The mind is slow in unlearning what it has been long in learning.[6]

The difficulty of selecting and using precise terms has been noted in trying to differentiate in this book among the various kinds of chemical publications. Stuart Chase elaborated on the problem in his book "The Tyranny of Words." Few can achieve complete success in their quest for certainty of expression, but writers and teachers can aim to communicate in a manner that cannot be misunderstood.

Inconsistency by an author, in addition to contributing to lack of clarity, may confuse the student and trouble the instructor using a text. Two familiar examples illustrate the problem.

One concerns the suffix "-ide." For a century or more in inorganic chemistry it meant a *binary compound,* such as carbide, halide, hydride, nitride, oxide, sulfide, and others. Suddenly the words "actinide" and "lanthanide" were used to designate *groups* of chemical elements closely related to actinium and lanthanum, but not to a binary compound of actinium or lanthanum, respectively.

Another example concerns terms used in analytical chemistry, such as "oxidimetry" and "precipitimetry" to denote the use as titrants of oxidants and precipitants, respectively. But in most cases, in the same book, acidimetry meant the determination *of* acids and not the *use* of acids as titrants. What does "aquametry" mean—the measurement *of* or *with* water?

Many similar cases might be cited, including naming analytical methods in terms of the unit of measurement used.[7] It is bad enough for mature chemists to have to deal with inaccurate, unclear, and inconsistent terms in primary publications, but in textbooks we are conditioning students to use them.

Perhaps the most important point is to invite our students to think about what they read and what they hear dispensed from behind a lecture table. Jeremy Bernstein[8] has written that "The most valuable commodity that we have in science is doubt." Even more concise is the quotation attributed to Cicero, "By doubting we come at the truth."

As some of the more important periodicals, treatises, and other kinds of publications have been mentioned, a short list of titles is included here for selected textbooks in several fields of chemistry. It should not be assumed that the ones cited are considered the best or latest in their areas. Rather they are

---

[6]Seneca, *Troades.*
[7]M. G. Mellon, *J. Chem. Educ.,* **50,** 690 (1973).
[8]*Am. Scholar,* **48,** 8 (1978).

intended only to be representative. For the most part the works are American, as these are the usual textbooks in this country.

General Chemistry

Lippincott, W. T., A. B. Garrett, and F. H. Verhoek: "Chemistry," John Wiley & Sons, Inc., New York, 1977.

Moore, J. W., W. G. Davies, and R. W. Collins: "General Chemistry," McGraw-Hill Book Company, New York, 1978.

Sienko, M. J., and R. A. Plane: "Chemical Principles and Properties," McGraw-Hill Book Company, New York, 1979.

Sisler, H. H., R. D. Dresdner, and W. T. Mooney, Jr.: "Chemistry: A Systematic Approach," Oxford University Press, New York, 1980.

Analytical Chemistry (Qualitative)

Cheng, H-S.: "Analytical Chemistry (Qualitative Analysis)," San Min Shu Chu, Taipei, Taiwan, 1977.

Clifford, A. F.: "Inorganic Chemistry of Qualitative Analysis," Prentice-Hall, Inc., Englewood Cliffs, NJ, 1961.

Lyalikov, Yu S., and Yu A. Klyachko: "Theoretical Principles of Modern Qualitative Analysis," Khimiya, Moscow, 1978.

Svehla, G.: "Vogel's Textbook of Macro and Semimicro Qualitative Inorganic Analysis," Longman, New York, 1979.

Note. Various properties of many chemical products depend upon the nature of the chemicals which the products contain. Chemists employed in industries buying, making, using, and selling these products must maintain the compositions desired. Just as important, perhaps, is to determine the composition of competitors' products.

To detect and identify the significant components in a product is to perform qualitative chemical analysis. The importance attached to this aspect of chemistry was noted for Gmelin's great inorganic treatise. Even more striking is the space allotted to qualitative analysis in the large analytical treatises.

For decades many teachers used texts based on the Fresenius system of separations of cations and anions. The best courses served as a framework for teaching inorganic chemistry. Such courses, and the textbooks used in them, seem largely to have disappeared.

Analytical Chemistry (Quantitative)

Bassett, J., R. C. Denny, J. Mendham, and G. H. Jeffrey: "Vogel's Textbook of Quantitative Inorganic Analysis," Longman, New York, 1978.

Crooks, J. E.: "The Spectrum in Chemistry," Academic Press, New York, 1978.

Kenner, C. T., and K. W. Busch: "Quantitative Analysis," Macmillan Publishing Company, New York, 1979.

McLafferty, F. W.: "Interpretation of Mass Spectra," University Science Books, Mill Valley, CA, 1980.

Biochemistry

Lehninger, A. I.: "Biochemistry," Worth Publishers, New York, 1975.

Metzler, D. E.: "Biochemistry," Academic Press, New York, 1977.

Stryer, L., "Biochemistry," W. H. Freeman and Company, San Francisco, CA, 1981.
White, W., P. Handler, E. L. Smith, R. L. Hill, and I. R. Lehman: "Principles of Biochemistry," McGraw-Hill Book Company, New York, 1978.

Industrial Chemistry (Chemical Engineering)

Cavaseno, V.: "Process Technology and Flowsheets," McGraw-Hill Book Company, New York, 1979.
Clausen, R. M., and C. Mattson: "Principles of Industrial Chemistry," Wiley-Interscience, New York, 1978.
Felder, R. M., and R. W. Rousseau: "Elementary Principles of Chemical Processes," John Wiley & Sons, New York, 1978.
Peters, M. S.: "Plant Design and Economics for Chemical Engineers," McGraw-Hill Book Company, New York, 1980.
Shreve, R. N., and J. Brink: "The Chemical Process Industries," McGraw-Hill Book Company, New York, 1977.

Inorganic Chemistry

Cotton, F. A., and G. Wilkinson: "Basic Inorganic Chemistry," Wiley-Interscience, New York, 1976.
Huheey, J. E.: "Inorganic Chemistry," Harper and Row, New York, 1978.
Jolly, W. L.: "The Principles of Inorganic Chemistry," Marcel Dekker, New York, 1976.
Purcell, K. F., and J. C. Kotz: "Inorganic Chemistry," W. B. Saunders and Company, Philadelphia, PA, 1977.

Organic Chemistry

Brown, R. F.: "Organic Chemistry," Wadsworth Publishing Company, Belmont, CA, 1975.
March, J.: "Advanced Organic Chemistry," McGraw-Hill Book Company, New York, 1977.
Morrison, R. T., and R. N. Boyd: "Organic Chemistry," Allyn and Bacon, Inc., Boston, MA, 1973.
Solomons, T. W. G.: "Organic Chemistry," John Wiley & Sons, Inc., New York, 1978.

Physical Chemistry

Alberty, R. A.: "Physical Chemistry," John Wiley & Sons, Inc., New York, 1980.
Barrow, G. M.: "Physical Chemistry," McGraw-Hill Book Company, New York, 1979.
Eyring, H.: "Basic Chemical Kinetics," John Wiley & Sons, Inc., New York, 1980.
Rosenberg, R. M.: "Principles of Physical Chemistry," Oxford University Press, New York, 1977.

INSTRUCTIONAL MEDIA IN NONPRINT FORM

Since the time of Lavoisier, printed textbooks on chemistry have been the principal means employed in teaching the science. Lectures and exercises in the laboratory have served to supplement and illustrate the presentations in the textbooks.

For several decades nonprint media have become available. Most librarians have catalogues and indexes of these media. Many educational institutions maintain audiovisual centers to facilitate the use of this means of transmitting chemical information.

Although usually such media have not been considered a part of the chemical literature, they have become important in various ways in supplementing textbooks. These nonprint media comprise audio, video, and audiovisual means.

Audio Media

Faced with the necessity of repeating the same lecture to several sections of the same course, some instructors have taped their lectures to avoid the repetition. The recordings may be made available for individual students to hear, if they missed the presentation or wish to hear it again.

An excellent example of such lectures is described in a brochure of the American Chemical Society, "Audio Courses." The 1981 catalogue lists a series of 52 professional-level audio courses by recognized authorities. The materials consist of two parts: (1) audiocassette recordings of the lecture(s) and (2) coordinated manuals for the material discussed. Two of these lecture series concern chemical literature as such. One is "Chemical Abstracts, An Introduction to Its Effective Use," and the other is "Use of the Chemical Literature." In the latter case the manual is essentially an outline of the items discussed in the lectures. It is accompanied by another manual containing illustrative library problems. To use these films there is an audiocassette player/recorder available.

The American Chemical Society also sponsors a series of lectures entitled, "Man and Molecules." These audio tapes feature talks by well-known scientists.

A third example of interest to literature chemists is a guide and cassette, "A Guide to Beilstein's Handbuch," by Martha M. Vestling.[9]

Such lectures, if separate entities, may vary widely in nature. They may amount to an updating of a subject discussed in an encyclopedia. In contrast, they may recount a late discovery. If given by a famous individual, such as a Nobel laureate, they feature his or her manner of presentation.

Video Media

This type of nonprint medium is of value to demonstrate how to perform some operation. No doubt it is most effective for an instructor personally to show the individual student how to carry out a manual assignment, but this alternative is impractical with large classes. A movie demonstration can be observed simultaneously by many students, and the film may be rerun if a student wants to see the demonstration again.

As an illustration, the author[10] prepared several such movies for a course in quantitative chemical analysis. One of them demonstrated the method of

[9] J. Hurley Associates, Boca Raton, FL, 1975.
[10] *Proc. Indiana Acad. Sci.*, **48**, 74 (1938).

operating the equal arm balance then available, and the process of weighing a factor weight of sample using counterpoised watch glasses. After this film was shown, a laboratory section was taken to the balance room. Following the example of the instructor, each student examined a set of weights and the movable parts of the balance, including the several means of adjustment. Finally, a small object was weighed and the mass value recorded.

Another film demonstrated dissolution of a sample of soluble chloride, adjustment of the pH of the solution, precipitation of the chloride ions as silver chloride, digestion of the precipitate, preparation of a Gooch filtering crucible, transfer of the precipitate to the crucible, and washing of the precipitate.

Four decades later A. A. Russell and B. L. Mitchell[11] reported the program descriptions for their series of 15 videotapes, "Experiments and Techniques for General Chemistry."

Most laboratory operations can be demonstrated in this way. Although they are described adequately in the student's manual, seeing the operation performed is helpful to many individuals.

A number of film/video courses are available from the American Chemical Society. An example is "Gas Chromatography" in two parts by H. M. McNair.

Audiovisual Media

Combining audio and video allows a speaker to be heard and manual operations observed. Using color film or videotape probably yields the best possible presentation.

One of the most extensive, and possibly one of the best, presentations of this kind was the series of broadcasts by the National Broadcasting Company entitled "Modern Chemistry" in 1959. It was one of three television courses of the program *The Continental Classroom*. Most of the chemistry lectures, with accompanying demonstrations, were conducted by Dr. John F. Baxter. Several visiting lecturers participated by giving single lectures. The audiovisual films were made available to those interested.

To accompany the course, a paperback text was prepared by Dr. Baxter and Dr. L. E. Steiner. It covered 39 topics, but not those of the visiting lecturers.[12]

Another example was the film "Curves of Color," prepared by the General Electric Company to illustrate the principles and use of their recording spectrophotometer. To supplement his talks on "Chemistry, Curves, and Color," the author used this film for some years.

Judith A. Douville and B. S. Schlessinger published an annotated list of 23 sources of audiovisual materials.[13]

[11]*J. Chem. Educ.*, **58**, 753 (1979).
[12]J. F. Baxter and L. E. Steiner, "Modern Chemistry," Prentice-Hall, Inc., Englewood Cliffs, NJ, 2 v 1959, 1960.
[13]*J. Chem. Educ.*, **57**, 796 (1980).

TERTIARY SOURCES—
GUIDES AND DIRECTORIES

If we could first know where we are, and whither we are tending, we could then better judge what to do, and how to do it.

Abraham Lincoln

There remains a variety of miscellaneous sources which might, perhaps, have been included as a subdivision of Chapter 9, since the publications are, in a sense, works of reference. However, for the most part they seem to be sufficiently different to justify separate consideration.

These works are designated here as "tertiary sources," inasmuch as they serve, in part at least, as guides to the secondary and primary sources. They are aids for searching all literature. Also, they provide facts about chemists and their work. The latter category includes plants, products, societies, and related items.

These publications often are of more use to nonchemists than to those actually practicing the profession. Reference librarians, for example, receive many questions for which answers are available in tertiary sources.

GENERAL GUIDES

A number of publications are devoted to the overall problem of chemical literature as such. From a broad viewpoint, they may be classified as books and periodicals.

Books about Chemical Literature

In recent decades the literature has grown to such an extent that books on its nature and use seem justified. At any rate, they have appeared and have found use both in specific courses devoted to the subject in departments of chemistry and in a more general way in library schools.

1919   Ostwald, W.: "Die Chemische Literatur und die Organisation der Wissenschaft," Akademische Verlagsgesellschaft m.b.H., Leipzig.

1921   Sparks, M. E.: "Chemical Literature and Its Use,"[1] privately published.

1924   Mason, F. A.: "An Introduction to the Literature of Chemistry," Oxford University Press, Fair Lawn, NJ.

1927   Crane, E. J., and A. M. Patterson: "A Guide to the Literature of Chemistry," John Wiley & Sons, Inc., New York.

1928   Mellon, M. G.: "Chemical Publications," McGraw-Hill Book Company, New York.

1938   Soule, B. A.: "Library Guide for the Chemist," McGraw-Hill Book Company, New York.

1946   Serrallach, M.: "Bibliographia Quimica," J. Bosch, Barcelona.

1951   Dyson, G. M.: "A Short Guide to the Literature of Chemistry," Longmans, Green & Co., Inc., New York.

1958   Singer, T. E. R.: "Information and Communication Practice in Industry," Reinhold Publishing Corporation, New York.

1960   Janiszewska-Drabarek, S., J. Surowinski, and A. Szuchnik: "Literatura chemiczna," Technika poslugiwania sie, Warsaw.

1961   Hanć, O., B. Hlavica, V. Hummel, and J. Jelinek[2]: "Chemická Literatura," Statní Nakladatelství Technické Literatury, Praha.

1961   Reid, E. E.: "Invitation to Chemical Research," Franklin Publishing Company, Inc., Palisades, NJ.

1962   Bottle, R. T. (ed.): "Use of the Chemical Literature," Butterworths, London.

1962   Novak, A.: "Fachliteratur des Chemikers," VEB Deutscher Verlag der Wissenschaften, Berlin.

1965   Burman, C. H.: "How to Find Out in Chemistry," Pergamon Press, New York.

1967   Terent'ev, A. P., and L. A. Yanovskaya: "Khimicheskaya Literatur i Pol'zovanic eyu," Khimiya, Moscow.

1969   Herner, S.: "A Brief Guide to Sources of Scientific and Technical Information," Information Resource Press, Washington, DC.

1970   Eberson, L.: "Introduktion till den Kemiska Literaturen," Studentlitteratur Lund, Sweden.

1973   Davis, C. H., and J. E. Rush: "Information Retrieval and Documentation in Chemistry," Greenwood Press, Westport, CT.

---

[1]This small book seems to be the first text for a course on chemical literature. Such a course was begun in 1913 at the University of Illinois.

[2]This book contains long lists of various kinds of publications, such as journals, abstract journals in specific libraries, reference works, monographs, and textbooks. Especially to be noted are those in languages of Southeastern and Eastern Europe.

1974   Woodburn, H. M.: "Using the Chemical Literature," Marcel Dekker, Inc., New York.
1975   Auger, C. P. (ed.): "Use of Report Literature," Shoestring Press, Hamden, CT.
1978   "CAS Printed Access Tools," American Chemical Society, Washington, DC.
1978   Wilen, S. H.: "Use of the Chemical Literature" and "Use of the Chemical Literature—Exercises," American Chemical Society, Washington, DC.
1979   Anthony, A.: "Guide to Basic Information Sources in Chemistry," John Wiley & Sons, Inc., New York.
1979   Maizell, R. E.: "How to Find Chemical Information," John Wiley & Sons, Inc., New York.
1981   Subramanyan, K.: "Scientific and Technical Information Resources," Marcel Dekker, Inc., New York.

The above list is arranged in the order of appearance of the first editions:

The following titles of closely related books are examples from a considerable number available:

Brunn, A. L.: "How to Find Out in Pharmacy," Pergamon Press, New York, 1969.

Chen, C-C.: "Scientific and Technical Information Sources," MIT Press, Cambridge, MA, 1979.

Coblans, H. (ed.): "Use of Physics Literature," Butterworths, London, 1975.

Gould, R. F. (ed.): "The Literature of Chemical Technology," *Adv. Chem. Ser.*, No. 78, American Chemical Society, Washington, DC, 1968.

Grogan, D. J.: "Science and Technology: An Introduction to the Literature," Linnet Books, Hamden, CT, 1976.

Holm, B. E.: "How to Manage Your Information," Reinhold Book Corporation, New York, 1968.

Jackson, E. B., and R. L. Jackson: "Industrial Information Systems," Dowden, Hutchison, and Ross, Inc., Stroudsburg, PA, 1978.

Malinowsky, H. R., and J. M. Richardson: "Science and Engineering Literature: A Guide to Reference Sources," Libraries Unlimited, Littleton, CO, 1980.

Mildren, K. W. (ed.): "Use of Engineering Literature," Butterworths, London, 1976.

Mount, E. (ed.): "Guide to Basic Information Sources in Engineering," John Wiley & Sons, Inc., New York, 1976.

Parker, C. C., and R. V. Turley: "Information Sources in Science and Technology," Butterworths, London, 1975.

Peck, T. P.: "Chemical Industries Information Sources," Gale Research Company, Detroit, 1979.

Rawles, B. A.: "A Guide to the Scientific and Technical Literature of Eastern Europe," National Science Foundation, Washington, DC, 1962.

Sewell, W.: "Guide to Drug Information," Drug Intelligence Publications, Inc., Hamilton, IL, 1976.

Stibic, V.: "Personal Documentation for Professionals," North Holland Publishing Company, New York, 1980.

Weiser, S.: "Guide to the Literature of Engineering, Mathematics, and the Physical Sciences," The Johns Hopkins University Press, Baltimore, 1972.

Whitford, R. H.: "Physics Literature," Scarecrow Press, Metuchen, NJ, 1968.

Periodicals on Documentation

In recent years the word "documentation" has come into use. One meaning is the production, dissemination, and use of publications.

A periodical dealing specifically with chemistry is the *Journal of Chemical Information and Computer Sciences.* More newsy is *Chemical Information Bulletin,* the quarterly publication of the Division of Chemical Information of the American Chemical Society.

More general periodicals of this kind are the *Journal of the American Society of Information Science* (United States); *Journal of Documentation* (Gt. Britain); *Revue de la documentation* (France); *Dokumentation, Dokumentation-Fachbibliothek-Werksbücherei,* and *Nachrichten für Dokumentation* (Germany); and *Tijdschrift voor Efficientie en Documentatie NIDER* (Netherlands). *Special Libraries* (United States) is also of interest.

Related secondary periodicals are *Annual Review of Information Science and Technology* and *Information Science Abstracts.*

BIOGRAPHICAL WORKS

Many times in office, study, or library one wants to know biographical data about chemists and workers in related fields. For individuals who are well known, such information generally can be found in works of the "Who's Who" type. Questions such as the following arise: date of birth, education, principal field of chemistry, experience, publications, and address of home or place of business.

Few such publications appear annually. As there may be some years between editions, one should note the date of publication whenever this is important

Some publications limit the information to mere biographical data. Others add the titles and dates of important publications by the author. The Poggendorff compilation also includes titles of published papers and their source.

Subject and area coverage vary widely. The following examples, which list the names of many chemists and chemical engineers, are arranged according to coverage:

International Listings

1. All Eligible Areas

"Authors and Writers Who's Who," Burke's Peerage, Ltd., London.

"Biographical Dictionaries Master Index," Gale Research Company, Detroit.

"Contemporary Authors," Gale Research Company, Detroit.

"Dictionary of International Biography," International Biographical Centre, Cambridge, England.

"International Who's Who," Europa Publications, Ltd., London.

"Who's Who," A. and C. Black, Ltd., London.

"Writer's Directory," St. Martin's Press, New York.

The names of those publishing in journals indexed by the Institute for Scientific Information are listed in its "Current Bibliographic Directory of the Arts and Sciences."

2. Science

"International Directory of Research and Development Scientists" and "Who's Publishing in Science," Institute for Scientific Information, Philadelphia.
"Poggendorff's Biographisch-Literarisches Handwörterbuch zur Geschichte der exacten Wissenschaften," Verlag Chemie, Leipzig.
"Dictionary of Scientific Biography," Charles Scribner's Sons, New York.
"World Who's Who in Science," A. N. Marquis Company, Chicago.

3. Chemistry

"Chemical Age Directory and Who's Who," Benn Brothers, Ltd., London.
"Chemical Who's Who," Lewis Historical Publishing Company, New York.
"Nobel Lectures: Chemistry," Elsevier Publishing Company, New York.
"Nobel Prize Winners in Chemistry," Abelard-Schumann, Ltd., New York.

National Listings

1. All Eligible areas

"Directory of American Scholars," R. R. Bowker Company, New York.
"Dictionary of American Biography," Charles Scribner's Sons, New York.
"Poor's Register of Directors and Executives," Standard and Poor's Corporation, New York.
"Who Knows and What" and "Who's Who in America," A. N. Marquis Company, Chicago.

2. Science

"American Men and Women of Science," R. R. Bowker Company, New York.
"Dictionary of British Scientists," St. Martin's Press, Inc., New York.
"Handbook of Scientific and Technical Awards in the United States and Canada," Special Libraries Association, New York.
"McGraw-Hill Modern Scientists and Engineers," McGraw-Hill Book Company, New York.
"Who's Who in Science in Europe," Francis Hodson, Ltd., Guernsey.

3. Chemistry

"Addressbuch deutscher Chemiker," Verlag Chemie, New York.
"Chemical Society's Memorial Lectures," The Chemical Society, London.

4. Education

"Who's Who in American Education," Who's Who in American Education, Nashville, TN.

State Listings

Science
"Indiana Scientists," Indiana Academy of Science, Indianapolis, IN.

There are "Who's Who" compilations (in English) for at least a dozen foreign countries, and also similar publications in the languages of the various countries. In the United States there are such works for various regions and many states.

SCIENTIFIC AND TECHNICAL SOCIETIES

In this category the information desired concerns the names of organizations that exist or did exist and people who belong to them.

Organizations

Useful directories to societies include the following:

Bates, R. S.: "Scientific Societies in the United States," MIT Press, Cambridge, MA, 1965.

Bolton, H. C.: "Chemical Societies in the 19th Century," Smithsonian Institute Publications Misc. Collections, No. 1314 (1902).

Colgate, C., Jr., and P. Broida (eds.): "National Trade and Professional Associations of the United States and Canada," Columbia Books, Washington, DC, 1981.

———: "Encyclopedia of Associations: National Organizations of the United States," 3 v., Gale Research Company, Detroit, 1978–1980.

———: "Scientific, Technical, and Related Societies of the United States," National Academy of Science, Washington, DC, 1971.

Many similar publications have appeared. The American Library Association publishes a "Guide to Reference Books," edited by E. P. Sheehy, which lists various categories of this kind of source. One section covers societies and congresses. The ninth edition is dated 1976, but there are supplements.

Membership Lists

Usually at irregular intervals societies publish lists of their members. Of most interest to chemists are those of the American Chemical Society, the American Institute of Chemical Engineers, the American Institute of Chemists, and the American Society for Testing and Materials. Included for the American Chemical Society would be the members of the various divisions of the society.

Examples of more specialized organizations are the Optical Society of America, the Society for Applied Spectroscopy, and the Coblentz Society. There are many others.

Physical chemists might be interested in lists for various physical societies, and biochemists in those for certain biological societies.

Other matters of interest might be the chemists elected to the National

Academy of Science, or the chemical engineers elected to the National Academy of Engineering. Similarly, knowing which chemists are fellows in the American Institute of Chemists and/or the New York Academy of Sciences might be important to some chemists.

No general source will provide all such information. Each organization has its own list. Occasionally one may be updated annually.

LABORATORY PLANNING AND ADMINISTRATION

Advances in chemical knowledge have greatly increased the complexity of facilities required for much current research. Technological developments have produced a comparable situation in plants producing chemical products. Laboratories for teaching, research, and development are a special architectural problem. The following books contain much valuable advice and information in this area:

Architectural Record: "Buildings for Research," F. W. Dodge Company, a Division of McGraw-Hill, Inc., New York, 1958.

Guy, K.: "Laboratory Organization and Administration," The Macmillan Company, New York, 1962.

Lewis, H. F. (ed.): "Laboratory Planning for Chemistry and Chemical Engineering," Reinhold Publishing Corporation, New York, 1962.

Munce, J. F.: "Laboratory Planning," Butterworths, London, 1962.

Palmer, R. R., and W. M. Rice: "Modern Physics Buildings: Design and Function," Reinhold Publishing Corporation, New York, 1961.

Schramm, W.: "Chemische und Biologische Laboratorien," Verlag Chemie GmbH, Weinheim, Germany, 1965.

For a discussion of some trends in planning chemical laboratories since 1930, see M. G. Mellon, *J. Chem. Educ.,* **52,** 345 (1975); **53,** 114, 454 (1976); **54,** 195 (1977); **55,** 194 (1978).

Tome I (1882) of E. Fremy's "Encyclopédia Chimique" contains photographs and plans of important laboratories.

NOMENCLATURE

Previous mention was made of the list of references on chemical nomenclature contained in Appendix IV of the "CAS Index Guide." For more ready reference, the following publications may be of interest:

Cahn, R. S., and O. C. Dermer: "Introduction to Chemical Nomenclature," Butterworths, London, 1979.

Fletcher, J. H., O. C. Dermer, and R. B. Fox: "Nomenclature of Organic Compounds: Principles and Practice," American Chemical Society, Washington, DC, 1974.

Several relevant bulletins have been issued by the International Union of Pure and Applied Chemistry:

"Biochemical Nomenclature and Related Documents," The Biochemical Society, London, 1978.

"Compendium of Analytical Nomenclature," Pergamon Press, New York, 1978.

"How to Name an Inorganic Substance," Pergamon Press, New York, 1977.

"Manual of Physico-Chemical Symbols and Terminology," Butterworths, London, 1975.

"Nomenclature of Inorganic Chemistry," Butterworths, London, 1971.

"Nomenclature of Organic Chemistry," Sections A, B, and C, Butterworths, London, 1969. An updating and extension of this work, under the same title, compiled by J. Rigaudy and S. P. Klesney and published by the Pergamon Press, New York, in 1979, covers Sections A, B, C, D, E, F, and H.

BOOK LISTS

As stated in the discussion of bibliographies, a librarian deals broadly with four kinds of publications: periodicals, bulletins, patents, and books. The books appear singly or as parts of sets.

Thousands of new books appear each year. There is a problem of knowing not only about the new ones but also about the ones already existing on a given subject. Perhaps the ideal solution would be access to a library which has everything. As there is no such collection, one must use other means.

New Books

One of the best means of trying to keep up with new books is to have one's name placed on the mailing lists of the chief publishers of technical books. Most publishers have catalogues listing their books in print. As new titles are published, current announcements are mailed. Journals carrying advertisements are sources of late information, as are book reviews. The September issue of the *Journal of Chemical Education* contains an extensive classified list of books available.

Chemical Abstracts lists new books in many of the sections. Among the many subjects listed in *CA SELECTS* is one entitled "New Books in Chemistry" (see Chapter 7). Included in the listings are all new books in chemistry and chemical engineering selected for citation in *Chemical Abstracts;* also all monographs, series publications, and conference proceedings dealing with any aspect of chemistry and chemical engineering.

Several publishers' periodicals deal with new books. Important examples are *Book Review Digest,* 1905–   ; *Publishers' Trade List Annual,* 1873–   ; *Publishers' Weekly,* 1872–   ; *Technical Book Review Index,* 1935–   [3]; *ASLIB Book List,* 1935–   ; and *New Technical Books,* 1915–   .[4]

"Biblioscan Q–Z" lists all new English language books classified in the Library of Congress categories Q through Z. A monthly cumulation of entries in

[3]Issued by the Special Libraries Association, New York.
[4]Issued by the New York Public Library.

Publishers' Weekly appears in *American Book Publishing Record.* The arrangement is by subject according to the Dewey Decimal Classification.

Started in 1980 by the Institute for Scientific Information, the *Index to Book Reviews in the Sciences* is multidisciplinary. The ten monthly issues and two semiannual cumulations cover the life sciences, clinical medicine, physical sciences, chemical sciences, agriculture, engineering, and technology. The arrangement is by author (or editor) and subject.

Old Books

Important general lists of books in English follow. There are similar lists for various foreign countries. The times of cumulation vary, there being monthly, annual, and longer periods. Very old or rare books should be sought through dealers in such publications.

United States

"Books in Print," 1948–   ; "Subject Guide to Books in Print," annual; "Publishers' Trade List Annual," 1873–   ; and "Scientific and Technical Books and Serials in Print," 1979–   [5]; all by the R. R. Bowker Company, New York.

"Cumulative Book Index," H. W. Wilson Company, New York, 1898–   .[6]

"Library of Congress Catalog—Books: Subjects," Pageant Books, Inc., Paterson, NJ, monthly and annual.

Great Britain

"British Books in Print," annual; "Reference Catalogue of Current Literature," 1874–   ; and "Whitaker's Cumulative Book List," 1924–   , J. Whitaker and Sons, Ltd., London.

"British Museum General Catalogue of Printed Books," British Museum, London.

"British National Bibliography," Council of British National Bibliography, British Museum, London, 1950–

"English Catalogue of Books," Publishers' Circular, London, 1801–

LANGUAGE DICTIONARIES

Useful chemical dictionaries follow.

Alford, M. H. T., and V. L. Alford: "Russian-English Scientific and Technical Dictionary," 2 v., Pergamon Press, New York, 1970.

Callaham, L. I.: "Russian-English Chemical and Polytechnical Dictionary," John Wiley & Sons, Inc., New York, 1975.

Czerni, S., and M. Skrzynskiej: "English-Polish Dictionary of Science and Technology," Wydawn Nauk-Tech., Warsaw, Poland, 1975.

Desov, H. E. (ed.): "Concise English-Russian Technical Dictionary," Soviet Entsiklopediya, Moscow, 1969.

[5]Contains list of micropublishers; abstracting and indexing services; and publishers and distributors.

[6]Formerly "The United States Catalog."

DeVries, L.: "French-English Science and Technical Dictionary," McGraw-Hill Book Company, New York, 1976.

———— and L. Jacolev: "German-English Science Dictionary," McGraw-Hill Book Company, New York, 1978.

———— and H. Kolb: "Dictionary of Chemistry and Chemical Engineering," vol. 1, German-English; vol. 2, English-German, Verlag Chemie International, New York, 1978.

Duncan, D. R.: "English-Esperanto Chemical Dictionary," British Esperanto Association, London, 1956.

Elmer, T. H.: "German-English Dictionary of Glass, Ceramics, and Allied Sciences," John Wiley & Sons, Inc., New York, 1963.

Emin, I.: "Russian-English Physics Dictionary," John Wiley & Sons, Inc., New York, 1963.

Fouchier, L., and F. Billet: "Chemical Dictionary in Three Languages: English-French-German," Werveries, G.m.b.H., Baden-Baden, Germany, 1953.

————: "Dictionnaire de Chimie," Netherlands University Press, The Hague, 1962.

Garfield, E.: "Transliterated Dictionary of the Russian Language," ISI Press, Philadelphia, 1979.

Goldberg, M.: "Spanish-English Chemical and Medical Dictionary," McGraw-Hill Book Company, New York, 1952.

Marolli, G.: "Dizionario Technico," Le Monnier, Firenze, 1961.

Neville, H. H., N. C. Johnston, and G. V. Boyd. "A New German-English Dictionary for Chemists," D. Van Nostrand Company, Inc., Princeton, NJ, 1964.

Patterson, A. M.: "French-English Dictionary for Chemists," John Wiley & Sons, Inc., New York, 1954.

————: "German-English Dictionary for Chemists," John Wiley & Sons, Inc., New York, 1950.

Renzo, D.: "Dizionario technico italiano-inglese, inglese-italiano," W. S. Heinman, New York, 1962.

Stephen, H., and T. Stephen: "Dictionary of Chemistry and Chemical Technology in Four Languages, English/German/Polish/Russian," Pergamon Press, New York, 1962.

For a more extensive list of dictionaries in non-English languages, see Hanć et al., "Chemická Literatura," p. 245.

DIRECTORIES AND TRADE CATALOGUES

Trade catalogues are published for the principal purpose of selling manufacturers' products.[7] To the student and engineer they offer useful information about processes, machines, and materials. Also they show the practical application of theories and principles of engineering. Catalogues of equipment are generally illustrated with photographs, diagrams, and drawings, and frequently detailed specifications of the design or properties of the product offered are included.

Usually from such sources one wants information about some kind of product, such as molecular stills, or about an organization. In the latter case,

---

[7]See R. T. Bottle, "Use of Chemical Literature," Butterworths, London, 1979, p. 227, for British trade catalogues.

annual reports and company histories are of interest, for example, to one making an economic or historical survey of a company or an industry.

Included here also are various kinds of directories. Examples are manufacturers of, or dealers in, a given chemical; testing and consulting laboratories; faculties of technical educational institutions; and sources of raw materials.

Some of these works contain a variety of information. Thus, for a given compound a buyers' guide may contain information on sources, producers, dealers, common properties, shipping containers and hazards, trade names, and related items.

The following unclassified selections are representative of this kind of publication. In many cases the titles indicate the general nature of the contents.

Some directories are annual publications, while others are irregular. The date of issue may be important to the user. Some examples of such compilations follow.

"Aldrich Catalog/Handbook," Aldrich Chemical Company, Milwaukee, WI.

"American Drug Index," J. B. Lippincott Company, Philadelphia, PA.

"Annual Buyers' Guide," Society of Chemical Industry, London.

"Buyers' Directory," Oil, Paint and Drug Reporter, New York.

"Buyers' Guide," Chemist and Druggist, London.

"Buyers' Guide," McGraw-Hill Book Company, New York.

"Chem BUYdirect—International Chemical Buyers' Directory," W. de Gruyter, Inc., New York.

"Chemical Engineering Catalog," Reinhold Publishing Corporation, New York.

"Chemical Engineering: Inventory Issue," McGraw-Hill Book Company, New York.

"Chemical Industry Directory and Who's Who," Chemical Age International, London.

"Directory of Graduate Research," American Chemical Society, Washington, DC.

"Directory of Testing Laboratories, Commercial and Institutional," American Society for Testing and Materials, Philadelphia, PA.

"Directories in Science and Technology," Library of Congress, Washington, DC.

"Handbook of Commercial Organic Chemicals," Society of Organic Chemicals Manufacturing Association, Washington, DC.

"Heat Treating Buyers' Guide," American Society for Metals, Metals Park, OH.

"Industrial Chemical Directory," W. A. Benjamin, New York.

"Industrial Research Laboratories of the United States," Jacques Cattell Press, Inc., New York.

"Laboratory Buyers' Guide Edition," American Laboratory, International Scientific Communications, Fairfield, CT.

"Laboratory Guide," Analytical Chemistry, American Chemical Society, Washington, DC. This guide includes Advertised Products, Laboratory Supply Houses, Instruments and Equipment, Laboratory Chemicals, Analytical and Research Services, New Books (as reviewed in ACS publications), Trade Names, and Company Directory.

"McGraw-Hill Directory of Chemicals and Producers," McGraw-Hill Book Company, New York.

"Modern Drug Encyclopedia and Therapeutic Index," Drug Publications, New York.

"Modern Plastics Encyclopedia and Engineers' Handbook," Plastics Catalogue Corporation, New York.

"Natural and Synthetic Fibers Yearbook," Interscience Publishers, Inc., New York.

"New and Official Drugs" (of American Medical Association), J. B. Lippincott Company, Philadelphia, PA.

"OPD Chemical Buyers' Directory," Chemical Marketing Reporter, New York.

"Pharm Index," PHARMINDEX, Portland, OR.

"Plastics Engineering Handbook," Reinhold Publishing Corporation, New York.

"Professional Services Directory," American Translators Association, Croton-on-Hudson, New York.

"Resins, Rubbers, Plastics Yearbook," Interscience Publishers, Inc., New York.

"Science Guide to Scientific Instruments," American Association for the Advancement of Science, Washington, DC.

"Sweet's File, Process Industries," Sweet's Cataloguing Service, New York.

"Thomas' Register of American Manufacturers," Thomas Publishing Company, Inc., New York.

————: "Chem Sources—U.S.A.," Directories Publishing Company, Inc., New York.

Most states have industrial directories. Examples are "The Alabama Book," "Arizona Industrial Buyers' Guide," "Industrial Directory of Arkansas," "Directory of Florida Manufacturers," "Idaho Industries," and "Guide Book to Minnesota Industry."

Manufacturers' or distributors catalogues are useful for information on the availability and comparative prices of many common chemicals. Generally specifications are included for "analytical" or "reagent grade" substances. A good example is "Eastman Organic Chemicals," from the Eastman Kodak Company, Rochester, NY.

An interesting smaller catalogue is "GFS Chemicals 50th Anniversary Catalog," distributed by the G. F. Smith Chemical Company, Columbus, OH. This small service company specializes in making and marketing analytically useful chemicals. It started from the founder's doctoral work with perchlorates and perchloric acid. Included in the catalog is a short history of the company and its founder, together with the following kinds of information: GFS Publications on Analytical Chemistry, Oxidation-Reduction Indicators (with literature references), Colorimetric Reagents for Iron and Copper (with literature references), and General Information on the Use of GFS Analytical Reagents.

FINANCIAL DATA

Financial data are of concern in several ways in the chemical industry. One of these is the cost of procuring suitable raw materials and producing and marketing the desired products. The one of interest here is the company itself. What is the price of its stock? What is the dividend yield? How do its assets compare with those of competitors? There are other related questions. Many libraries have, or can recommend, a variety of sources for such information.

Two well-known financial newspapers, published by Dow Jones and Company, New York, are the *Wall Street Journal* and *Barron's National Business and Financial Weekly*. Many metropolitan general newspapers, such as the *New York*

Times and the *Chicago Tribune,* carry daily stock reports. Of a different nature are the *Chemical Marketing Reporter* and *European Chemical News.*

Fortune magazine issues an annual publication, "The Fortune Directory," which ranks the 500 largest United States industrial corporations. This report is followed by the second-ranking 500 corporations. The lists include chemical companies. Each entry lists data on sales, assets, net income, and several other items of financial interest.

Similar in nature, but restricted to chemical firms, are the special reports in *Chemical & Engineering News.* The issues of May 4 and June 30, 1981, listed the "top" 50 products and producers, respectively.

The July 27, 1981, issue of the same journal contains a comprehensive survey of the data available on chemical research and development in the United States. Separate sections cover the Federal government, industry, and colleges and universities.

The following more general works may be of interest:

Daniels, M. M.: "Business Information Sources," University of California Press, Berkeley, 1976.

Hanson, R. P.: "Moody's OTC Industrial Manual," Moody's Investors Service, Inc., New York, 1978.

Wasserman, P., et al. (eds.): "Encyclopedia of Business Information Sources," Gale Research Company, Detroit, 1976.

————: "Directory of Corporate Affiliations," 1979, and "Consumer Register of American Business," National Register Publishing Company, Inc., New York.

————: "Register of Corporations, Directors, and Executives," 3 v., Standard and Poor's Corporation, New York, 1979.

PART

PUBLICATIONS:
STORAGE AND USE[1]

[1]See Lucille J. Strauss, Irene M. Shreve, and Alberta L. Brown, "Scientific and Technical Libraries," Wiley, Becker, and Hayes, New York, 1972; Jean K. Gates, "Guide to the Use of Books and Libraries," McGraw-Hill Book Company, New York, 1979; and Allen Kent et al. (eds.), "Use of Library Materials," Marcel Dekker, Inc., New York, 1979.

CHAPTER

LIBRARIES AND INFORMATION CENTERS

Knowledge is of two kinds. We know a subject ourselves, or we know where we can find information upon it.

Samuel Johnson

If one wants published information on some chemical subject, the first question is where to go to find publications discussed in the preceding chapters. In general, it may be assumed that some publications will be found wherever progressive chemists are, or where there is a demand for chemical facts.

Private collections began to disappear several decades ago. In the early 1900s it was not uncommon to find small collections in the home libraries of the more affluent college professors. A bound set of the *Journal of the American Chemical Society* and the *Berichte der Deutschen Chemischen Gesellschaft* lent prestige to the owner, especially if a set of *Chemical Abstracts* was included. Occasionally there were thousands of volumes, as in the personal libraries of E. C. Worden and S. C. Hooker. Now such collections have practically disappeared because of lack of funds to maintain them and of space to house them.

LIBRARIES

Whether called information centers, resource centers, or libraries, such institutions house the primary and secondary publications, along with appropriate means for using them. Both general and specialized types are found.

259

General Libraries

A general library contains publications in several subject areas, usually including at least some on science and technology. Such libraries may be grouped as public, academic, and government.

Public libraries are found practically everywhere, from small villages to the largest metropolitan centers. Except in unusual cases, publications for serious chemical study are unlikely to be found in centers with a population under 100,000. Generally in such situations there is neither need for the publications nor funds to provide them.

In large cities the need may be varied and extensive, as illustrated by the questions listed in Chapter 1. Here one may expect to find a science and technology division. Perhaps the outstanding example is the New York Public Library.

Academic libraries are a part of the hundreds of colleges and universities throughout the country. Collections of much importance can hardly be expected in the community and junior colleges, with their curricula limited in most cases to 2 years. To these, unfortunately, must be added many of the smaller colleges having 4-year curricula. There is an occasional well-endowed small college.

Few large institutions with an extensive graduate research program, have one central library covering all subject areas. Administratively, librarians probably prefer a central arrangement, as it is most efficient and least costly to manage.

The general practice in large universities is to have departmental libraries for specific subjects or areas, such as chemistry, physics, biological sciences, and mathematics and computer sciences. There may be combinations, such as a joint engineering library for the various schools of engineering, a science and mathematics library, or other arrangements for particular situations. These libraries are comparable to the special libraries, discussed below.

Government general libraries are not common. Probably the unequaled example is the great Library of Congress, which may have the world's largest collection. Strictly, it is for members of Congress, but others may arrange to work there. It is designed for in-house research.

State libraries are generally collections of more or less provincial interest, with little specialization.

Special Libraries

In nature these libraries are like departmental libraries in large universities, but they are not part of a larger unit. They are now so common that there is a society, the Special Libraries Association, with its own journal, *Special Libraries*. Many different subject areas are represented, and special libraries vary considerably. The interest here is the ones devoted to science and technology.[2]

---

[2]See M. L. Young, and H. C. Young, "Subject Directory of Special Libraries and Information Centers," Gale Research Company, Detroit, MI, 1981. Also "Directory of Special Libraries and Information Centers," 1977. See G. Lavendel, *Chem. Eng. News,* **55** (16), 34–36, 38, 40, 42 (1977), on the capability of special libraries to supply varied technical information.

In the United Kingdom there is the Association of Special Libraries and Information Bureaux (ASLIB). The ASLIB Library is the nation's documentation center. In addition to several other publications, there is the *Journal of Documentation.*

Examples in the public domain are the John Crerar Library in Chicago and the Linda Hall Library in Kansas City, MO. Such libraries are likely to have originated from a private endowment.

Industrial special libraries are now reasonably common. As might be predicted, they are located in large cities and in or near centers of chemical industry. Organizations having them vary somewhat in nature. For example, Arthur D. Little is a consulting firm, the Battelle Memorial Institute is an endowed industrial research organization, and the American Society for Metals is a professional society.

Most common among such libraries are those associated with large companies, such as the General Electric Company and General Motors. All the large chemical companies have them, well-known examples being E. I. du Pont de Nemours and Company, Hercules, Inc., the Dow Chemical Company, and Monsanto.

In Appendix C of their book, "Industrial Information Systems," Jackson and Jackson[3] indicate which of *Fortune's* list of 500 corporations have such libraries. Included is the nature of the library holdings in each case.

Private special libraries are not common, but a number are important. In New York City alone are the Chemists' Club,[4] the American Association of Engineering Societies, and the New York Academy of Medicine. The Library of the Franklin Institute is in Philadelphia.

There are a number of admirable governmental collections, such as those of the National Library of Medicine, the National Library of Agriculture, the Patent and Trademark Office, and the National Bureau of Standards. Many other Federal agencies have similar, but generally smaller, collections.

The present period is one of transition in various ways for libraries. Primarily the changes are occurring in large institutions, and especially in industrial research laboratories. "Information centers" has become a designation commonly used instead of "libraries."

As already noted, the kinds of chemical publications and their general nature have not changed significantly in several decades. The problems of handling them and the means developed to do it have.

An outstanding example of development has been reported by Herman Skolnik for the Hercules Technical Information Division.[5] The account covers the following information groups and services: the open literature library, the proprietary files (internal records), translating and editing, technical reports, and patent documents, chemical and market data, literature research, and

[3]Dowden, Hutchinson, and Ross, Stroudsburg, PA, 1978, p. 246.

[4]See P. B. Slawter, Jr., *CHEMTECH,* **3,** 714 (1973), for a history of the library of the Chemists' Club.

[5]*J. Chem. Inf. Comput. Sci.,* **14,** 123 (1974).

computerized information systems design and programming. Several special systems are described.

Library Holdings

Although numbers do not guarantee quality, in general one is likely to rate a library in terms of its holdings of publications. But even a very large collection might not include specialized journals one needs. Chapter 7, under CASSI, shows the small number of libraries holding the analytical journal *Quimica Analitica*.

The holdings of periodicals are part of the data in CASSI. The entries cover 300 libraries in the United States and 69 in foreign countries. These 369 libraries are rated the most important by Chemical Abstracts Service.

Holdings of technical reports are likely to be spotty. Although Federal reports are widely distributed to more than a thousand libraries, probably few good collections are maintained outside the Regional Depository Libraries listed in Chapter 3.

For patent documents the prime holding is at the United States Patent and Trademark Office at Arlington, VA. There are at least partial collections of United States patents in the depository libraries listed in Chapter 4. Many industrial firms manufacturing chemicals and chemical equipment undoubtedly maintain collections relating to their processes and/or products.

Holdings of the great reference works, such as those of Beilstein, Gmelin, and Landolt-Börnstein, are difficult to document. Many small colleges can hardly afford them now, even if the staff were interested. They should be available in the large, research-oriented institutions and industrial laboratories. The reported global distribution of the Beilstein "Handbuch" is some 2,200 copies, split about equally between industries and universities.

Centers for Shared Resources

Millions of titles of publications have been published since Lavoisier laid the foundation for modern chemistry. Chemical Abstracts Service is monitoring more than 14,000 periodicals, and thousands of books are published annually. It seems unlikely that any one library can ever acquire all the published works in science and technology. Establishing a single library limited to chemistry and chemical engineering seems an unattainable goal.

Cooperation in acquisitions and sharing of resources has been suggested as a reasonable and possible solution. As early as 1964 S. L. Warren[6] proposed a National Library Science System. He envisaged a computer-based pool of all the scientific literature. Included also was a suggested distribution network through which the information could be made available upon request to any scientist who wanted it.

---

[6]*Chem. Eng. News,* **42**, (24), 21 (1964).

A decade later a conference was held to discuss various aspects of sharing library resources. Edited by Allen Kent, the papers constitute the book "Resource Sharing in Libraries."[7]

Although implementation of the proposed national system seems to have made little progress, a number of regional developments are under way. Probably all librarians realize that no library can own all the material that its clientele may need. Several of the cooperative arrangements are of interest.

Founded in 1949 as the Midwest Interlibrary Center, the name was changed in 1965 to the Center for Research Libraries to reflect its developing national importance. Now this library has some 170 academic, government, and industrial libraries as members. The holdings of over 3 million volumes are primarily research materials not readily available in the members' own collections.

A detailed inventory of the more than 50 collections in the center is contained in the "Handbook." Most items of chemical interest are listed under science and technology. Since 1956, the Center has aimed to acquire all journals and serials abstracted in *Chemical Abstracts* and *Biological Abstracts* that are not widely held.

Only member libraries may borrow materials or obtain photocopies. Details are outlined in the "Handbook." There is a printed catalogue of the collections.[8]

In 1974 the Research Libraries Group began operation. It started with the research libraries of Columbia, Harvard, and Yale Universities and the New York Public Library as initial members.

More recently the nine campus libraries of the University of California arranged to combine their resources in support of graduate teaching and research.

A shared catalogue project was begun by 50 Ohio college libraries in 1968 by the Ohio College Learning Center in Columbus, OH. This is now the center in the United States for sharing computerized cataloguing.

Other developments in this general direction are noted in two recent papers.[9]

Renewed interest in the general problem is shown in the publication of a 178-page book, "Scholarly Communication," a report of the National Enquiry, published by the Johns Hopkins University Press in 1979. This study contains 12 recommendations relating to publication, dissemination, storage, and availability of scholarly materials. Although directed especially to the humanities and social sciences, the conclusions are relevant for science and technology.

The first three recommendations follow: (1) a national bibliographic system, (2) a national periodicals center, and (3) a national library agency.

[7]Marcel Dekker, Inc., New York, 1974.
[8]5721 South Cottage Avenue, Chicago, IL 60637.
[9]I. T. Littleton, *The Torch*, Spring 1978, p. 17; J. B. Smith and H. B. Schleifer, *AAUP Bull.*, p. 78 (1978).

Interlibrary Loans and Photocopying Service

In Chapter 2 mention was made of the copying services available at the Library of Chemical Abstracts Service. As only a limited period of recent time can be covered, there is a long gap back to the start of Crell's *Chemisches Journal* in 1778, or the issuance of the first British patent in 1617.

What can be done for the items not held, and for the thousands of books (not bound journals) that have been published, now for more than four centuries (see Chapter 1)? In general, the answer is an interlibrary loan of the desired work, or a photocopy of the section of interest.

Librarians are unlikely to risk losing a rare book in the mail. Sending full-size photostats or a microfilm copy is the safe way. Anyway, photocopies are desirable any time one needs an accurate copy of complicated features, such as formulas, diagrams, and tables.

It is the task of the librarian to find out where such service is available for the publication in question.

CLASSIFICATION SYSTEMS

For the efficient use of the chemical literature, it is necessary to have an orderly arrangement of the publications in the library, together with some means for finding them easily. Fortunately, librarians have devised systems of arrangement which are relatively easy to understand and use.

Arrangement of Publications

Publications are arranged according to some definite scheme, so that those possessing more or less common characteristics are grouped together. Although various bases of classification may be used, for general work the material is most satisfactorily arranged according to subjects. In carrying out this scheme the Dewey Decimal Classification is most common.[10] In accordance with this system, the works covering all branches of knowledge are divided into 10 main divisions, designated with Arabic numerals, as shown below.

000	General works (bibliography, encyclopedias, general periodicals)
100	Philosophy (psychology, ethics)
200	Religion (Bible, church history)
300	Sociology (economics, education)
400	Philology
500	Natural science (mathematics, physics, chemistry, biology)
600	Useful arts (architecture, engineering, agriculture)
700	Fine arts
800	Literature
900	History (travel, biography)

[10]An elaboration of the Dewey system, known as the "Universal Decimal Classification" (U.D.C.), provides for more detailed classification. See J. Weston, A Century of Dewey's Decimals, *Chem. Ind., (London),* **1976,** 842.

For convenience these classes are usually referred to as hundreds rather than as units; e.g., the 500s mean the sciences.

Each of these classes may be divided and subdivided, by the addition of other figures, to an almost unlimited extent. For example, 500 is natural science; 540 is chemistry; 547 is organic chemistry; 547.2 is acyclic, aliphatic compounds; and 547.28 is carbonyl compounds. Thus, C. W. Shoppee's book, "Chemistry of the Steroids," is classified 547.28. As a subdivision of carbonyl compounds, 547.284 is ketones and 547.2843 is acetone.

Each book, as it comes to the library, is given a number that corresponds as closely as possible to its subject, and all books having the same number are shelved together.

To distinguish books written on the same subject but by different authors, another symbol is used, called the author number or book number. This consists of a letter followed by one or more numerals. The book by Shoppee cited above has the designation Sh 7c, which is for the surname Shoppee and the location of the book on the shelf. In addition, the book number may contain other entries to indicate title, edition, language, or translator.

Books are arranged on the shelves from left to right, first according to their class numbers, and then books with the same class numbers according to their author numbers.

The classification number 547.28 and the book number Sh 7, taken together, are spoken of as the "call number," since by means of it the book can be located on the shelf. This call number is written on and in each volume on the shelves. It appears also in the upper left-hand corner of the card in the card catalogue, which is discussed later. When the call number is written down for reference or when a book is called for, the number should be copied accurately and completely.

The following abridged classification indicates the chief classes of publications that contain information of chemical interest, the Dewey[11] number being shown on the left. Many changes are made with time.

ABRIDGED SYSTEMS OF CLASSIFICATION

Dewey		Library of Congress	
000	General works	A	
016.54	Bibliography of chemistry		Z 5521-5526
301	Sociology	H	
310	Statistics	HA	
317.3	U.S. Bureau of the Census		HA 201
340	Law	K	
340.6	Legal chemistry		KA 1001-1171
370	Education	L	
371.623	Laboratory equipment		Q 183
389	Weights and measures		QC 81-114
500	General (pure) sciences	Q	

[11]M. Dewey, "Decimal Classification and Relative Index," Forest Press, Essex County, NY, 1942.

ABRIDGED SYSTEMS OF CLASSIFICATION

Dewey		Library of Congress
510	Mathematics	QA
530	Physics	QC
530.85	U.S. Bureau of Standards	QC 100
535	Light	QC 350-467
535.84	Spectroscopy	QC 450-467
536	Heat	QC 251-338
537	Electricity	QC 501-718
539	Atomic physics	QC 171-2
540	Chemistry	QD
541	Theoretical and physical	QD 450-655
542	Practical and experimental	QD 40-65
542.1	Chemical laboratories	QD 51
543	Analytical, general	QD 71-142
544	Analytical, qualitative	QD 81-100
545	Analytical, quantitative	QD 101-121
546	Inorganic	QD 146-199
547	Organic	QD 241-441
548	Crystallography	QD 901-999
549	Mineralogy	QE 351-399
550	Earth sciences	QE
551.9	Geochemistry	QE 515-516
557.3	U.S. Geological Survey	QE 75-76
570	Life sciences	QH 301-705
578	Microscopy	QH 201-278.5
600	Technology	T
608.7	Patents	T 201-342; TP 210
610	Medicine	R
612	Physiology, human	QP
612.015	Biochemistry	QP 501-801
614	Public health	RA 421-790
614.09	U.S. Public Health Service	RA 11; 815-819
615	Therapeutics and pharmacology	RS
620	Engineering, general	TA
621.35	Electrochemistry	QC 601-607
628	Sanitary engineering	TD
630	Agriculture	S
630.24	Agricultural chemistry	S 583-588
631.4	Soil science	S 631-669
637	Dairy and dairy products	SF 221-275
640	Domestic arts and sciences	TX
641.1	Applied nutrition	TX 501-612
660	Chemical technology	TP
661	Chemicals, industrial	TP 200-248
662	Explosives and fuels	TP 268-301
663	Beverages	TP 500-659
664	Foods	TP 368-465
665	Oils, fats, waxes	TP 343-360
666	Ceramics, glass	TP 785-889
667	Bleaching, dyeing, cleaning	TP 890-929
668	Other organic industries	TP 930-997

ABRIDGED SYSTEMS OF CLASSIFICATION

Dewey		*Library of Congress*
669	Metallurgy and assaying	TN 550-799
669.0284	Electrometallurgy	TN 681-687
670	Manufactures, articles made of	TS
671	Metals	TS 200-770
672	Ferrous metals	TS 300-360
673	Nonferrous metals	TS 564-589
674	Wood, lumber, cork	TS 800-937
675	Leather and fur	TS 940-1047
676	Paper and pulp	TS 1080-1220
677	Textiles	TS 1300-1865
678	Rubber (elastomers)	TS 1870-1935
679	Other products	TP 1101-1185
690	Buildings	TA; TT
691	Materials, processes	TA 401-495
698	Detail finishing	TT 300-385
700	Fine arts	N
770	Photography	TR
900	History and geography	D; E; F
920	Biography	QD 21-22
925.4	Chemists only	

Each Dewey class may have subdivisions, as .03 for dictionaries, .05 for serials, and .06 for society publications. For further details concerning this system, the reader is referred to Dewey's "Decimal Classification and Relative Index for Libraries."

Another system of classification particularly suited for large libraries is that used in the Library of Congress, the Chemists' Club in New York, and others. "LC" call numbers are shown above. According to this scheme, the general fields of knowledge are designated by letters, class Q being science. A second letter is used to differentiate the divisions of each class, QD being chemistry; QP, physiological chemistry; TP, chemical technology; RS, medical and pharmaceutical chemistry; and S, agricultural chemistry. Details regarding this system may be found in the "Library of Congress Classification," available in libraries or obtainable from the Superintendent of Documents, Washington, DC.

In addition to the two foregoing systems, some librarians have devised special systems of their own or modified one of the more generally used schemes in order to adapt it to special situations.

There are many cases of seemingly wrong and/or inconsistent classification by cataloguers. A book entitled "Metallurgical Analysis" is very likely to be catalogued on the basis of the adjective "metallurgical," rather than the noun "analysis." If so, the number will be 669 instead of 545.

In the author's institution, his two books "Colorimetry for Chemists" and "Analytical Absorption Spectroscopy" have the respective numbers 545.8

(chemistry) and 535.3 (physics), although the latter book is substantially a fivefold expansion of the former.

Other examples of scattered cataloguing follow:

539.1	"Principles of Ultraviolet Spectroscopy"
543.01	"Official Methods of Analysis" (Association of Official Analytical Chemists)
615	"United States Pharmacopoeia" (American Pharmaceutical Association)
628.161	"Standard Methods for the Examination of Water and Waste Water" (American Public Health Association)
663.3	"Methods of Analysis" (American Society of Brewing Chemists)
665	"Official and Tentative Methods (of Sampling and Analysis)" (American Oil Chemists Society)

Obviously, all these works deal with analytical chemistry, and mostly with quantitative chemical analysis (Dewey 545).

In general, scattered cataloguing means that one cannot browse in a given shelf number sequence, or in the card catalogue, to find related publications. Similarly, analytical chemists cannot restrict their reading in *Chemical Abstracts* to sections 79 and 80, which deal with analytical chemistry. In every issue there are many cross references to other sections, because the document abstracted seemed less analytical than something else.[12]

Citing four different books on polymer chemistry as an example, H. Skolnik[13] illustrates the dispersion arising from classifying these works by the Dewey or Library of Congress systems. He believes that ease of browsing among books in a library is the primary requirement from a user's viewpoint. Consequently, he designed a different numerical classification for the extensive literature on polymers in the library of the Hercules Company. These new Hercules call numbers bring together works of the same general nature. In addition, they have the advantage of being short. Thus, the number for adhesives is 9.100.

The Card Catalogue

The card catalogue serves as an index or guide to the publications contained in a library, just as the index of a book enables one to find what is contained therein. In making such catalogues, the practice of librarians varies somewhat. In many cases the catalogue is a single file consisting of a list of the publications by author, by title, and by subject, all in one alphabetical arrangement. In such a case it is known as a "dictionary card catalogue." Under the names of individuals will be found lists of the works they have written or works written

---

[12]See Martha L. Manheimer, "Cataloging and Classification," Marcel Dekker, Inc., New York, 1980.

[13]*J. Chem. Inf. Comput. Sci.*, **19**, 76 (1979).

about them, which the library possesses. Again, for each publication dealing with a specific subject, a card is placed in the catalogue under the name of that subject, thus bringing together in one place all references dealing with that subject. The following example indicates the method of indexing a publication when the headings on the cards are for (1) author, (2) title, and (3) subject:

1 541.39   Sabatier, P.
   Sal   Catalysis in organic chemistry.
2 541.39   Catalysis in organic chemistry.
   Sal   Sabatier, P.
3 541.39   Organic chemistry, catalysis in.
   Sal   Sabatier, P.

If no author is given or if the title is a significant one, a card is included for the title. From the card catalogue one can ascertain at once the publications in the library on a given subject with a certain title or by a particular author. By means of the letters on the front of the drawers and the guide cards inside, one searches in the proper alphabetical place for the author's name or the title or the subject of the publication desired. Each catalogue card, whether its heading is for author, for title, or for subject, furnishes information such as the following:

Author's full name (Government, society, or institution)
Title of publication
Call number (in upper left-hand corner)
Number of pages, size of book, illustrations
Publisher, date, and place of publication
Descriptive and bibliographical notes, table of contents

The example of a catalogue card on page 270 was prepared for the fourth edition of this book by the Library of Congress.[14] On the lower half of the card are shown (1) each of the more important subjects under which the book should be catalogued, *A;* (2) the number which would be assigned to the book in accordance with the Library of Congress classification, *B;* and (3) the order number by which the card is obtained from the Library of Congress, *C.* The *D* entry is the call number for the Dewey system. It is typed in the upper left-hand corner of the card, if this system is used.

Some librarians include more on the cards than the bare bibliographical data shown here. Thus, this card might show the titles of the 12 chapters. Such additional information is likely to be found for monographs in which a number of authors contributed different sections. In such cases each entry includes the author's name and the title of the contribution. To make the card catalogue even more useful, separate cards may be entered for each author and subject. Similar practice could apply to a series of papers published as a separate symposium volume.

---

[14]In addition to preparing such printed cards for every book receiving a United States copyright, the Card Division of the Library of Congress is prepared to sell collections of cards for special fields.

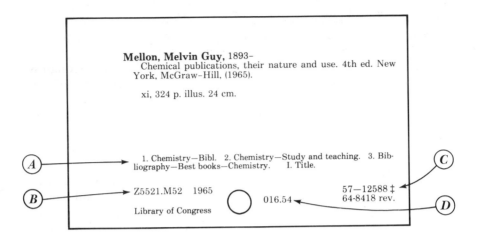

> **Mellon, Melvin Guy,** 1893–
> Chemical publications, their nature and use. 4th ed. New York, McGraw–Hill, (1965).
>
> xi, 324 p. illus. 24 cm.
>
> 1. Chemistry—Bibl.   2. Chemistry—Study and teaching.   3. Bibliography—Best books—Chemistry.   I. Title.
>
> Z5521.M52   1965   016.54   57—12588 ‡
>   64-8418 rev.
> Library of Congress

(A)   *(B)*   *(C)*   *(D)*

In contrast to the foregoing scheme, another system consists in maintaining three card files as follows:

1 Author-title catalogue—arranged alphabetically, containing names of all authors and societies; also titles of all periodicals, of anonymous works, and of important sets or series of works.

2 Classified subject catalogue—based on the Dewey, or other, system already described. The arrangement of the cards in the classified catalogue corresponds to the actual arrangement of the publications on the shelves.

3 Subject index to the classified catalogue—arranged alphabetically. To find all the works on a given subject, one consults this index under the name of the subject. The number at the right-hand corner of the card gives the class number under which the work desired may be found in the classified catalogue. For example, if one is interested in electric lighting, the following will be found in the index:

Electric lighting	621.32
Illumination, electric	621.32
Lighting, electric	621.32

The subject is thus fully indexed. By consulting the classified catalogue under the number 621.32 all works on the subject of electric lighting will be found.

A subject index similar to the one just mentioned has been devised on a decimal system by the U.S. Office of Experiment Stations for the articles in the reports and bulletins issued by the agricultural experiment stations of the various states.[15] In using this index, it is necessary to locate the desired number for the subject in the key to the index and then turn to the cards with this number. Works on analytical chemistry, for example, bear the number 1.26, 1 being for general science, 0.2 for chemistry, and 0.06 for analytical chemistry.

[15]*States Relations Service Doc.* 37, U.S. Department of Agriculture, States Relations Service.

Computerized Card Catalogue

In an interesting discussion, "The Machine in the Library," Ruth Gay[16] has pointed out that large libraries are finding card catalogues increasingly cumbersome and expensive to maintain. Her chief example is the Library of Congress. The trend is to mechanize operations and abandon the present type of card catalogue.

Most significant, probably, is the change at the Library of Congress, where more than 20 million cards had been accumulated by 1979. The card catalogue was frozen December 31, 1980. A second catalogue, computerized for accessing by computer terminal, lists subsequent acquisitions.

This cathode ray terminal, or computer console, consists of a keyboard with a television screen. The user inputs author, title, subject heading, or call number for a desired publication. The screen will show the available information about a publication in stock. The document may then be requested or found in the stacks.

Future developments in other libraries seem likely to be adaptations and extensions of adoptions at the Library of Congress.

Working in a Library

Although a knowledge of the actual layout of the library in which one is to work is of value in making efficient use of one's time, no general statement can be made covering what will be found in all libraries.

Usually in a general library there will be one or more reading or reference rooms, depending upon the size of the library, in which are shelved the reference works, periodicals, and other publications most often used. Current numbers of periodicals may be segregated in a separate periodical reading room. If the library is of sufficient size, there may be several major divisions. Then the chemical publications will probably be found in a technological division having its own reading room and library equipment. But whether the library is of a general or a specialized nature, the usual practice of librarians is to shelve in the stacks the publications not found in the reading and reference rooms.

In shelving publications, librarians may differ in making certain special arrangements, even though they use the same system of classification, such as the collection in one place of special types of publications, dissertations, governmental bulletins, and odd sizes of books, for example, regardless of the nature of the subject matter in them.

Although one may obtain a given publication with certainty by looking up its call number in the card catalogue and presenting this to a competent member of the library staff, it is frequently much more advantageous, especially if a considerable number of publications must be consulted, to familiarize oneself with the layout, the system of classification, and the rules of the library and then

[16]*Am. Scholar,* **49,** 66 (1979).

to obtain permission to go directly to the stacks to use the various works there. In working in a periodical set, for example, one frequently wants to refer to various volumes of the set which were not indicated in connection with the first reference sought.

Two uncommon arrangements in a library are convenient for the user. One is the distribution of study carrels in the stack area. This provides a place to study and write while using volumes taken from the shelves.

A second item concerns arrangement of sets of publications, such as *Chemical Abstracts,* the comprehensive treatises, and compilations of tabular data. It is very convenient to sit at a narrow table with shelves at the back on which are rows of the publications, preferably only two shelves high. One can then easily reach from volume to volume over some range without getting up. Better yet is the rotating circular table used at Chemical Abstracts Service. Turning the table readily brings a desired volume to the user.

SEARCHES I— MANUAL SEARCHING OF PRINTED SOURCES[1]

Anyone who wants to use a library effectively must already have some knowledge of the same nature as that which he hopes to find there. If not, he does not know where or how to look; nor can he grasp fully what he finds, even if he happens to hit upon the right book.

Sir William L. Bragg

The preceding chapters show that the chemical literature is varied and vast. Once one has become acquainted with the nature of the different kinds of these publications, with the manner of arranging them in a library, and with the means afforded for finding them, there remains the question of what constitutes a reliable and efficient method of ascertaining the specific information available upon a given subject. One must be able not only to find the proper works but also to find specific data which they contain, or to determine that desired facts are unavailable.

The kinds of questions presented in one large technical library were noted in Chapter 1. The concern now is the searcher and the nature of the searches necessary to find answers to such inquiries.

An efficient searcher should possess the following attributes: (1) a clear concept of the kind of information desired, (2) a general familiarity with the nature of what the different kinds of publications contain, (3) a knowledge of how to find information in the publications at hand, and (4) an awareness of what to do if a publication outside the local collection is needed. Perhaps, too,

[1]Searching patent literature was discussed in Chapter 4.

an accomplished searcher possesses an intuitive sense of what to do in specific situations. Searching is an art for which experience develops skill and efficiency.

The proper procedure to follow is more or less dependent upon the nature of the information desired. As F. E. Barrows stated[2]:

> Searches or investigations of the chemical literature . . . may be made from various standpoints—for example, by the research chemist, to familiarize himself with the available published information along the lines of his research; by the student, as a part of his studies or research; by the writer or author who, in his articles or publications, desires to give credit to the work of others, or to review the prior literature along the lines of his own publication; by the bibliographer, as the basis of his bibliography; by the manufacturer, to obtain information of interest along the lines of his manufacture or along new lines of development; by the patent investigator, in connection with questions relating to the novelty and patentability of inventions, and the validity and infringement of patents.

The volume of material which has to be consulted will vary widely, depending on the object the searcher has in view. To discover the solubility of sodium chloride in water under ordinary conditions, for example, it is sufficient merely to turn to a simple handbook, such as that by Lange or Weast, for the data. If knowledge of the details of the procedure employed by the investigator in determining these data is required, larger works of reference must be consulted where the original reference is given. Again, if a complete summary is required covering all the work that has been done on this determination, one must make a much more extended search of all portions of the chemical literature likely to contain such facts.

It is obvious that a searcher entirely unfamiliar with the chemical literature might experience much difficulty in finding even the solubility of sodium chloride in water. Such a person would lack the general knowledge or perspective which provides one with a sense of direction regarding the proper course to follow. This sense of direction, acquired only by experience with the literature, is of immense value in locating facts. Without it, one is greatly handicapped; with it, not only is much time saved but also the thing desired is much more likely to be found. It is a common experience to have senior students, when given a subject to look up in *Chemical Abstracts*, return shortly with the statement that they looked all around but could find little or nothing on their subject.

Given this sense of direction, or perspective, in handling chemical literature, able searchers will know where to look for certain kinds of information. Journals of physical chemistry will not contain papers on the synthesis of new organic compounds; details for the quantitative determination of lead will be found in works on analytical rather than inorganic chemistry; and commercial statistics relating to the chemical industry are to be found in industrial and trade journals and government publications rather than review serials. Such knowledge helps. But the sense of direction is probably of most value in connection with the use of

[2]*Chem. Met. Eng.,* **24**:517 (1921).

indexes, such as those of reference works, card catalogues, and abstracting journals. So important is this point that the indexing portion of chemical publications deserves special consideration.

INDEXES

In general, an index may be considered as a device which indicates the existence, location, or means of finding definite information. In an ordinary book, for example, the subject index differs from the table of contents in that the latter indicates, more or less in detail, the main divisions of the material in the order in which they are treated, while the former indicates, in alphabetical order, each significant item in the text. Since indexes of publications should be adequate directories for the purpose intended, it is of the first importance to know something about the nature and use of these facilities for locating information.

E. J. Crane, long-time editor of *Chemical Abstracts,* presented a valuable discussion of this subject from the viewpoint of one who had vast experience in both making and using certain kinds of indexes. The quoted paragraphs in the following portion of this chapter are taken from his paper.[3]

> The main problem, of course, in using the journal literature is in finding references, all that are pertinent to the subject in hand, in order that one may learn what the literature contains and all that it contains relating to this subject. This is often difficult. It is doubtful if there is a more important problem for students, in college or out, to learn how to solve. Its solution involves to some extent a familiarity with the more important journals, particularly the abstract journals; but above all it involves a knowledge of indexes and how to use them. References to the journal literature are often obtained from books or from one paper to others, but mostly they are obtained from indexes, usually of abstract journals. It is frequently assumed that the use of indexes in making literature searches is a simple matter requiring no special experience or ability. This is a mistake. The making of indexes is an art in itself, involving more than a comprehensive knowledge of the general subject being covered, and the use of indexes is not less an art. This deserves emphasis. It is true, because existing indexes vary greatly in kind, thoroughness, and quality. Even in the use of the best subject indexes the user must meet the indexer part way for good results. Conscious effort to become a good index user will well repay any scientist. Many a day has been spent in the laboratory seeking information by experiment which might have been obtained in a few minutes, or hours at the most, in the library, had the literature search been efficient.
>
> What constitutes a good index? The test is to determine whether or not an index will serve as a reliable means for the location, with a minimum of effort, of every bit of information in the source covered which, according to the indexing basis, that source contains. To meet this test an index must be accurate, complete, sufficiently precise in the information supplied, and so planned and arranged as to be convenient to use. Existing indexes fall far short of this ideal in many cases, and of course somewhat short of it in all cases.

[3]*Ind. Eng. Chem.,* **14:**901 (1922).

The main purpose in indexing is partially lost sight of through an effort to bring some sort of classification into it. Classification in connection with indexing frequently detracts from, rather than enhances, the efficiency and usefulness of an index, and is beside the main purpose.

Scope

Indexes range from a small number of entries in the back of a small book to sets of large volumes. Each individual reference book, bulletin, or monograph will ordinarily contain its own subject index. In the case of a set of reference works consisting of a number of volumes, each volume may or may not have its own index. If not, and occasionally even if it does, there will be a general index to the whole set. For periodicals, it is the usual practice to have an annual index in the final issue of the year, in case there are two or more issues during the year. In addition, many periodicals have cumulative or collective indexes covering a period of years, frequently 5, 10, or more. These are very desirable for making searches, as in the case of the cumulative indexes of *Chemical Abstracts*. They not only save time but also provide the indexer with an opportunity for correcting and improving the material taken from the volume indexes to make up the cumulative work. Indexes to patents may be volume or cumulative. It has already been mentioned that the index serials are annual publications.

Types

Depending chiefly on the nature of the information to be indexed, one finds in more or less common use at least five following bases of indexing:

1. By Patent Number This type is considered in Chapter 4, which deals with the literature on patents. In most cases one would not have to rely on patent-number indexes only, since other information, such as the title of the patent or the patentee's name, would be available for searching in the other types of indexes. *Chemical Abstracts* contains the best-known patent-number indexes.

2. By Formula Two variations of formula indexes are in use: one consists of empirical formulas and serves for all types of compounds; the other consists of ring formulas for complex organic compounds. A formula index may be used for any chemical compound whose empirical or ring formula is known. It avoids introducing the uncertainty of name which may exist for some compounds, particularly those with the more complicated structures.

Permuted formula indexes would facilitate searching for organic compounds containing elements other than C, H, O, and N. The Rotaform index of *Current Abstracts of Chemistry* is a fairly recent example (see Chapter 6B). Presumably this scarcity indicates a low cost-effective evaluation by publishers.

3. By Author In practically every printed article relating to chemistry, whether abstract, book, bulletin, periodical, or patent, the name of the author (or patentee) is mentioned. Also in treatises frequent mention is made of the names of investigators. Whenever the number of these names becomes fairly large and it is of importance to be able to find readily the mention made of them, separate author indexes are issued. This practice is more or less general for periodicals.

Although the alphabetical arrangement used in such indexes is simple, the user should keep in mind a number of variations which may be encountered.[4] The following examples are not uncommon[5]:

1 Names with variant spellings, such as Mellan, Mellen, Mellin, and Mellon.

2 Names beginning with 'Mc' and 'Mac'; also 'St.'

3 Names and prefixes such as 'de,' 'de la,' 'du,' 'van,' 'van der,' and 'von,' which may precede or follow the surname.

4 Hyphenated names, if not consistently entered under the first part of the name.

5 German names with umlauts. If replaced by the letter 'e' or letters 'oe,' the alphabetical sequence may be altered.

6 Transliteration from languages not using roman characters.

4. By Subject Although at times patent-number and author indexes, and particularly formula indexes, are very useful, their general value is not comparable to that of an index of subjects alphabetically arranged. Subject indexes are not only the most generally useful type but are, undoubtedly for this reason, the most common.

It is comparatively easy to make a complete index of the other types, but a complete subject index is an ideal much more difficult to attain. Those available vary widely in this respect, a state of affairs for which the maker is primarily responsible.

5. By Citation A citation index is a directory of cited references, each accompanied by a list of citing source documents. It provides a means for tracing the publications which have appeared since specific papers and which have used such papers as references. Thus, one goes forward in time with a citation index, but backward with conventional indexes.

Other Indexes

The possibility of using, at least for specific purposes, certain other kinds of indexes has been discussed briefly by C. L. Bernier.[6] He noted especially group, notation, reaction, and citation indexes.

---

[4]See introduction to author sections of cumulative indexes of *Chemical Abstracts*.

[5]See T. E. R. Singer, *Adv. Chem. Ser.*, No. 30, p. 75 (1961), for a discussion of language problems in literature searching.

[6]*J. Chem. Doc.*, **1**:62 (1961).

The Use of Indexes

The chief problem with indexes, after one learns the types available, where they are likely to be found, and the kind of information they contain, is to learn how to use them. For indexes of patent numbers and of authors the procedure is obvious; for formula indexes one should know the Hill (Chapters 4 and 7) and the Richter (Chapter 10) systems.

Subject indexes, owing to their nature, require special attention. The following discussion is from the previously mentioned paper by E. J. Crane[7]:

> The first step in learning how to use subject indexes with maximum effectiveness is to become familiar with the characteristics and peculiarities of existing chemical indexes.[8] The most significant point to note is whether or not a so-called subject index is really an index of subjects or an index of words. The tendency to index words instead of thoroughly to enter subjects constitutes the greatest weakness in the literature of chemistry. There is a vast difference. Words are of course necessary in the make-up of a subject index, but it is important for an indexer to remember that the words used in the text of a publication are not necessarily the words suitable for index headings or even modifying phrases. Word indexing leads to omissions, scattering, and unnecessary entries. After the most suitable word or group of words from the indexing point of view has been chosen for a heading, it should of course be used consistently no matter what the wording of the text may be. To illustrate a kind of scattering of entries which may result from word indexing, let us consider such a series of article titles as follows: "An apparatus for the determination of carbon dioxide"; "A new absorption apparatus"; "Apparatus for use in the analysis of baking powder"; "An improved potash bulb"; and "Flue-gas analysis." Word indexes would no doubt contain an entry under the heading "Carbon dioxide" for the first title, one under "Absorption apparatus" for the second, under "Baking powder" for the third, under "Potash bulb" for the fourth, and under "Flue gas" for the fifth, and probably no others. These entries seem reasonable enough if the titles are considered separately without thought of the others. And yet the articles may all be descriptive of the same sort of apparatus. As a matter of fact, all these titles might conceivably be used for the same article; if the author happened to be working on baking powder or on flue-gas analysis when he conceived the idea for his novel piece of apparatus, or had it in mind particularly for one purpose or the other, he might choose one of the more specific titles for his article rather than one of the more general ones.
>
> In an index entirely based on subjects rather than words, it would be the task of the indexer to see that all these articles got indexed under one heading, or under each of more than one heading, best with cross references pointing from the other possible headings to the one or more headings used. Or, if there seems to be some justification for scattering owing to differences in point of view (word indexing cannot be gotten away from entirely), he would make sure that the necessary cross references are supplied to lead the index user about from heading to heading so that all entries can be readily located. It is not hard to determine whether or not an index is a word index;

[7]For other advice on the use of indexes, see J. P. Smith, *Adv. Chem. Ser.,* No. 4, pp. 19–23 (1951), and T. E. R. Singer, *Adv. Chem. Ser.,* No. 4, pp. 24–25 (1951).

[8]For example, the key or explanatory matter given in both the author and subject indexes of the cumulative indexes of *Chemical Abstracts.*

when this is suspected or noted, one should look around pretty thoroughly in its use instead of being satisfied that the entries found under the obvious heading are all that the index contains on a subject.

It is important to note the approximate degree of completeness of an index in use. There is perhaps no definite point at which a subject index may be said to be complete. The indexing basis is too indefinite. A great many subject indexes are not as full as they ought to be. Aside from word indexing the indexing merely of titles is the most common reason for incompleteness. Titles cannot be depended upon to furnish the information necessary for adequate subject indexing. An index may be reasonably complete from one point of view and not from others. For example, a publication devoted to bacteriology may not reasonably be expected to be indexed fully from the chemical point of view. Completeness in the information supplied in modifying phrases, as well as completeness in index headings, needs to be taken into consideration. It is necessary, of course, to call forth one's resourcefulness to a special degree if a relatively incomplete index is to be used.

Cross references play an important role in subject indexing and in the use of subject indexes. Word indexing is really hard to avoid and cross references are the great preventive. It is a good sign if a subject index has a plentiful supply of cross references, both of the "See" kind and the "See also" kind.[9] They make for uniformity and proper correlation. "See also" cross references are of just as much importance as the "See" kind, though not as much used. The service which they render in directing the index user to related headings or to headings which, though dissimilar for the most part, have entries under them likely to be of interest to the investigator who refers to the original heading, is often the chief means of making a search complete. It is not reasonable to expect an index user, or an indexer, as a matter of fact, to think of all the headings representing related or significant subjects under which headings he may find valuable references that otherwise might be missed. Nevertheless, in the careful indexing, year after year, of a periodical devoted to a more or less definite field, as an abstract journal for example, subjects are met in such a variety of connections and from so many angles that it is possible for a truly comprehensive list of cross references to be built up. The suitability of a given "See also" cross reference may not be clear, much less suggest itself, until a specific case in which it is helpful is observed. It often pays to follow up such a cross reference even when it does not look as if it applies in a given case. The indexer, in surveying the whole field year after year, is in a position to make valuable suggestions in the form of cross references calculated to lead the index user from place to place in the index, so that the chances that his search will be really exhaustive as far as that particular index is concerned are much increased.

Persistence is a good qualification for index searching. It is desirable to avoid being too soon satisfied. There is no task in which thoroughness is more important. It involves first a knowledge of the index system and of the characteristics of the index. Then one needs to be resourceful, exhausting all possibilities, if he is to avoid some futile searches or incomplete findings. One's fund of general knowledge can usually be brought into service to good advantage.

On account of the necessity of drawing on one's general knowledge in making a

[9]Cross references which refer from a possible heading under which no page references are given to the chosen heading where they may be found are called "See" references, as "Mineral oils. See Petroleum." Those which connect headings representing allied subjects or containing related entries are called "See also" references, as "Iron alloys. See also Steel."

literature search in any field, it is in many instances important that one should make his own searches. It is not always safe to let someone less informed in a certain filed make an index search when a complete survey is desired, even though his familiarity with indexes and the literature in general may be better than one's own. Just as some tasks in the laboratory can be turned over to another to advantage, but not the more important determinations and experiments, so some tasks in the library can be delegated to an assistant, but not all such tasks. Knowledge, skill, and power of observation are factors fruitful of important results in the library as well as in the laboratory. Literature searching is a dignified pursuit, and it cannot with impunity be assigned to a lower level than that of the laboratory side of the problem, as far as the attention it receives is concerned.

With a given problem at hand the first step, of course, is to think out the most likely places to look in the indexes to be used. This may be a simple matter or it may be a very difficult one, depending on the nature of the problem. Difficulty increases with indefiniteness. Experience is necessary. In fact the beginner is often completely at a loss to know what to do at this very first stage of his search. This point is stressed in the Report of the Subcommittee on "Research in Chemical Laboratories" presented to the Committee of One Hundred on Scientific Research in December, 1916.[10] In this report, which commends and recommends courses in chemical literature searching in the universities, it is pointed out that the average graduate "fails to analyze the subject" in which he is interested "into its factors, and, hence, generally looks for topics which are too general. Because he does not find any references to the problem as a whole as he had it in mind, he assumes that nothing has been done upon it and that there is nothing in the literature which will be of aid to him in the investigation. Were he to separate his subject into its essential parts and then to consult the literature on each factor, he would find considerable information which he otherwise would miss." Even though some index headings to which to turn, perhaps the more important ones, may be brought to mind without ingenuity, the completeness of a search may be marred by a failure properly to analyze the problem. Indexes with cross references, particularly "See also" ones, help.

Too much dependence on cross references is not advisable. They may not be available at all and they are never complete. With a given heading in mind it is well to cudgel one's brain for synonymous words or phrases to try, as well as for variously related subjects, and it is advisable to try these even though entries as expected are found in the first place to which one has turned. Words or phrases with an opposite meaning to the one in mind may serve as subject headings under which desired entries may be found. For example, the searcher interested in viscosity may find significant entries under the heading "Fluidity" in addition to those under "Viscosity." Incidentally, it may be noted that the word "consistency" may serve as a heading for still other related entries. Or, some entries under "Electric resistance" may interest the searcher whose thought on turning to an index was of "Electric conductivity." If such related subjects are not suggested by cross references and have not been thought of in advance, they may be suggested by the nature of some entry under the heading first turned to if one is on the lookout for them.[11]

[10]*Science,* **45:**34 (1917).
[11]See C. H. O'Donohue, *J. Chem. Doc.,* **14,** 29 (1974), on the importance of profiling as the key to successful searching; also T. E. R. Singer, *Rec. Chem. Prog.,* **18,** 11 (1957), on the need for imagination and skepticism when making literature searches.

The resourcefulness required in making a thorough search through subject indexes can best be discussed by considering an example. Suppose one were interested in looking up all possible references on vitamins. The first place to which to turn naturally would be "Vitamins" in the indexes to the various reference sources to be used. This would rarely, if ever, be far enough to go. If only one of the indexes contained "See also" cross references, these might be helpful in the use of the other indexes. This playing of one index against another, so to speak, is always a possible means of helping out. Cross references should be looked for. Since it is not always possible to find such cross references and it is not safe to depend too much on them, to be complete one might follow out a line of thought as follows: Vitamins are constituents of foods. It may be worth while to look under "Foods." Entries may be found there with some such modifying phrase as "Accessory constituents of." Vitamins are a factor in health and the effect of foods on health involves the idea of diet or ration. These headings, or this heading if they are combined under "Diet," for example, as would seem best in a true subject index, would no doubt prove fruitful of significant references. Studies of proper diet or of an adequate ration for an army would beyond doubt involve the vitamin theory. Experiments to determine the nutritional value of foods are frequently called feeding experiments, so a heading "Feeding experiments" may be looked for to advantage. Food is taken for the purpose of nutrition and the vitamin problem is a nutrition problem. Therefore, the general subject "Nutrition" needs to be examined in the indexes. There is, of course, such a thing as plant nutrition analogous to the vitamin theory in animal nutrition, so the heading "Plant nutrition" or the heading "Plants" would be suggested. If he did not know it, he would likely learn there are substances supposed to be factors in plant life, called "auximones" by Bottomley, which are analogous to the vitamins in animal nutrition. The heading "Auximones" would, of course, then be suggested for reference. The lack of vitamins in the diet is considered by some to be the cause of certain diseases (beriberi, pellagra, polyneuritis, scurvy, xerophthalmia). These ought, therefore, to be referred to as index headings. The general heading "Diseases" should be tried also, such a modifying phrase as "Deficiency" being looked for. Perhaps the next thing for the index user to do would be to ask himself, or someone else, whether or not there is a definite name for this general type of disease; he would find that there is and that the name is "Avitaminosis," which should then be turned to as a heading. Certain specific foods have been used and studied, particularly with reference to the vitamin theory, as, for example, polished rice, milk, butter, orange juice, yeast, tomatoes, etc. It seems unreasonable to be expected to think of these, or at least all of them, and yet an article entitled, say, "The effect on pigeons of eating polished rice," may be word indexed only in some index, and therefore only get under the headings "Rice" and perhaps "Pigeons." Vitamins have been differentiated as "fat-soluble A," "water-soluble B," etc. and are sometimes spoken of merely in these ways. It is conceivable that some indexes may have these names as headings. In the earlier literature studies resembling the modern vitamin studies are to be found in which other names for the accessory food constituents are used, as nutramines (Abderhalden), bios (Wildiers) and oryzanin (Suzuki). The text referred to from any one of the above-mentioned headings may suggest still other headings, as the names of specific foods supposed to be rich in vitamins.

The principle of referring to the general as well as to the specific subject, as exemplified in the preceding paragraph by the subject "Avitaminosis" for the general and by the individual deficiency diseases (beriberi, etc.) for the specific, is a good one to keep constantly in mind in using subject indexes. This principle applies aptly in searches for information regarding compounds. Group names for compounds may serve as index headings under which entries of interest to the searcher interested in an individual compound may be found.

The resourcefulness necessary in the location of information by means of the great variety of subject indexes in existence may seem to be little more than clever guessing at times. A paper on glass, so-called and indexed only under "Glass," may reveal a principle governing the action of metals or other undercooled melts. Authors fail to see the full significance of their experimental results, and it is not often that the indexer will go further than the author in bringing out this significance for attention. The kind of flexible ingenuity necessary for the location of information in this way is perhaps only to be acquired by experience. It is really more than guessing that results in the location of information in this way, and yet it seems as if a little more than reasoning power, something like intuition, is sometimes necessary.

Chemical publications present a special problem, both to the subject indexer and to the index user, in that many headings must consist of the names of chemical compounds. The difficulties encountered are to be attributed (1) to the fact that many compounds have, or may have, more than one name, (2) the names, or at least the best names, of the more complex compounds may be difficult to ascertain, and (3) new compounds are constantly being prepared, which, if named at all, may receive more than one name which is justified from one point of view or another, and the possibilities of incorrect names are great.

It is not feasible to enter into a detailed discussion of the best procedure in building or using indexes of chemical compounds. The difficulties increase with increasing complexity of compounds. Some indexes are based on systematic nomenclature, irrespective of names used by authors; others are not. Cross references within an index and introductions thereto, and the use by index searchers of dictionaries, chemical encyclopedias, handbooks, and other sources of information, leading to a knowledge of the names, sometimes numerous, of compounds, are helps to be utilized. As mentioned above, a knowledge of what constitutes good nomenclature is a great aid in the location of compounds in name indexes. This is particularly important for the organic chemist. It is on account of the almost insurmountable difficulties due to the complexities of chemical nomenclature and because of language differences, that a basis other than their names—namely, their empirical formulas—has been sought and, to a limited extent, used, in the indexing of compounds. A formula index provides a certain means for the location of individual compounds; it is very doubtful if the average chemist can locate compounds in all cases in name indexes even though systematic nomenclature may have been consistently followed in the indexing. In name indexes it is possible, by appropriate devices, to group related compounds to good advantage.

The subject index searcher is confronted with nomenclature problems relating to fields other than that of chemical compounds. For example, the chemist interested in plants must contend with the fact that some indexes use the scientific names (genus and species) of plants as headings and others use common names, of which there are frequently several for the same plant.

The use of indexes in foreign languages presents obvious difficulties. It is one thing to be able to read a foreign language and another to translate one's thoughts into that language. The use of an English-French, English-German, or other like dictionary, depending on the language involved, is about the only help available. The introductions to Patterson's "German-English Dictionary for Chemists" and "French-English Dictionary for Chemists" contain helpful suggestions for determining German and French names of chemical compounds.[12]

Locating Names of Compounds

The attainment of an efficient working knowledge of chemical nomenclature, as mentioned above, is a formidable task. Yet one who tried to locate in an index the more complex compounds, now comparatively common, especially in organic chemistry, without an idea of the system used in naming them, would be almost hopelessly lost.

Unfortunately, chemists have been slow to adopt an international basis of nomenclature, although progress has been made. But even if that highly desirable goal were now achieved and everyone used the system, preceding the time of its realization there are some 175 years of literature of modern chemistry more or less confused in this respect. This factor almost necessitates one's learning something about each subject index to be used.

Only the indexing practice of *Chemical Abstracts* is considered here, as it probably represents the best there is now. The procedure being followed was adopted only after years of study of, and experience with, the problems involved.

The previous edition of this book contained an adaptation of excerpts from (1) the section on nomenclature in "Directions for Abstractors" for *Chemical Abstracts* and (2) the introduction to the subject index for the current volume of this periodical. In this new edition, however, the searcher is referred to the following sources for the latest usage by Chemical Abstracts Service:

1 The Introduction to the "Chemical Substance Index" to *Chemical Abstracts.* Beginning with 1972, this index has been separate from the "General Subject Index."

2 "Directions for Abstractors," 1975. These instructions relate to the selection of names for the "Chemical Substance Index."

3 "1977 Index Guide." This latest guide is for the tenth collective index, which covers volumes 86 to 94. Some 1,200 pages include headings and cross references. Appendix IV deals with rules followed in deriving systematic names for chemical substances.

These sources must be studied for present practice. However, as one may have to search backward for decades, the indexes for *Chemical Abstracts* are much less elaborate and so necessitate adjusting to their nature. In general, it seems likely that indexes for most publications will be found increasingly unsatisfactory in earlier years.

In using any subject index involving the name of a chemical compound there

---

[12]See English–foreign language dictionaries listed in Chapter 12.

may be considerable difficulty, especially with foreign languages. Even English indexes may be troublesome if the searcher is not familiar with the system used.

The problem is worse in German, French, and other foreign languages, as even the IUPAC name varies somewhat because of the peculiarities of the languages.[13]

CURRENT-AWARENESS SEARCHES

This type of search deals with the present and involves finding what is new in order to be up to date. It seeks the current facts, along with their interpretations and applications.

Usually the goal is not to learn some single, isolated fact, such as the final patent number in the list of patents granted last Tuesday in the *Official Gazette* (Patents). Rather it is trends, situations, developments, as revealed in papers, reports, patents, regulations, prices.

Developments may concern the research program of an individual or an industry. There may be new processes, products, uses, or raw materials.

Sources of such information are periodicals, such as *Current Meetings, Current Contents, Chemical Titles, Chemical Industry Notes,* and similar publications, as noted in Chapters 6 and 7. An alert individual watches selected *CA SELECTS,* as well as the appropriate section(s) of *Chemical Abstracts.* Specialized abstracting journals are published in some fields.

Chapter 15 deals with computer means of searching these sources. Of course, an individual may examine manually the ones issued in printed form.

Late spot news, as well as general items of interest, appear in periodicals such as *Chemical & Engineering News, Chemical Engineering, Chemical Week, Chemistry and Industry, Chemical Marketing Reporter, European Chemical News,* and the *Journal of Chemical Education.*

Current-awareness questions are endless. A few examples follow: What are the latest U.S. patents on lasers? What fire-retardent chemicals are in current use? What article appeared last month on sequencing proteins? What is the current price of sulfuric acid? Which company produces the most nylon?

RETROSPECTIVE SEARCHES

This type of search concerns the past. Except for relatively late information, the desired facts appearing first in primary publications will have been collected, screened, and organized in several kinds of sources for ready reference.

Almost any kind of publication might serve for ascertaining some facts. The following kinds may be the most generally useful: abstracting journals, bibliographies, biographical works, book lists (old and new), book reviews, catalogues,

---

[13]See E. E. Reid, "Chemistry through the Language Barrier," The Johns Hopkins Press, Baltimore, MD, 1970.

dictionaries (language and chemical terms), directories, encyclopedias, formularies, indexing journals, monographs, reviewing journals, tabular compilations, textbooks, treatises, and perhaps others. These publications are probably the best example of where an experienced searcher knows the sources most likely to yield the desired information.

Such sources vary widely in usefulness. Abstracting journals are invaluable for extended searches on many different subjects. Textbooks serve only occasionally for isolated facts.

There are two kinds of retrospective searches. One may be designated as short, or spot-check, in nature, and the other as more or less comprehensive. They are considered separately.

Short Searches[14]

This type of very common search ordinarily involves a single entity, such as the crystal structure of sodium chloride, or a method for determining carbon in an alloyed steel. The time may be last year or the last century (possibly alchemical times). In general, this kind of question stands pretty much by itself.

As examples of the simplest kind, the following questions serve: What is cupellation? What is the systematic name for valium? What is the half-life for strontium 90? When does U.S. patent 4,126,578 expire? What is the approved method for determining residual chlorine in water? What is the density of durallium? What halogen substitution products of cyclopropane are known?

To become a little more involved, suppose that one wishes to know the physical and chemical properties of n-butanol. This compound has been known for many years, and its properties determined and recorded in periodicals for a sufficient length of time that one should be able to find the desired information in the general works of reference, excepting the possibility of some fact recently discovered. Accordingly, a search would be made in tabular compilations for data such as boiling point and refractive index. The chemical properties should be sought in a treatise on organic chemistry.

The identification of organic compounds is a different kind of problem. Having determined the empirical formula, one may then turn to formula indexes to find compounds having this formula. It then remains to compare properties of the compound with those of any others found in this way. There are other approaches.

Again, it may be necessary to find a method of preparing some compound, such as lithium perchlorate, or to introduce a nitro group into some compound containing the benzene ring. In cases such as these, the desired information may generally be found by consulting the large, general works of reference covering the field to which the problem belongs. These works include the treatises and encyclopedias mentioned in Chapter 10. It is important to bear in mind that not

---

[14]See Lucy O. Lewton, *J. Chem. Educ.*, **28**, 487 (1951), for suggestions on searching for on-the-spot information; also P. Fenner and M. C. Armstrong, "Research: A Practical Guide to Finding Information," W. Kaufmann, Inc., Los Altos, CA, 1981.

all methods of preparation are equally satisfactory for all purposes. One may have to produce a very nearly pure product and not be concerned with obtaining a yield of 95 per cent. On the other hand, a method giving a yield of 95 per cent may be the salvation of an industry if it is competing with another using a method giving a yield of 85 per cent. If it is a problem involving the necessity of finding the best method or of devising a special kind of method, a more detailed search, as described below, is generally required.

What has been stated regarding the above-mentioned specific examples applies, in general, to similar problems, whether the problem is general, inorganic, analytical, organic, physical, or some other division of chemistry. Use is made primarily, of course, of any reference works dealing especially with the field under consideration, and following this, if necessary, with consultation of works on related phases of the subject. If the method desired should be of recent development, the works of reference are usually too old to include it, and one should then go directly to the abstracting journals, as indicated below. However, if the method is not so new, it may not be necessary to go to the original literature unless it seems desirable to compare the statements found in the works of reference with those of the original.

A consideration of sources of information bearing on some of the questions a chemist takes to the library, as outlined in Chapter 1, may be of value as further examples of the question and the publications likely to be of value in finding information thereon.

1 *References on the physiological action of plutonium 239:* Published bibliographies, biochemical reviews, governmental reports, *Chemical Abstracts*.[15]

2 *Life of T. W. Richards:* Histories of chemistry, biographical articles in periodicals, biographical compilations (see Chapter 12), if available for the period concerned.

3 *Bromine substitution products of cyclopropane:* Formula Indexes of Beilstein and *Chemical Abstracts* (or *Chemisches Zentralblatt*) for all the compounds theoretically possible.

4 *Formulas for mothproofing textiles:* Formularies and recipe compilations, abstract journals.

5 *Fluorination of organic compounds:* Treatises on organic syntheses, "Organic Reactions," abstracting journals.

6 *Physiological effect of dihydrogen sulfide:* Encyclopedias, monographs on physiological action of chemicals, compilations on toxic agents, treatises on biological chemistry.

7 *Employment of $SOCl_2$ in synthetic chemistry:* "Organic Reactions," Theilheimer's treatise, abstracting journals (see various possible alkyl halides).

8 *Determination of PCB in milk:* "Official Methods" of Association of Official Analytical Chemists, works on food analysis, *Chemical Abstracts*.

9 *Claims for a process for making tetracycline:* In *Chemical Abstracts* and/or

---

[15]Use *Chemical Titles* for articles not yet abstracted, and keyword indexes in each issue of *Chemical Abstracts* until the annual indexes appear.

Official Patent Gazette of the U.S. Patent and Trademark Office locate a patent (United States); obtain a copy and consult the claims.

10 *Statistics for the uranium industry:* "Minerals Yearbook," census reports, circulars from U.S. Department of Commerce, current market reports, abstracting journals.

COMPREHENSIVE AND EXHAUSTIVE SEARCHES

In many instances, having defined the nature, purpose, and scope of a problem, one must gather all the available information relating to the whole subject involved or to certain phases of it, either for the entire period covered by chemical literature or for some definite part of this period. Searches of this kind will usually involve the use of both primary and secondary sources of information.

Before starting a comprehensive search one should have some idea of the approximate degree of completeness desired. In some cases one wants everything known; in others only the significant facts from a given date to the present time are sufficient. As stated by E. E. Reid[16]:

> It is well to realize that, at present, few searches of scientific and technical literature are ever complete. The practical question, then, is merely how far one can or ought to go in a particular case. According to the nature of the subject and the object with which the search is undertaken a point will always be reached, eventually, where all competent judges must agree that the probability of finding a reference and its possible value if or when found do not warrant the time, trouble or expense involved in continuing.

It would be a waste of time, for example, to search for the application of the vacuum tube in chemical measurements before the time of the invention of the tube.

Different individuals will begin such a search in different ways, although each seeks all the relevant facts in the case.[17] Whatever the method used, it should meet the tests of accuracy, dependability, and efficiency. Some individuals recommend starting with the earliest publications available relating to the problem in hand. By means of indexes, reviews, and abstracts they work forward to date. Others practically reverse the procedure, beginning with the latest indexes and working backward as far as seems profitable. The procedure described below is essentially the practice followed by the author.

Assuming that all the pertinent facts on a given subject are to be collected, it is probably a question whether one should start on the works of reference or in the indexes of the abstracting journals. Doubtless the former should be consulted first in many cases. A concise, recent summary of information on the topic under consideration, such as those in *Chemical Reviews,* general reviewing

---

[16]"Invitation to Chemical Research," p. 295, Franklin Publishing Company, Inc., Palisades, NJ, 1961.
[17]See Lewton, op. cit., on the art of searching the literature; also Singer, op. cit.

journals, encyclopedias, and monographs, may be of much value in providing a background for beginning the more extended search. More frequently, one begins with the most recent indexes of the abstracting journals. An important advantage is that one is likely to find in recent articles a bibliography bearing directly upon the subject in question. Even before the indexes are searched, any relevant bibliographies are consulted for compilations promising to be of value.

It is desirable at this point, or before the search of the index has progressed far, to make an outline of the possible headings which may furnish information of value. A systematic search under each of these should then be made, and all references which offer any promise of containing desired material should be noted. In general, the only safe way is to include all uncertain references until abstracts, or the original, indicate they are irrelevant. After all references seeming to have any bearing on the subject have been selected from the index, the abstracts are then read. Whenever the abstract indicates that the reference is obviously without value it is discarded. All others are recorded, each on a separate card, as indicated later, and filed.

Usually alphabetical filing by authors will be found most desirable. Often nothing more is needed; but in some cases, especially when many references are involved, it is preferable to adopt some system of subdividing the cards. This may be by years, countries, periodicals, natural divisions of the subject, or other suitable classification. In each of these divisions the filing would then ordinarily be alphabetical by authors.

Although one may expect abstracts to furnish a general statement of the contents of the publication abstracted, it is always well to maintain a skeptical attitude toward them. In many cases—probably most—they do not furnish the desired details. Unless it is reasonably certain from an inspection of the title of the publication, and of abstracts of it, either that nothing of interest is contained therein or that all desired information is provided thereby, the original publication should be consulted if possible. In case the original is unobtainable, several abstracting journals giving abstracts for the same article may help.

In some instances all the indexes of *Chemical Abstracts* and other abstracting journals are examined in this way before one starts to read original articles (collective indexes for periodicals should always be used if available). At other times some of the readily available articles, books, and treatises are read after the collection of only part of the references in the abstracts. Just as soon as the reading is begun, the references branch out in all directions, since all the authors will have included those references they considered significant. All those of value for the problem being investigated are recorded and put in the same file. Perhaps the better way is to divide the file into two sections, putting all references which have been examined and checked in one list and those to be examined in the other. The advantage of such filing appears soon after the search begins. If a reference is found when an article is being read, no record is made until the alphabetical list is examined to ascertain whether such an entry has not already been included, for one will often find the same reference in a number of different papers. In this way all references are put in the list as found, and duplication of cards is avoided.

The main abstracting journal for general searches is now *Chemical Abstracts. Chemishes Zentralblatt,* the *Journal of the Chemical Society,* and the *Journal of the Society of Chemical Industry*[18] were of value before being discontinued. The Russian abstracting journal, *Referativnyi Zhurnal Khimiya,* is general in coverage but began in 1953. A number of others, having less significance or covering a specialized area, are mentioned in Chapter 6.

Whether one begins with the abstracting journals or with other types of publications, ultimately both will be used in making a comprehensive search. Each individual will reach her or his own decision regarding the most efficient practice for the particular type of problem. In any case, a systematic handling of notes on the references is desirable.

Although this fairly systematic method of searching is, as a rule, most efficient and should generally be followed, simply looking about, more or less at random, is frequently productive of valuable information. In the case of general books one may look here and there, selecting for examination whatever works offer promise of interest; for periodicals it may be worthwhile, as a last resort, to leaf through the volumes, page by page.[19] This is a tiresome, time-consuming task, but one may find just the formula, equation, or other readily apparent item which is desired, when an examination of the titles of articles and the indexes indicated nothing of value.

As far as possible, one should develop the ability to go through the original publications rapidly, selecting at the same time the significant points for the search in hand. This is not an easy task, and the development of efficiency requires considerable practice. The individual familiar with the problem involved knows better than anyone else what is significant and thus is the best fitted to conduct this part of the search.[20] One not thinking along the line of investigation involved can hardly be expected to sense the importance of the various points encountered or to be alert in following what may seem at first only unpromising side issues.

Abstracting and reviewing journals, bibliographies, various kinds of reference works, and monographs are among the important sources for retrospective searches.

The general nature of the problems involved in such searches may be illustrated by the following subjects: analytical spectometry prior to 1930, dimethylsulfoxide as an extractant, disposal of nuclear waste, metal carbonyls, methane from organic wastes, physiological action of ethanol, precipitation of alkaloids with heteropoly compounds, synthetic tooth enamels, gasoline antiknock additives, and perfluorination of heterocyclic rings.

Hidden Facts An occasional comprehensive search involves finding what

---

[18]The latter two were combined in 1926 as *British Chemical Abstracts,* and subsequently as *British Abstracts,* which ceased publication in 1953.

[19]See H. S. Booth and D. R. Martin, "Boron Trifluoride and Its Derivatives," Preface, John Wiley & Sons, Inc., New York, 1949.

[20]Because of his knowledge and experience, one well-known director of research insisted on checking the literature himself in critical cases.

may be considered hidden facts. By this is meant facts the existence of which ordinarily will not be indicated by any of the usual indexes. This omission occurs because the facts are not considered important enough to mention in the title and/or the abstract of the publication containing them. Also they would not rate in a list of keywords.

An example concerned the use of ethanol in analytical chemistry. The problem arose when ethanol was not readily obtainable and isopropanol came on the market. The question was where the new reagent might be substituted for the old one. A long search revealed many incidental uses of ethanol, but unconventional approaches to searching were necessary.[21]

Retrospective searching for facts and ideas in sources published prior to 1850 may be difficult. Many early books and periodicals are listed in Chapters 1 and 2, respectively. Unless one has specific references to such sources, dependence must be placed on using their indexes and tables of contents. The indexes are unlikely to be detailed.

As stated in Chapter 6, secondary sources did not get under way until the first half of the nineteenth century. By the end of this period development of indexing, abstracting, and reviewing was well under way. In 1800 it was practically nonexistent.

Two papers in *Advances of Chemistry Series,* No. 30 (1961) may be of help. G. M. Dyson (p. 83) discussed searching the older literature, and M. G. Mellon and R. T. Power (p. 92) considered less familiar periodicals.

Three other topics involved similar problems of finding references containing the desired information: methods of determining the composition of amalgams,[22] the analytical uses of color,[23] and heteropoly compounds in analytical chemistry.[24]

Deficiencies in the Literature

The searcher's problem may not end with finding the desired reference source. Some years ago the author, as an analytical chemist, reported some personal difficulties with the literature. No doubt others have had similar, and probably more important, problems.[25]

Briefly, the nature of these problems follows: titles, summaries, definitions, symbols, abbreviations, editorial conventions, spelling, inadequate data, nomenclature, abstract deficiencies, and subject indexes.

Two examples illustrate what one may encounter. In checking papers over a 10-year period, 270 were found dealing with the determination of nickel in iron and steel. The concern was the method used. For 10 percent of the papers neither the title nor the abstract indicated the method.

[21]G. W. Ferner and M. G. Mellon, *J. Chem. Educ.,* **10,** 243 (1933).
[22]M. G. Mellon, *Proc. Indiana Acad. Sci.,* **34,** 157 (1924).
[23]M. G. Mellon and G. W. Ferner, *J. Chem. Educ.,* **10,** 691 (1933).
[24]E. R. Wright and M. G. Mellon, *Proc. Indiana Acad. Sci.,* **50,** 110 (1941).
[25]M. G. Mellon, *Adv. Chem. Ser.,* No. 30, 67 (1961).

Important details may not be included. Thus, the absorbance of a solution of cobalt(II) chloride may depend upon the wavelength of measurement, the spectral bandwidth used, the temperature, the solvent, the concentration, and the cell thickness. To give a molar absorbance index without specifying these parameters is to give inadequate information. Such details will not be in the title. If not in the abstract, one must consult the primary source (which may itself be deficient in one or more details).

THE PLACE OF MANUAL SEARCHING

In every current issue of the *Journal of Chemical Information and Computer Science* most of the papers deal with some aspect of searching by means of computer-readable data bases. Chapter 7 notes the extent to which Chemical Abstracts Service has mechanized its operations to facilitate the production of the various services now available. Chapter 15 presents an overview of computerized means for searching and the services available based on their use. There is little question about what computers can do with a given data base.

In view of all of these developments, what, then, is the place of manual searching? Several publications, in one way or another, seem relevant.

P. S. Davison[26] described a cost-effective manual system which provides a wide range of information services by the use of a tri-level classification.

E. Kiehlmann[27] reported the organization and efficient manual searching of *Chemical Abstracts Indexes, Science Citation Index, Index Chemicus, Chemical Titles,* and *Chemischer Informationsdienst.* He concluded that efficient manual search procedures remain essential for both current-awareness and retrospective literature work.

J. Santodonato[28] compared on-line and manual modes of searching *Chemical Abstracts* for specific compounds.

A fundamental question was raised earlier by D. R. Swanson[29] in the statement:

> The matter of impracticality notwithstanding, to suppose that fully successful catalog search is a possibility implies that all human knowledge can be organized by means of some immutable and unambiguous classification or indexing scheme. Apart from the cardinal disadvantage of being nonexistent, such a hypothetical scheme has the even more unfortunate disadvantage of requiring that information be cataloged with precognition to insure recoverability in response to all conceivably relevant future requests. Both human knowledge and language change in the course of time, and the basic function of the library—to serve as a communication link between past and future—cannot, within the framework of present indexing and searching concepts, accommodate such change.

[26]*Inform. Sci.* (London), **6,** 15 (1972).
[27]*J. Chem. Doc.,***13,** 78 (1973).
[28]*J. Chem. Inf. Comput. Sci.,* **16,** 135 (1976).
[29]*Science,* **132,** 1099 (1960).

Thus, Swanson's skepticism centers on the coder, the one who compiles the facts and sets the systems to make cards, film, or tape. A specific example is the chemistry of the platinum elements. Howe's superb bibliography[30] covers the literature back from 1896 to 1750. Who is going to read all the original sources cited, in addition to all of those published since 1896, and extract all the relevant items for coding? Then, who will code all of them so that anyone may search for any present or future problem involving the facts?

To be even more specific, do arsenic, germanium, phosphorus, or silicon ever appear in the platinum elements as impurities? If so, has heteropoly chemistry been applied in the determination of these impurities? Unless the reader of the original publication notes all such facts, they will not be coded. Obviously, then, the machine can yield no references to these facts. In the reference cited by Kiehlmann, this same general idea is noted as a principal difficulty in achieving a largely automated system.

F. A. Tate[31] made an important contribution in noting that data bases have not existed long enough to permit conducting most comprehensive searches. Few go back more than 15 years. Thus, a search of the primary literature prior to the early 1960s usually must depend upon printed tools.

This conclusion is supported by R. Solkot[32] of the Medical Library of the Bureau of Drugs (Food and Drug Administration). He states, "Literature searches support all drug program activities. . . . Many surveys must be carried out, in part or entirely, by manual retrieval since they require delving into pre-1965 literature."

The following quotation emphasizes the importance of printed sources in the form of the great treatises[33]:

> Computers have no intelligence, in the sense that they cannot classify problems by search probabilities or resolve ambiguities. Generally these machines can only solve explicit problems which are within their capacity, providing one has developed the rules and strategies and tactics involved and expressed them in a program. As a rule, this approach can provide no help at all when one is looking for and seeing new problems. . . . The machine merely transmits a knowledge of documents while the "Handbuch" transmits information.

As a final word, it seems likely that librarians will continue to consult Lange's (or other) small handbook, for example, for the density of rhodium, and the alcoholimetric tables in the "Methods of the Association of Official Analytical Chemists" for the percentage (v/v) of ethanol in a distillate having a specific gravity of 0.9600 at 20°C.

---

[30]J. L. Howe, *Smithsonian Inst. Publ.*, No. 1084 (1897).
[31]*Pure Appt. Chem.*, **49**, 1897 (1977).
[32]*J. Chem. Inf. Comput. Sci.*, **19**, 11 (1979).
[33]Introductory brochure of the Gmelin Institute for Inorganic Chemistry for "Gmelin's Handbuch der anorganischen Chemie," 1974.

Selected References on Searching

Many papers on searching have been presented before the Division of Chemical Information of the American Chemical Society. Of those listed herewith, some are general in nature and some deal with specific problems.

Advances in Chemistry Series, No. 10 (1951), "Literature Resources for Chemical Process Industries," contains 59 papers on many different subjects. They are classified under market research, resins and plastics, textile chemistry, food industry, petroleum, and general. The last group has the widest interest, especially the following two papers: "Basic Principles of Literature Searching," by J. F. Smith and C. M. Schaler (p. 438), and "Sound and Unsound Short Cuts in Searching the Literature," by C. M. Schaler and J. F. Smith (p. 441).

Advances in Chemistry Series, No. 30 (1961), contains four papers on the resources in specific libraries. In general, the others deal with a variety of searching problems, as indicated by the titles. There have been changes and advances in some cases, but most of the papers and subjects remain relevant.

Advances in Chemistry Series, No. 78 (1968), "Literature of Chemical Technology," has 40 chapters which cover a broad range of chemical technology. Each general résumé of the subject ends with an extensive bibliography. In the discussions reference is made to practically all the kinds of primary and secondary publications, with citation of the most important examples in each case.

Altogether, this report is an impressive bibliographic assemblage of references.

For searching on many subjects in the near past, probably the most helpful means are the present-day printed indexes of *Chemical Abstracts*. They are described and illustrated in the "CAS Printed Access Tools 1981." This problem/workbook shows efficient approaches to using the volume and other indexes.

A new overview of the problems of retrieving information appears in the Kirk-Othmer "Encyclopedia of Chemical Technology," vol. 13, pp. 278–336 (1981). Contributed by four members of three Exxon Information Centers, the various subdivisions of the subject contain a wealth of references to printed sources.

Recording Information

The method to be used in recording the information obtained during a search is of some importance also. A useful scheme is to employ a card file of convenient size, using cards of bond paper, cut to the desired dimensions, from a stock of the desired weight. Sheets of the 3- by 5-inch size may be used, although the 4- by 6-inch size affords more room for writing. Such paper is thinner than the ordinary cards.[34] On each one is recorded the required information, as indicated

[34]Library Bureau cards, no. 1192 (3 by 5 inches), unpunched and without vertical lines, are very satisfactory.

in the form shown on page 169. The reference contains the standard abbreviation, if a periodical, or the title, if a book, including volume, page, and year. In the upper right-hand corner is a word to indicate the nature of the bibliography so that different ones can be easily separated if the references become mixed. The citation to the abstracting journal enables one to return quickly to the periodical if desired. The notes include any facts likely to be of value.

When the reference is first taken down from the abstracting journal, or other source, and filed, only the items at the top are recorded. If, on consulting the original source, nothing of value is found, a check is placed in one corner to indicate such examination, but no further entry is made. Keeping the card in the file saves repeating the examination at some later time, if the reference is not recognized when encountered again. If the material is of value, the title is taken down, along with the notes, and the card is checked.

Dependability

Only the completely credulous individual uses the scientific literature without questioning the reliability of the information recorded therein. Undoubtedly not all the material printed is correct or always the best obtainable; but in general the achievement in this respect is believed to be high. Nevertheless, in examining every contribution a critical searcher will maintain a skeptical attitude.

To judge the reliability of another's work may be difficult. Wide experience and knowledge on the part of the critic are of great value in providing perspective. For writers of reputation one will be influenced by the quality of their previous contributions. In experimental work the technique used is an indication of general quality. The conclusions reached by an author should be examined in light of the facts upon which they are based. All references should be verified, as many otherwise careful writers are negligent in checking the accuracy of their citations.

Reports on Information Found

One who merely collects all the references bearing on a given problem has at hand the fundamental material for a bibliography on this problem. To make this material readily usable for selecting and looking up the references desired, there remains only the problem of arranging the references in some one of the ways suggested in Chapter 8. If the various references have been examined sufficiently to provide for the inclusion of carefully prepared annotations for each citation, the bibliography as such may constitute a satisfactory report of the search.

But if the bibliography, when properly prepared, does not in itself constitute a sufficient presentation of the material covered, it becomes necessary to prepare a more general type of report. In such a case it will take one of two forms: noncritical or critical. If it is noncritical, there is included for the work of each investigator a statement of what was done (including how it was done, if important experimentally) and the conclusions reached. A critical report, on the other hand, is distinguished by its emphasis on the relative merit or value of each contribution. In this connection one should keep in mind, of course, that, while one report may be entirely noncritical, the extent to which another is critical is a question of degree. Probably in few cases is the best possible approached.

For instructions regarding the preparation of reports the reader should consult Library Problem 20, pages 389–393, including the references mentioned there.

SEARCHES II—
COMPUTER SEARCHING
OF DATA BASES[1]

A knowledge of where and how to find the record of a fact is often of more practical use than a knowledge of the fact itself.

John Shaw Billings

INTRODUCTION

Today, computers assist the scientific, technical, and social science communities in obtaining information and data quickly and efficiently from files stored in the computer.

Computer technology in the last two decades has revolutionized accounting and other business data functions. Computers to record, process, and report scientific and technical data have made possible many of the technological advances during the same period, including operation of many critical chemical processing operations, and even getting to and back from the moon. The growth of computers in the processing of alphabetical material has been in parallel with that of data processing and computation, except not quite so rapid nor so extensive.

Among the earliest high-volume alphabetical applications of computers was their integration into the publishing process. In the early 1960s, the processing of typesetting for the indexing and abstracting publications was first moved to

[1]Contributed by Edward P. Bartkus, Ph.D., and David F. Krentz, B.S., M.S.L.S., Information Systems Department, E. I. du Pont de Nemours and Company, Inc., Wilmington, DE.

computer. Thereafter, the ability to produce the printed word by computer influenced the increase in use of computers for information supply, storage, and retrieval applications. In effect, the computerized publishing process itself helped to accelerate the growth of scientific and technical literature. Publishers recognized that not only could the computer tapes be used to produce print, but also procedures could be developed to search those computer tapes in much the same way that individuals manually search for information.

Before the onset of computers as search tools, the choice of a resource to search was generally limited to those references in the library at the user's location. Manual searching was laborious since pages in printed volumes had to be turned by hand and checked visually. The selected material had to be copied by hand. Because of time constraints, and the impatience of the user, the common practice was to search until a few requisite references were found.

As more and more word text material becomes available in computerized form, the behavior pattern of the information user is changing. Computers are affecting how an individual searches for information and how a scientist or engineer keeps up to date. It is now possible for one individual to search over a hundred different reference or data files by computer in less time than had been previously taken to search one reference manually.

The exponential growth of the written record, coupled with the increase in college graduates and the increase in numbers of subject specialities, has led to demands by the users for a variety of information (often of an interdisciplinary nature), and for increased speed with which the information requested is to be delivered.

Individuals have also started to use computers to keep files, prepare indexes, organize bibliographies, generate reports, and prepare papers for publication.

This chapter is aimed at instruction in the value and use of computer-searchable files as an aid to literature and data searching. At the end of the chapter are two tables to assist in selection of computerized files to be searched to answer specific subject inquiries. One table describes the files, the other is a subject index to the files.

The sections in this chapter cover the following elements essential to searching computerized files:

Basic Equipment and Terms for Computer Searching
Types of Computer-Based Information Services
Value of Computer Searching
Types of Computer-Searchable Data Bases
Basic Makeup of Data Bases
Types of Computer Searching
Data-Base Searching Procedures
On-Line Computer Searching
Keeping Up to Date
Computer Searching for Patent Information
Developments and Trends in Computer-Readable Data Bases

BASIC EQUIPMENT AND TERMS FOR COMPUTER SEARCHING

Just as in other areas, a jargon has evolved in computer technology. It is essential that an individual understand these new jargon terms in order to be able to use the computer efficiently and effectively to handle literature and information problems, and to communicate with the information supply community to obtain the information services needed. Several key terms used in this chapter are described below.

Computer-Searchable Data Bases The files of information and data processible by computer are called data bases. The term "computer-searchable" is interchangeable with "computer-processible" and "machine-readable." Sometimes a shorter phrase, "computerized files," is used.

Data bases are computerized files of any kind, whether the content of the file is numerical, alphabetical, or graphical. The term "data base" applies equally to files of accounting data, mass spectral reference numerical data, chemical formulas, bibliographical references, abstracts, and full-text material. To say it differently, a data base is an organization of files or a bank of material in computerized, machine-readable, or computer-searchable, form.

The term "data base," although it is jargon, has become accepted essentially as a standard phrase, worldwide. Some personnel in the information supply community (librarians, information specialists, information services vendors) at times stretch the term to apply it to manual subject files which are organized and indexed. Occasionally the term "data bank" is used to mean a data base containing numerical data.

Computer Input Information or data which is to be processed by the internal storage mechanism of the computer is the computer input. In information searching, computer input comprises the data base to be searched and the searching strategy translated to computer-searchable format.

Central Processing Unit That part of a computer system which manipulates the data and does searching, matching, and calculating is the central processing unit (CPU). Computer-use charges specify the amount of CPU time used.

Computer Output Information processed by the computer and transferred from internal storage to some output device, i.e., printed on paper, converted to punched cards, or displayed on a TV-type screen, is the computer output.

On-line Any activity taking place while a piece of auxiliary equipment is connected to the computer and both are operating is on-line. *Off-line* means that the auxiliary equipment is operated separately from the computer.

Computer Terminal The computer terminal (or simply the *terminal*) is a typewriterlike device that permits interacting or conversing with the computer.

The simpler and least expensive models are standard teletype terminals, but more expensive and more sophisticated terminals operating at faster speeds are available. The fact that the terminal can be used where there is a telephone and a place to plug into an electric circuit means that a computer search does not have to be done in a library. The terminal can be in any office of a scientist, engineer, or professor; in a laboratory; at a plant site; or at home.

Cathode Ray Tube Terminals   The most practical devices for information searching are the terminals equipped with a TV-type screen or cathode ray tube, commonly referred to as a CRT. These permit the user or searcher to see the material typed for search input, and/or the material which the computer has extracted from the data base (output).

Printer   The printer is the device used to reproduce the material extracted from the computer *(computer output)* in printed form *(hard copy)*. If the computer output is printed on-line at the searcher's terminal immediately after the computer search, the procedure is called *on-line printing.* If the material is held for printing later, at the computer facility, at a time when computer and printing costs are lower, the procedure is called *off-line printing.*

Hard Copy   The printed computer output, i.e., the results of a search, is called hard copy.

Computer Output Microfilm   Computer output on microfiche, the sheet form of microfilm, is often abbreviated as COM.

Software   The program which "tells" the computer the sequence of step-by-step procedures by which material has to be processed, and by which the data base can be searched, is the software. The input software and the output software are different for each data base, and for each search strategy.

Queuing   The computer job priority which orders the sequence in which computer search jobs have been received and the sequence in which they are to be processed is called queuing.

Time-Sharing   Time-sharing involves a combination of hardware, communications, and software, in which numerous terminals can utilize a central computer concurrently for input, processing, and output functions.

Telecommunication System   The electronic communication configuration which ties the user and terminal (and printer) to the computer and to the data bases to be searched is the telecommunication system. The elements of this communication system include the telephone lines, a coding-decoding device called a *modem,* and a telephone. The user, in a searching sequence, dials the computer's telephone number and connects the terminal through the modem to the computer.

TYPES OF COMPUTER-BASED INFORMATION SERVICES

Traditionally, information searching meant going to the library. With the accelerating growth in the use of computers for searching for information and data, new types of information services began to provide an alternative. Many of these information resources came into being because some libraries had difficulty accepting the different modes of operation required to search computerized files effectively. Computer searching is not inexpensive, but this technique is being used because costs can be justified. These search costs have often been outside the normal budget of the library, and outside the normal concept of "free services" by the traditional librarian. However, the libraries have begun to respond to requests by potential users who have used the service elsewhere, or who have read about the new computer-searching techniques and their benefits, and have heard about the new approaches at technical meetings. Many libraries and information centers are now modifying their normal procedures to train information services librarians in computer-searching capabilities.

Among the computer-based information sources and services that have come into being are the following:

Computer Service Bureaus   These organizations buy, lease, or operate data bases that already exist in computerized form, and provide searching service of the data base at a cost. Depending on the needs of a client, a service bureau will accept the search strategy developed by the client, and make the computer search; or the service bureau will accept the *question* as stated by a client, structure the search strategy needed to obtain the results desired from the computer search, and then make a computer-search run. At a client's request, the service bureau can screen the computer-search results; remove redundant citations; provide abstracts (if not already provided by the computer); provide copies of many of the original articles, reports, or other original source documents; and manipulate the computer-search results to provide an organized bibliography, an annotated bibliography, or some other kind of information report.

The computer service bureau may be a commercial money-earning company, may be associated with an academic institution, or may actually be a function in an industrial company (sometimes part of library activities, but often combined with the libraries into what is called an information center). Several government departments or agencies operate computer service centers.

Among university-associated computer service bureaus are those at the University of Georgia, Indiana University, the University of Pittsburgh, the University of Connecticut, the University of New Mexico, the University of North Carolina, the University of Southern California, Ohio State University, Lehigh University, the University of California at Los Angeles, and Illinois Institute of Technology Research Institute. Among industrial companies that have in-house information computer service centers are Bell Laboratories, E. I. du Pont de Nemours & Co., and IBM Corporation. Commercial service companies include the New York Times, Chemical Abstracts Service, and

Predicasts, Inc. An example of a U.S. government computer service center is the National Library of Medicine.

Data-Base Vendors Information-supply companies which purchase or lease a number of data bases, develop time-saving procedures for searching the data bases, arrange for telecommunication networks to make the search services readily accessible, develop billing mechanisms, promote the use of these search services, and train potential users are called data-base vendors. Companies may provide services of but one data base, or they may offer access to multiple data bases (as many as 100). Examples of data-base vendors are Lockheed Information Systems Company, System Development Corporation (SDC) Search Services, National Library of Medicine (NLM), and Bibliographic Retrieval Services, Inc. (BRS).

Information Finders These commercial businesses have evolved from a recognition that (1) the computer searching field was increasing so rapidly that many large companies could not adjust their organizations fast enough to develop searching capability; (2) many small business concerns likely would never be able to afford their own search centers; and (3) the number of data bases was growing so fast that only specialists could keep up with each of the new data bases and use them effectively.

Information finders offer both manual and computer-search services on a unit cost or contract basis. Many offer high-speed document delivery. One factor which contributes significantly to the success of these information finders is the preservation of the anonymity of the original client. Since these companies must satisfy their clients to remain viable, their services often readily justify their contracts with a client.

Examples of this type of growing business are FIND, Inc. (New York City), Information Unlimited (Berkeley, CA), Information Specialists, Inc. (Cleveland, OH), Warner Eddison Associates, Inc. (Lexington, MA), Information Resources (Toronto, Ontario), and InfoSource, Inc. (Pittsburgh, PA).

Data-Base Generators Organizations which develop and maintain the data bases are called data-base generators or producers. The initial data bases were converted from hard-copy (printed) indexing and abstracting services publications (such as *Chemical Abstracts* and *Engineering Index*), with the material becoming available as a result of using computers to produce the publications. With the potential power of the computer becoming an accepted bonus in information supply, more and more organizations (industry, publishers, academia, government, research institutes) are developing computer-searchable files as really the only practical means by which vast amounts of reference material, numerical data, and full texts can be searched and the desired material retrieved. Some organizations produce data bases for in-house search only. Some commercial organizations develop, purchase and/or lease, and convert files to machine-readable form and offer search services for sale.

E. I. du Pont de Nemours & Co., Inc., Hercules Inc., Dow Chemical USA, Allied Chemical Corporation, Olin Corporation, and many other companies generate and operate in-house, computer-searchable research and development (R&D) report information systems. The National Institutes of Health (NIH)/ Environmental Protection Agency (EPA) Chemical Information System (CIS) contains a number of data bases obtained from several sources which are made available for search through a commercial vendor. IFI/Plenum Data Corporation produces and leases a computerized data file called *World Hydrocarbons*, which contains raw material and plant capacities, worldwide, for about 80 commodities through several product derivative levels.

VALUE OF COMPUTER SEARCHING

Regardless of whether the computer search is for the purpose of finding what is available in earlier literature files or to keep aware of current developments through search of the file updates, computer searching does have a number of significant advantages over manual searching.

Among the more significant advantages of machine search over manual search are the following:

1 Deeper indexing aids accessibility to needed contents in each reference. In many data bases, to aid in computer searching, the indexing and abstracting services have added supplemental search terms not found in the printed volumes.

2 Computer manipulation permits faster, more useful answers to complex, multiconcept questions, particularly from very large data bases.

3 Computers allow multiyear coverage of major indexing and abstracting services in one computer search, thereby reducing the need for multiple-volume handling as in manual searching.

4 Searches of a number of different data bases or reference files can be made during one search session. In many instances, the searcher's library has only a small percentage of the series of reference documents available in printed form, compared to the resources available from vendors of machine-searching services.

5 Since the output of the computer is already in printed form, time-consuming hand copying of citations can be avoided. The computer can organize the material into bibliographic format as an added procedure to save time.

6 An experienced intermediary or computer-search specialist, who handles many inquiries per year, develops an in-depth knowledge of *(a)* what can and what cannot be retrieved from computer-readable data bases, and *(b)* the various search strategies required to solve a particular problem. This means a greater likelihood of achieving search results which satisfy the user, as compared to manual searches by or for the client. Further, important citations are generally more likely to be retrieved.

The arguments in favor of computer searching do not mean that manual searching or scanning is not worthwhile, even with current machine accessibility. Computer searching is not a total replacement for manual searching. But for most situations, where the data base provides coverage of the field and of the time period from which information is needed, machine or computer searching has been found to be beneficial as compared to manual review and search of the literature. The search strategy must be right, however, in order to retrieve needed material.

If a substantial amount of information about a particular subject is needed, a combined computer and manual search of the literature, carried out under a carefully planned search strategy, is the only procedure with a high probability of obtaining the maximum number of pertinent references. Obviously, for literature not in a computer-stored file, manual search is the only procedure.

There are some *disadvantages* to computer searching as compared to a manual approach:

1 Costs for searches vary and can be substantial. Costs depend on such factors as the number of data bases searched, the search strategy itself, the amount of time during which the computer and its auxiliaries are used, and the number of citations obtained. The costs of using the computer for searching, the communications mechanism, and access to the data bases are "out-of-pocket," or additional, costs to an organization's budget. In many libraries operating under a committed annual budget, manual search assistance often is not charged to a client. Increasingly, however, computer search assistance is becoming a direct charge to the user client, and is an addition to the normal library expenditures.

An understanding of the costs of information searching is gradually evolving. The "free" library's traditional willingness to supply the references for the user, and/or the assistance of the searching "reference librarian" without a line item charge to the user, makes charges for computer searches very visible by contrast. Actually, if manual searches were fully billed to the user, such searches could cost several times the charge for a computer search, because of the value of personnel time needed to achieve equivalent results.

2 Development of the right search strategy to obtain desired search results is a difficult exercise, and requires insight into both what is and what is not wanted or needed.

Since computer searches often increase the number of relevant citations to be reviewed, the number of requests for the original documents increases. This means that the use of conventional library services (e.g., photocopies, interlibrary loans, purchases, microfiche copies) increases. The computer search is often followed by a request for a supplemental manual search in subject areas not available via machine-readable data bases.

TYPES OF COMPUTER-SEARCHABLE DATA BASES

Among the types of chemical-related data bases (see Tables X and XI) now available for computer searching are:

1 Bibliographical/reference
2 Referral
3 Thesaurus/word list/description/indexing term/classification
4 Chemical synonym
5 Graphical
6 Full texts
7 Numerical/statistical data

1. Bibliographical/Reference Files The bibliographical/reference computer-searchable files are the most common and the most universally used. They are also the fastest growing with respect to use and the number being developed for computer searching.

This type of data base in many instances corresponds to the printed copy of an indexing and abstracting service, such as *Chemical Abstracts* and *Biological Abstracts*. The bibliographical data base contains the elements in a literature citation or reference to a particular document. The citation includes the author(s), often the title, the name of the publication, the volume and/or date of publication, pages, and publisher (if a book). In some instances, the references cited may also include keywords or descriptors, and in others, the full abstract. In summary, the bibliographical data base is alphabetical in nature, and contains the reference citations, with or without abstracts. Examples are CA SEARCH (Chemical Abstracts Service) and EI-COMPENDEX (Engineering Index).

2. Referral Data Base The referral data base is a mechanism by which sources, computer-searchable and/or manual, are matched against a subject of interest. Such a data base is, in fact, an index to other data bases or manual information sources, indicating where to go, and where not to go, for information or data requested. Examples are: (1) the Chemical Data Base Directory (alternately called Chemical Information Resources Directory) being developed as part of the U.S. Government Interagency Toxic Substances Data Committee's Chemical Substances Information Network (CSIN), and (2) the System Development Corporation's (SDC) Data Base Index (DBI), which contains the keywords in SDC's data bases and identifies which data bases may contain the keywords.

3. Thesaurus/Word List/Descriptor/Indexing Term/Classification This type of data base is generally limited to a major subject area such as physics or electronic engineering in which the volume and complexity of vocabulary present problems to the searchers. The objective is to aid an individual in development of a computer search strategy through use of suggested and related indexing terms, concepts, or calssifications. The procedure is first to review this type of data base, select the various related terms, broaden or narrow the search strategy, and then use the selected terms to search the computer.

4. Chemical Synonyms The chemical dictionary or synonym data base is a helpful first step by which the appropriate synonyms can be retrieved and used for searching of bibliographical and other data bases for material needed. Such a chemical synonym data base may also provide such identifiers as the Chemical Abstracts Registry number for a material. Examples include System Development Corporation's CHEMDEX, Lockheed's CHEMNAME, and the National Library of Medicine's CHEMLINE (all described in Table X).

5. Graphical Representations A pictorial representation often is the clearest way to present the attributes of a chemical. Two current graphical data bases are *(a)* a structural representation of chemicals (NIH/EPA Chemical Information System's Chemical Structure and Nomenclature System) and *(b)* phase diagrams (Man-Labs National Physical Laboratory Materials Databank). (Both are described in Table X.) These data bases are capable of displaying (printing out) graphical representations, points, and lines on the CRT, and on hard copy via a printer.

6. Full-Text Full-text data bases provide the full text of the original source document. In the area of chemistry, there are no publicly available full-text data bases. However, Meade Data Central offers its LEXIS full-text-searchable file of legal statutes of states.

7. Numerical Data Bases Numerical data refers to information expressed in numerical form, such as physical and chemical properties, quantities manufactured, and mass spectra data points. Identification numbers which refer to literature articles are not generally called numerical data. Numerical data bases also normally do not include code numbers, although this classification may vary, depending on the scope and the designation by the data-base generator. An example of a chemical-related numerical data base is the Physical Properties Data Service offered by the Institution of Chemical Engineers (London). (See Table X.)

Numerical data bases are special-purpose files. They are unlike bibliographical data bases in that each data file of specific subject matter differs in format from the data file for any other subject. They have a different basic content, as described below. Most of the time, the numerical data bases are too specific for the scope of work of the general library or information center.

Numerical data bases may contain data in any one or more of four representation modes:

1 *Point*, which refers to single data points, or individual measurements, e.g., a vapor pressure or viscosity measurement.

2 *Tabular*, in which several data points are stored, requiring interpretation for intermediate data points.

3 *Equation*, with equations for calculating the desired data points. If equations were not used, the required computer capacity for storing many groups of data would be so expensive as to be limited to highly critical material (e.g., oceanographic, defense).

4 *Graphical*, with representation of the data measurement in image form, either on a TV-type CRT screen or printed out as computer output on paper.

In the same way that the bibliographical data bases require computer software for the desired material to be retrieved in useful format, each set of numerical data must have software for data to be extracted by the user. Input to the files generally has to be carefully structured to limit the amount of computer storage required, and to facilitate the extraction of selected data from the data base or data bank.

Because it is not economical or even possible to regenerate much of the data upon which many journal articles and books are based, a means must be established to find the data which have been published. More and more of the bibliographical data bases are providing *data indexing,* which means that the presence of data in a published document is indicated. The original document then may be reviewed to ascertain whether the data therein are useful.

Two types of data indexing are used: (1) *data flagging* indicates that numerical data are contained in the document but does not necessarily identify the type of data; (2) *data tagging* is more descriptive and characterizes the data in the document. Both Chemical Abstracts Service and American Institute of Physics have recently established procedures for citing data in their abstracts. Review of an abstract can often give a searcher a signal that data are contained in the source document. Without the data flag or data tag, the searcher would have to scan the original document to see whether any data were included in the text.

BASIC MAKEUP OF DATA BASES

Each computer-searchable data base is different from the others in subject matter (see Tables X and XI). Each data-base producer or generator has different philosophies and policies. It is not likely that the information elements in any two data bases have the same format for retrieval. There is no standard data-base search software for use by all information services vendors.

In order to know which data base to search for which information, in how much detail, and covering which time period, many aspects of the contents of the data bases and their original sources must be identified:

1 Type of data base (e.g., bibliography, bibliographical with abstract, numerical, full-text)

2 Contents policy (e.g., all articles of a core group of journals included, or only selected articles)

3 Types of sources (e.g., periodicals, patents, reports, books, theses, proceedings)

4 Number of different sources covered (e.g., number of periodicals)

5 Subject areas covered (e.g., chemistry, physics, engineering, electronics, environment, toxicity)

6 Geographical coverage (single country, region, worldwide)

7 Time period covered (*Note:* many data bases cover only the last few years.)

8 Total cumulative number of different reference or other items currently in the data base

9 Number of items added per day, week, month, or year

10 Indexing techniques to assure required retrieval (e.g., keywords, descriptors, special identifiers)

All the material placed on the computer tape which describes an indexed document is referred to as a *record* or *unit record* for that document. Each individual record consists of a complete citation including document title, author, journal title, volume, date, pagination, index terms (free-text, descriptors, and identifiers), plus other information such as work location, abstract number of secondary publication (e.g., *Chemical Abstracts*), contract numbers, report numbers, language of original publication, corporate author, patent number, patent classification, abstract.

Each of the individual items is called a *field* or *data element* and is assigned a specific location within each record. Each field always has the same position in each record, i.e., if the author is the second item in the first record, it is the second in all the other records. The sum of all unit records on the same tape is called a *file*. A file usually refers to a segment of the total number of searchable files which is called a *data base*.

Each of the fields described above is available for searching on the computer. However, the choice of what will be made searchable depends on data-base vendors (e.g., Lockheed Information Systems Company, System Development Corporation, Bibliographic Research Service, National Library of Medicine).

It is the vendor who, generally with the aid of users, decides which data elements are best for searching and which give the most useful information in the most economical manner. Once chosen, computer software is written to allow searchers access to the data base by these data elements. Data elements differ with each data base. Not even data elements that are the same among different data bases are searchable in each use.

To assure reliable search results (assure effective retrieval), many data-base producers, such as Chemical Abstracts Service, supplement their citation entries with additional indexing terms. These index terms, called *keywords* or *descriptors,* reflect the terminology of the material in the data base.

Many similarities exist between keywords in one data base and those in others. There are subtle and distinct differences which may not affect manual searching, but which can make the difference between a successful and an unsuccessful computer search.

Keywords or descriptors may either stand alone as single words, called *unit terms* (e.g., pollution, fuel), or are coupled or bound to others, and known as *bound terms* (e.g., air pollution, fossil fuel).

The indexing terms may be added under two different indexing philosophies:

1 *Controlled vocabulary,* a method by which the index terms must be taken from a special subject thesaurus or a preselected vocabulary list

2 *Uncontrolled vocabulary,* or "free-text indexing," which permits the author of a source document and/or a data-base producer's indexer to supply individually chosen index terms

Controlled vocabularies are intended to assure indexing consistency of the material (in source documents) being placed into the file, and to act as a tool to aid searchers to get material out of the file. Many times, these vocabularies are structured in a sequence which classifies relationships between words and technologies. This structure incorporates *scope notes*, such as "see," "see also," "use," and "used for," to assure that the preferred indexing terms are used to describe the concept in each indexed document (e.g., "Chemical kinetics, see also Reaction kinetics").

Additional scope notes, such as "broad term," "narrow term," and "related term," are used to guide users/searchers to hierarchical relationships among the subject terms of interest (e.g., "For keyword "ytterbium," use broader term "rare earth elements").

Some data bases include additional identifiers to aid the retrieval process, such as:

1 *Codes,* which identify particular search terms, or special attributes, and describe the material being added to the data base [In Chemical Abstracts Service computerized services, the CA Number refers to the CA abstract, the CA Registry Number to a specific chemical, and the Section Number to the content of a particular subject section of CA. In Engineering Index (EI) Compendex, the computerized version, a Card-Alert code refers to the subject coverage of a discontinued manual selective dissemination of information service.]

2 *Free-text terms* which are not in a thesaurus nor in the controlled vocabulary list (e.g., words in the title, words in an abstract, specified chemical names)

The differences in both spelling and terminology between British and American use of the English language are reflected in the data bases produced in various countries. The data bases produced in United Kingdom, for instance, will use keywords, descriptors, or search terms with British spellings; while those from the United States will use keywords, descriptors, or search terms with American spellings. For instance, note the British suffix "-re" versus the American "-er," as in "centre" versus "center" and "metre" versus "meter." The British lift the bonnet of a lorry, but Americans lift the hood of a truck. "Sulfur" will never be found if it is entered as "sulphur."

In the same way, data bases differ in the types and numbers of abbreviations used. *Chemical Abstracts* provides a list of standard abbreviations used as keywords (e.g., "prepn" for "preparation," "manuf" for "manufacture," "manufd" for "manufactured," and "manufg" for "manufacturing"). *Rubber and Plastics Research Abstracts* allows the author to use "PU" for "polyurethanes," "PE" for "polyethylenes," and "HDPE" for "high-density polyethylenes."

These differences complicate searching immeasurably. In computer searching of a data base, it is important to develop a search strategy which considers these differences. Leaving out one of the spelling variations may limit the number of potential references retrieved. Even when the country of origin of the data base is known, it is more than likely that many of the titles, and perhaps abstracts, may have come from another country. So, when searching is done. all variations should be identified and listed as part of the search strategy.

Software to search entire abstracts or the full computerized text of a source document is now possible. However, for many situations, the process is too complex, and the cost too expensive. Further, many information service organizations do not provide *free-text searching* with their information search services. In the near future, it can be expected that the combination of improved search strategy concepts and search software, and lower unit computer and telecommunication costs, will make free-text searching commonplace.

TYPES OF COMPUTER SEARCHING

Computer searching replicates the types of searches done manually. An information search is conducted for one of two (or both) purposes:

1 To retrieve references to the full-text document or abstract, relative to a specific subject already in the file being searched. The procedure is called *retrospective searching,* and is equivalent to a manual search which produces a bibliography.

2 To keep up to date by providing material as soon as possible after it becomes available for searching. This technique is called *current awareness*. The term often used to describe current awareness via computer searching of new updating data-base increments is *selective dissemination of information,* or SDI. SDI can be a continuing service equivalent to periodic updating a manual bibliography.

Both retrospective and SDI machine searching can be done in one of two time-related procedures, batch mode or on-line:

1 *Batch mode* is the method in which the search strategy for a proposed information search is first structured, and the material then sent to a computer center for processing along with other similar searches. In the batch mode, individual search requests are accumulated as they come to a computer center. On a regular schedule, all the searches in the batch are placed in sequence (called *queuing*), and processed against the selected data base(s) in a single pass through the computer.

Retrospective batch searching of data bases according to a schedule is the only practical way that a number of information sources can provide search services because of costs or computer service limitations. The batch-mode approach can also be used for current-awareness or SDI searches. Periodic running of profiles of interest against data-base updates will provide the SDI service. The computer search results are generally printed out on paper (as *hard copy*), and sent to clients.

For retrospective searching, the batch-mode operations are at a disadvantage when compared to on-line operations described below. In batch mode, if the first search results are unsatisfactory, revisions must be made to the search strategy and the search rerun through the computer. Each revision and each computer iteration may involve an overnight, a several-day, or even a week time lapse, and must be separately paid for. Such time delay may hurt a project or be otherwise unacceptable.

2 *On-line mode* is a procedure in which a searcher interacts directly with the computerized data files via a computer terminal, adjusts the search strategy until the desired level of search results is achieved, and then gets the search answer. The searcher is able to get search results, or at least an indication of the success of the search, within a few minutes after the communication with the computer is started.

Although the primary and most frequent use of on-line searching is to answer questions of interest to a client, on-line searches can produce periodic retrieval of file updates. In other words, the on-line approach for current awareness or SDI offers periodic searches with the initially developed search strategy.

If the on-line search produces only a few references, then the search results are often printed out at the terminal *(on-line printing)*. If the search has determined that a substantial number of references has been obtained, the references are printed *off line* to reduce time and associated increase in costs, but at a time-delay penalty.

DATA-BASE SEARCHING PROCEDURES

Learning about Computer Searching

An individual can become familiar enough with the procedures to search a single base through:

1 Training sessions offered by vendors, libraries, professors, consultants, or experienced colleagues
2 Vendor-supplied instruction manuals and user aids
3 Computer-assisted instruction programs
4 Multimedia instructional programs (e.g., cassette-slides)
5 User groups which meet to exchange searching experiences

A combination of the above is probably the best solution. The most difficult and time-consuming single approach is the review of vendors' user manuals together with self-learning and practice.

What often happens is that an individual who develops proficiency in searching becomes known to associates and is often asked to do all computer searching for the group.

Learning to search one data base effectively and efficiently is difficult enough for the occasional searcher. Attaining search capabilities across the whole range of complicated data bases is mind-boggling. The rapidly increasing number of available data bases adds to the difficulty.

Every data-base vendor offers different searchable data elements, requires different search strategies for retrieval, and provides different information system search software. It can be expected, though, that standard software will eventually evolve into user-oriented search systems which should ease the problem of learning to search.

Computer Presearch Procedures

As mentioned before, computer searching of data bases can be used for both retrospective searching and for current awareness or SDI. When developing a search strategy for either of the searching purposes, whether batch or on-line mode, the same strategy development procedure generally applies.

A written presearch record of both the administrative details and technical needs will increase the efficiency and effectiveness of the search and reduce its costs, whether performed by an intermediary or the user. A standard form or questionnaire is often available to assist in organization of needed information.

Administrative items are required before a search can be made, regardless of whether the searcher is in industry, at a college or university, in a government or research institute, or dealing with a commercial supplier of information:

1 Date
2 Search identification number
3 Name of the user or client
4 Organization or institution and location (address)
5 Telephone number
6 Payment reference number (invoice, project, grant, purchase order, department fund, location fund)
7 Date the search is wanted or needed
8 Scope and purpose of the search (orientation; researching new technology or a new project or subject; making a bibliography; paper preparation; teaching; basic research; preparing a thesis or dissertation; getting answers to a specific question; getting background material; keeping up to date; "brainstorming"; substantiating or refuting
9 Time span of the literature to be searched

Technical specifications of the search needs usually include:

1 Title of the project
2 Narrative, specific description of the problem to be searched

3 Keywords, descriptors, search phrases, synonyms, identifiers, related terms, alternate spellings

4 Similar statements of aspects of the problem to be excluded from the search strategy, and for which *no* reference citations are wanted

5 Several authors and references known to the user and pertinent to the problem

6 Names of several journals or other source documents thought to have pertinent material

7 Abstracting and indexing services of potential value

Search Strategy Preparation

The search specialist, or the user, with the technical search needs described above, translates these needs into a searching strategy. Without a written, planned search strategy, the search is not likely to be successful.

A search strategy is composed of keywords and other index terms which describe the search question. These are combined into logic statements which define what is to be retrieved and what is not to be retrieved. The logic statements are specific for use with a particular data base and with the data-base vendor's search software.

Search strategy preparation generally has the following sequence:

1 Formulate the query by describing the question. Sample question: "What methods are used to prepare or make pyrocathechol (1,2-benzenediol)?"

This question will have terms which define two concepts: *(A)* the manufacture or preparation of, and *(B)* the chemical itself. Each concept will be considered a set, and each set will be represented by a Boolean circle (also called a Venn diagram), one *A* and the other *B*. Terms which make up set *A* are words such as manufacture, preparation, and synthesis; those in set *B,* pyrocathechol, its CA Registry Number, 120-80-9, and other identifiers.

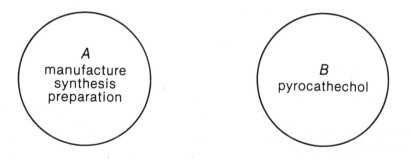

2 Obtain the data-base vendor's manual. Review the questionnaire prepared by the inquirer. Obtain a subject area thesaurus and indexes to potentially relevant publications. Ascertain whether the data-base vendor has a search term data base available.

For set *A,* dealing with preparation or manufacture, develop a list of synonyms, taking into account any special features of the vendor's search software.

Since computers retrieve only what they are "asked" to find, it may be necessary, to assure retrieval, to add keywords expressing the concept in other forms, e.g., plural, past tense (such as manufacturing, prepared, synthesized, syntheses). But in doing so, it soon becomes evident that listing all forms of each keyword is a lengthy and time-consuming process.

From some information vendors, the computer software allows term searching by a shortcut method called *truncation,* or *term-stem* searching. When this technique is used, all forms of a word stem can be found without fully spelling the work. A special symbol (e.g., ? or *, designated by the vendor), is placed after the last letter in the word stem (e.g., manuf?). The question mark in this case tells the computer to search for each word having the word stem "manuf" without regard to the letters following the "f." Word forms such as "manufacturing," "manufacture," and "manufactured" are then retrieved as search answers. As a result, maximum word and answer retrieval is achieved.

Below are two keyword lists for set *A, manufacture,* needed to search the sample question, one for use with a data-base vendor or computer service bureau which provides truncation, and the other for a vendor or service without this searching feature:

Without truncation	With truncation	
Manuf	Prepg	manuf?
Manufacture	Preparation	prep?
Manufactures	Prepn	ppn
Manufactured	Synthesis	synthes?
Manufacturing	Synthesizing	prodn?
Prepare	Produce	
Prep	Produced	
Prepared	Prodn	
Prepd	Producing	
Preparing	Ppn	

For set *B,* which deals with the chemical itself, obtain the CA Registry Number and synonyms from a chemical dictionary data base (e.g., CHEMNAME, CHEMLINE, CHEMDEX) or CA indices, (e.g., CA Registry Name File), the TSCA Inventory List, or other hard-copy references or data bases which are known to contain such material. Look at pertinent journal articles, proceedings of symposia, and patents.

Prepare a list of search terms, again taking into account whether the data-base vendor uses truncation in the search software. Consider abbreviations, trade names, chemical names, chemical class names, processes, and methods of use (e.g., reactant, preparation, and catalyst). Two lists for set B follow.

Without truncation		With truncation
RN = 120-80-9 (CA Registry Number)		RN = 120-80-9
1,2-benzenediol	Pyrocathechin	Benzenediol
Benzendiol	Ortho-dihdroxybenzene	Cathechol
Cathechol	O-dihydroxybenzene	Pyrocatech
Pyrocathechol	1,2-dihydroxybenzene	Dihydroxybenzene
Pyrocatechinic acid	Dihydroxybenzene	Oxyphenic acid
	Oxyphenic acid	

Within the search strategy, the keywords used are grouped into categories or sets which define the major subject matter or concepts of the question, and are linked together by logical connectors. AND, OR, and AND NOT determine the relationships between the keywords and concept sets.

3 Develop the search strategy, using the keywords, descriptors, and identifiers. Prepare the logic statements which define what is to be retrieved and what is not to be retrieved.

For set *A, manufacture,* to obtain all documents indexed for each keyword within each set, all documents assigned to the term "manufacture" are added to all those assigned to the term "synthesis" and so on. To do this, the Boolean operator OR is used, represented here by +:

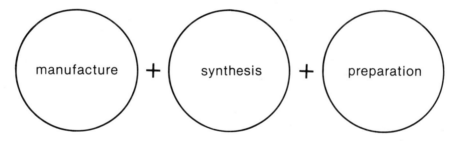

Each circle represents all documents containing their respective words. Since these three sets represent a single concept, they can be represented as a single circle *A:*

For set *B,* the chemical, to obtain all documents within each word set, all documents assigned to catechol are added to all of those assigned to the CA

Registry Number (120-80-9), and to dihydroxybenzene, and so on. Again, the Boolean operator OR, represented by +, is used:

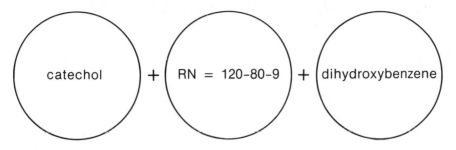

Since these three sets and others represent a single concept, they can be represented as a single circle B:

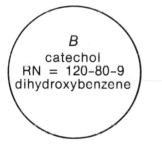

In the two separate sets, each set contains all documents in a particular data base for all keywords used to describe the concepts chosen from the original question. If the concepts were searched separately, the search result may contain nonrelevant documents. What is needed are all documents which contain a combination of one or more keywords from each set, i.e., *preparation* of *catechol*. Using the Boolean operator AND, designated by *, the terms of set *A* are combined or intersected with those of set *B*. The search results will then contain only those references which satisfy the combination:

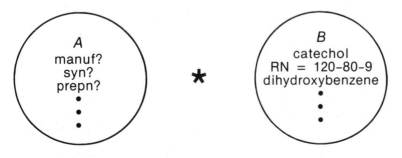

Logic statement: (A * B).

A third Boolean operator AND NOT is also available. This operator negates all documents with keywords specifically not wanted. For example, if the use of catalysts in the above question is not wanted, a third set of terms is created, such as catalytic, catalysts, and catalysis, to form concept *C*. All *C* documents are then subtracted from the answer (A * B), e.g., (A * B)−C, to give a final answer set that has all documents for the noncatalytic preparation, manufacture, or synthesis of catechol.

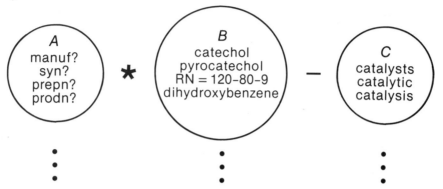

4 Review the search strategy against the original query formulation and revise as needed.

5 Select the data bases to be searched (see Table X, Contents of Chemistry-Related Data Bases, and Table XI, Acronym Index to Chemistry-Related Data Bases). List the priority in which the data bases are to be searched.

6 Select the data-base vendor(s) and the order in which their services are to be used (*Note:* For some data bases, only one vendor provides search services). Modify the search strategy to meet the constraints or special features of the information service vendor's instructions.

7 Conduct the computer search (batch, on-line, or SDI). The search mode used will decide when the results of the search can be first seen.

8 Review the results for relevancy. If the search result is unsatisfactory, modify the search strategy, and do another computer search. The on-line search mode permits essentially immediate opportunity to review the results and to adjust the search strategy.

9 Print the search results either on-line or off-line.

10 After final review of the results, record the data bases searched, the vendor(s) used, the query logic used, and the results of the search, including an estimate of the value of the search results.

Below are three items retrieved from search of CA CONDENSATES (references to *Chemical Abstracts*) according to the above search strategy. One is a patent, and the others are periodical citations.

CA08625189487D
  Pyrocatechol and hydroquinone from phenol
  Author: Kuwata, Hiroshi
  Location: Japan
  Section: CA025010   Publ. Class: PAT
  Journal: Japan, Kokai   Coden: JKXXAF   Publ: 76118   Pages: 2

Patent No: 76133240   Applic No: 75/57247   Date: 750516   Class: C07C39/08

Assignee: Ube Industries, Ltd.

Identifiers: phenol oxidn photochem catalyst, iron chelate oxidn catalyst, pyrocatechol, hydroquinone

CA08624189379V

Descriptors: dehydrogenation; ring cleavage; hydrocarbons, preparation; ketones, preparation; phenols, preparation; epoxides

Identifiers: Monoepoxide deriv. formation vinylcyclohexene monoxide reactions uses miscellaneous formation piperylene dimer limonene pyrocatechol prepn hydroxycyclohexanone alkenylcyclohexene monoxides alkenylcyclohexenes

CAS Registry Numbers: 106-86-5 17410-45-6D 138-86-3 620-17-7 41911-60-8 297-95-0 132-07-9 71-43-2 100-42-5 62744-64-3 99-87-6 499-75-2 533-60-8 120-80-9PP 100-41-4

CA08613089303X

Application of enzymes to the synthesis of catechol

Author: Nishimura, Tsutomu; Ishikawa, Toshio

Location: Natl. Chem. Lab. Ind., Tokyo, Japan

Section: CA025010, CA007XXX   Publ. Class: JOURNAL

Journal: Yuki Gosei Kagaku Kyokai Chi   Coden: YGKKAE Publ: 76 Series: 34   Issue: 9   Pages: 652-6   Language: Japan

Identifiers: pyrocatechol, benzenediol, phenol hydroxylation, tyrosinase, benzoquinone redn

ON-LINE COMPUTER SEARCHING

The most important development in computer retrieval of information, and the fastest-growing use, has been searching in an on-line mode. As highlighted earlier, on-line searching is a procedure by which a searcher communicates, via a terminal, directly with the computer which contains the data base(s) to be searched.

The equipment system for on-line searching includes: (1) a terminal, with or without a CRT, and with or without a printer; (2) at least one telephone; (3) a modem which connects the terminal to the telephone; (4) a telecommunications network; and (5) the computer at the data-base vendor.

The most effective search combination consists of a terminal with a CRT, a printer, and two telephones. With a CRT, the searcher can screen preliminary search results, and then can quickly adjust the search strategy, if necessary, to meet the specific needs of the query. With a printer, hard-copy samples of the search output from each of the search strategy adjustments can be made for later review, or the entire search can be printed out at once if not too long a time is involved. The second telephone permits communicating with the computer service vendor to get help to resolve searching problems, or to discuss intermediate search results with the inquirer.

Emphasis has been placed earlier on the need to know the details about the data bases, associated searching software, and the various user procedures of the computer service vendors, in order to be able to achieve satisfactory search results.

The searching strategy involved in the on-line searching process is outlined above, under Data-Base Searching Procedures.

Following are the actual specific steps to be taken to make an on-line search at the terminal:

1 Obtain the local identification and user number.
2 Check all terminal settings to assure communication with the computer.
3 Telephone the computer.
4 Listen for the computer signal or whistle.
5 Place the telephone headset in the modem.
6 Identify the terminal being used.
7 Give the user ID and user number (called *log-on* step).
8 Request the data base you wish to search.
9 Enter the search terms (search strategy).
10 Combine the search terms in the proper sequence.
11 Review the answer by printing some citations on-line.

Note: During the course of on-line searching, it may be necessary to broaden or narrow a search because either too many or too few answers are obtained for the strategy in use. For example, if the strategy prepared for the query "preparation of catechol" gave 3,000 answers, it may be desirable to reduce the search answer to 50 or 100 answers. This is usually done by adding specific terms which define reactions, reaction conditions, or processes. Conversely, if a strategy is too limiting, the search may be broadened by deleting terms which are highly specific.

12 Adjust the strategy if necessary.
13 Print answers on-line or off-line.
14 Request and search other files in a similar manner.
15 Log off the system, using proper protocol designated by the vendor.
16 Hang up the telephone and turn off the terminal.
17 Record the search time and results in a logbook. (A logbook should always be kept to record the time each data base was used, the number of citations printed, and the anticipated cost. The log-on, log-off records are a check against monthly vendor invoices, and the basis for client charges).

The many reports in the literature, and comments by data-base vendors and information supply personnel about their experiences, indicate the following significant advantages to on-line searching over batch mode and manual searching:

1 Fast turnaround in obtaining search results
2 Single fast access to accumulated years in a data base
3 Fast access to many extensive files/data bases
4 Access to files without having to go to more than one location
5 Greater recall in retrieval

6 Ease in asking a search question

7 Control over the search, the ability to change the direction of the search as it is being made

8 Ease in switching from one data base to another data base

9 No need for the organization to operate a computer service or develop computer programs/software

10 Access to material not available in printed form

KEEPING UP TO DATE

Keeping up to date with the latest scientific and technological developments can be called selective dissemination of information (SDI), keeping informed, maintaining current awareness, subject monitoring, subject alerting, and maintaining subject intelligence.

The great increase in the amount of scientific and technical know-how developed in the past several decades has multiplied the number of subject areas, study disciplines, publications, and scientific and technical societies required for intercommunication about the latest developments. The result is an increase in the need for several-subject coverage by individuals.

Each individual, over a period of years, must develop a personal mix of keeping-up-to-date techniques, including reading journals, attending society meetings, reviewing abstract journals, and exchanges with peers. Because of work demands, an individual has only a limited amount of time, either at work or off the job, to spend on screening and reading publications to keep informed, to go to meetings, or to interact with associates.

The problem is compounded as more and more multidisciplinary material becomes essential to a project or task. Therefore, the scientist and engineer have had, progressively, to reduce personal breadth of coverage of subjects.

The availability of computerized reference files has begun to have a significant impact on the mechanics by which scientists and engineers do keep up to date about the latest technology. Access to computerized data bases covering thousands of publications has made it possible for individuals to keep current in not only their own subject areas but also those of peripheral interest.

Computer selection of new material, covering personal-interest profiles, has added a major dimension to the individual's information mix and, indeed, has displaced some individual information-gathering habits.

The computer selectively provides information ranging in specificity from browsing or initial new subject orientation to explicit answers about individual chemicals or techniques. Via SDI, the computer can, on a periodic basis, give to an individual or group the material from the latest data-base updates which represents their specific interests as defined in the search strategy used to query the file. Essentially, the same search strategy that is used for retrospective searching (see Data-Base Searching Procedures, above) can be adopted for SDI, adjusted to meet a specific vendor's data-base specifications for software for SDI.

The search strategy represents an ongoing profile of interest for a person or group. If the update material for SDI is to be distributed to a number of persons or groups, the profile must be broad in scope and is called a *general-interest profile* (GIB). If a search strategy is tailored to a specific individual or group, and can be frequently changed, the profile is called a *custom-interest profile* (CIP), or *special-interest profile* (SIP).

Because of customization, the cost of the CIP is higher than that of the GIP. However, the difference in cost may be justified because of the reduced number of references to be reviewed via the CIP, and because the interest profile can be adjusted as needed. A GIP can only be adjusted at the convenience of the computer service center, or by consensus of the groups being served.

As in the case of retrospective searching, no search strategy is so nearly perfect as to select every pertinent reference. Therefore, if it is absolutely critical that no significant reference be missed, a planned combination of periodic manual and computer searching is necessary. Unfortunately, some data bases do not have corresponding hard-copy service. Computerized SDI then remains the sole mechanism to keep up-to-date in those situations.

SDI, as is retrospective searching, is available via two modes:

1 *Batch*. Search strategies for specified data bases previously sent to computer centers are run periodically against the new data-base updates. The computer output representing the interest profile is printed and delivered to the user.

Computer centers offering computerized SDI may be operated in-house in an organization (e.g., Bell Laboratories and IBM Corporation lease tapes from data-base generators and provide SDI service for company employees using company computers). Many organizations cannot justify such an operation and use a service center, such as the information-dissemination centers located at universities (e.g., the University of Pittsburgh, Indiana University, the University of Connecticut, and the University of Georgia).

2 *On-line*. Search strategies, or interest profiles, may be run in an on-line mode against update segments to obtain SDI service. Many vendors provide special procedures and codes for such services. The searcher can query the data base on-line each update period, or can arrange to have the update search made automatically. The results, depending on the volume of "hits" (citations in the output), are printed either on-line or off-line and delivered to the user. The major on-line service vendors which have such SDI capability, in addition to retrospective searching, are Bibliographic Research Services, Lockheed Information Systems, System Development Corporation, and National Library of Medicine.

E. I. Du Pont de Nemours & Company has been providing an internal SDI service for over 5 years, covering several hundred profiles of interest. Du Pont users repeatedly reported that computerized SDI:

1 Generated ideas for future work.
2 Identified citations which would otherwise be missed.
3 Forced habit changes to assure review of computer output.
4 Increased scope of coverage beyond manual capability.

Just as importantly, SDI users reduced the time previously spent reviewing journals to keep up to date, and were able to divert this time to other work.

One of the disadvantages of computerized SDI is that the user might be overwhelmed with irrelevant material, if care is not taken to limit the profile of interest or to update it as the work changes.

SDI Example

An example will illustrate the value and simplicity of batch-mode computerized SDI. If an individual were interested in being alerted only to *review articles about reactions* cited in *Chemical Abstracts,* but with the exception of biochemical reactions, manual search of all of the 61 CA (nonbiochemical) sections would take several hours weekly. The weekly computer printout, however, from CA SEARCH takes only a few minutes to scan. Below is a sample printout of a citation from one university computer center:

CA vol. 091 No. 01 CA Sec. & Subsec 023000 Abstract No. 004901w
　　Title: Oxidation of Paraffin Wax to Fatty Acids—A Review
　　Author: Lahikri, C. R.　　Chaudhuri, K. K.
　　Location of Work: Calcutta Univ., Univ. Coll. Technol., Calcutta 700 009, India
　　Source: Journal—Indian Chem., J., 1979, Vol. 13, Iss. 8, PP. 18–25 (Coden: ICLJAG)
　　Language: English
　　Keywords: Review Oxidn Paraffin Wax Fatty Acid

The search, of course, could have been broadened to include each reference about specific reactions; but the expected large number of citations would likely take an hour or two to scan. The CA SEARCH printout does not provide abstracts. If the scan suggests potential value to the citation, the individual has the choice of obtaining either the abstract or the full document.

The search strategy to achieve the limited search for reaction review articles is listed below (using the search strategy of the New England Research Applications Center at the University of Connecticut):

Section terms	Search strategy	Explanatory notes
00*	Deletes sections	1 through 09 (biological)
01*	Deletes sections	00 through 19 (biological)
Review*	Asks for review articles only	
X000	Asks for reviews in each of 61 sections, nos. 20 through 80, inclusive	
.		
.		
.		

*Refers to left and right truncation, which means prefixes and suffixes that apply.

Keyword Terms

The terms listed below indicate that review articles for the following reactions are to be searched. Note the use of both American and British spellings.

Nitration	*Alkylation*	Fluorination*
Chlorination	Acetylation*	Acylation*
Amidation*	Amination*	Ammonolysis*
Bromination*	Carbonylation*	Carboxylation*
Cyclisation*	Cyclization*	Dehydrohalogenation*
Decarboxylation*	Diazotization*	Diazotiation*
Diels-Alder*	Diene synthesis*	Dimerization*
Dimerisation*	Dissociation*	Esterification*
Etherification*	Ethylation*	Halogenation*
Hydration*	Hydrogenation*	Hydrolysis*
Iodination*	Isomerization*	Isomerisation*
Methalation*	Methylation*	Nitrogen fixation*
Oxidation	*Oxidn*	Ozonization*
Ozonisation*	Phosphorylation*	Photosynthesis*
Reduction*	Solvolysis*	Olefinat*
Chemical reactor*	Sulfation*	Sulfonation*
Sulfuration	Sulphurization*	Sulfurization*
	Hydroboration	
	(Purdue's Nobel Prize reaction)	

Note that the search strategy enables designation of which CA sections are to be searched. The list is prefaced by one word, *review*. The numbers are added to assure that no section between 20 and 80 is missed. On the other hand, the two terms 00* and 01* delete the searching of sections 1 to 19, covering biochemical reactions. The search strategy essentially says, "In sequence, combine each section term with each type of reaction term, and then search—but leave out biochemical reactions." Note also that the keywords list includes both American and British spellings of a reaction and abbreviations. These inclusions are necessary to assure retrieval of articles published worldwide.

COMPUTER SEARCHING FOR PATENT INFORMATION

An earlier chapter on patents emphasized that patents have always been a valuable source of chemical-related information. However, many individuals have not had the opportunity to appreciate just how much useful information is contained in the patent literature. Reasons given are lack of patent files in most libraries, the need for searching in Washington, DC, at the U.S. Patent and Trademark Office, and the difficulty of understanding patent jargon.

Searching for pertinent patents was a time-consuming effort until the development of computerized file systems. However, manual searching is still the best, or only, way to search some patent classes.

The machine-readable patent data bases have greatly aided research and development and product design personnel, and those associated with filing and patent litigation, in identifying pertinent patents for a number of purposes. Unfortunately, the patent abstracts and bibliographical references have been available for only a few years. Manual searching is still mandatory for those years in which the patent information is not in computer-readable form, and for many patent classes.

Even with computerized searching of recent patents, there are times when both manual and computer searching are required to meet a specific technical or legal need. But with more than 100,000 new patents issued worldwide per year, the more sensible approach is to try computer searching first, if the years of coverage needed are expected to be in the data base.

In the ideal situation, there would be a single computerized file of patent information. Such a single file does not exist. The various computerized patent files or data bases differ considerably in their content and in their capability to retrieve desired, specific information. The files cover different time spans. Within a single data base, differences exist, depending on the time period covered by any section of the file.

Some of the data bases consist entirely of patent information, such as IFI-Plenum Data Company's CLAIMS-Chem file and Derwent's World Patents Index (WPI). Yet, assured retrieval of needed patent information may require the searching of the general chemical literature data bases since they also contain patent references. Some of the data bases which contain patent references are *Chemical Abstracts,* National Technical Information Service, *Pollution Abstracts,* and *INSPEC.* It has been reported that over 30 data bases contain patent information.

Each patent-containing file has certain strengths in the time span it covers, in the currency of updating and breadth of coverage of subjects, and in the number of countries included. Some files are stronger for bibliographical needs, others emphasize subject. The files complement each other. Search of all or many files may be necessary to satisfy a specific need.

In some countries, patent applications are also published. One fact normally not understood is that patents and patent applications are normally published within 18 months to 4 years after an invention takes place. This publication may take place well in advance of publication of the same information as an article in a technical journal.

Not all patent-containing data bases are on-line. In fact, an important chemical patent file, the IFI/Plenum Comprehensive Data Base, is a computerized file which is not yet completely available on-line. An annual subscription costs several thousand dollars and requires special search software for its effective use. The on-line files which are available are not generally the tools for doing a complete search on a subject in the U.S. patents. The Comprehensive Data Base is designed to do that for the chemical field. However, many answers can be found via on-line searching.

To do an effective job of searching patent-containing data bases, the sequence of searching involves several steps:

1 List the specific coverage needed: subject, classes of patents, country of origin, whether a U.S. patent is involved, subject coverage, assignees, and inventors.

2 Obtain synonyms and trade names for chemicals from CHEMLINE, CHEMDEX, or other source.

3 Develop a search strategy.

4 Computer-match the prospective data bases against the questions to be answered.

If one data base meets the need, then only one need be searched, preferably on-line. If several data bases must be searched, then the applicable data bases must be searched, whether on-line or off-line. One major advantage of the on-line search continues to be the ability to narrow or broaden the search terms (and such variations as spelling of names) in interaction with the terminal. This is a significant time saving over off-line batch searches (except in the cases where many off-line batch searches are made at the same time). Yet even here, the need to adjust the search strategy on-line may offset the on-line cost.

DEVELOPMENTS AND TRENDS IN COMPUTER-READABLE DATA BASES

Some trends in machine-readable data bases are readily perceivable, and the implementation of some of these developments and improvements is under way. Changes are being made, largely due to lower costs of computers, terminals, telecommunications, and data-base searching. Among these trends and developments are:

1. More Data Bases   Governments, academia, research institutes, technical societies, trade associations, publishers, service contractors, and industrial manufacturers have learned of the cost benefits of generating data bases for internal use, and of achieving further benefits through making the data bases available commercially. About 100 new data bases of all kinds have been generated during each of the last several years.

2. Standards   The complications and cost of learning the individual software characteristics of each data base and of each data-base vendor tends to restrict, significantly, the use of the less well-known data bases. It can be expected that software will evolve, together with standard input and output procedures, which will simplify the ultimate user-searcher's interaction with the computer-based files.

3. Full-Text Availability   Some data-base vendors now supply, on request, copies of original full-text documents cited in pertinent references. This trend will continue, together with the developments of the capability for the searcher to call up on the CRT the full text of the document for immediate review, prior to making a copy of selected pages or requesting a copy of the document. Search software is approaching the point where full-text searching, complemented by keyword and other identifier searching, will be practical and economical. Much of the text will likely be delivered on computer output microfilm.

4. Computer-Produced By-Products Computer-readable tapes are now available as a by-product of publishing in print. Machine-to-machine transfer of bibliographies and abstracts is starting to take place between the journal and book publishers and the abstracting and indexing services. The full text used to produce the printed document will soon be available over telecommunication systems from remote sources, whether in response to requests for documents from results of searches of the bibliographical files, or from full-text searching.

5. Networks Several time-sharing communications networks now operating in the United States permit on-line interactive computer access by individuals to data-base vendors almost anywhere in the country. Such communications capabilities should grow at an accelerating rate as the following increase: *(a)* the number of data bases, *(b)* the number of data-base vendors, and *(c)* the number of time-sharing networks.

The European information network called EURONET has started up. EURONET is expected to make available, on-line, computer access to a number of data bases in each of the subscribing countries.

6. Encyclopedias, Catalogues, Orientation, Education On-Line Information searching is a subset of the education process, the steps of acquisition and dissemination of knowledge already available.

The educational process of transmitting available knowledge, or reinforcing what is already known, has for years utilized various modes of electronic media, including records, telephone, radio, and television. At many universities, students can pick up telephones, at the end of a day and hear a missed lecture, or review a lecture they have heard, to reinforce their own notes. Videotape and videodisk recordings are adding the visual dimension. Computerized lectures will eventually be incorporated into widely available data bases for call-up, on demand, from almost anywhere.

Search information services are not limited to universities. Individuals can now telephone special numbers and hear selected medical lectures on audiotape. The most-used on-line, interactive, computerized visual system today is the airline reservation system.

Two major systems under development, aided by microcomputerization, are called "teletext" and "viewdata." They are the beginnings of encyclopedic information services over a TV (CRT) set.

Under *teletext,* a computer transmits all kinds of print messages (e.g., supermarket items and prices, weather reports, recreation schedules). The data remain invisible until the subscriber dials or punches the correct buttons on a decoding device, which activates the desired printed material on the reading screen.

In *viewdata,* telephone lines link the subscriber to several information banks. The subscriber dials or punches appropriate codes which select the material from the various information banks and transmit that data to the viewing screen. The system operates interactively so that the subscriber can order material, and even transfer funds from bank accounts.

Computer-aided education techniques will be extended to the early steps in project orientation. For instance, an individual starting on a new project, or in a new area of work, will first develop an orientation searching strategy which defines the needed areas of information and data not known or on hand. Feeding such needs to an information center will set up a specially tailored information supply sequence, bringing selected material in from many information sources at intervals to meet the absorption rate of the individual, and in the detail required to meet individual current knowledge requirements.

7. Computerized Directories and Information Centers   Printed information source directories and libraries of books and journals and pamphlets will continue to be a major source of information and data. The fact remains, however, that for special subject areas, major computerized information centers will evolve for the purpose of facilitation of access, on demand, to thousands of computer-readable, or computer-searchable, data bases.

8. Copyright   A major deterrant to growth in data bases and their widespread use will continue to be the constraints of copyright. The problems still to be resolved are many, such as *(a)* conversion of computer output to other uses, *(b)* transmittal across country boundaries, *(c)* reformatting, and *(d)* use by government agencies.

A last note: The time is fast approaching when the combination of accessibility to several hundred data bases, improved telecommunications, simplified search procedures, and low-cost computers and auxiliary equipment will significantly affect the ways one works. The need then will be for a master computerized directory to lead the search to the appropriate data bases.

CONTENTS OF CHEMISTRY-RELATED DATA BASES

Table X summarizes the capabilities of a number of machine-readable data bases which contain chemical-related information. New data bases continue to be added to the list of those already available for searching.

The coverage for each data base was developed over a 10-year period, primarily from personal computer searching, SDI experience, and literature scanning. The original sources for the subject area coverage terms cannot be specifically identified, but include several editions of vendor manuals and data-base generator descriptive literature and indexes, user group discussions, thesauri, and articles from the open literature.

For an index, see Table XI, Acronym Index to Chemistry-Related Data Bases.

For each data base, the following items are noted:

Data Base

Acronym, or abbreviation
Full name (if different)
Data-base generator or vendor

Subject Coverage

Major subject areas

Data Base Content

Input: Number of references accumulated since start of a data base
Number of items added per period indicated
Source material
Relationship to printed information services
Output: Citations to references, indication of abstracts in the data base, and use of keywords and additional identifiers

Note: Unless otherwise stated, each data base can be searched on-line. With respect to the others, some can be searched only in retrospective batch mode, and some can be used only for SDI. Some data bases can be searched only via a single data-base vendor. Many of the data bases (such as *Chemical Abstracts, EI Compendex*, and National Technical Information Service) can be searched via several data-base vendors.

TABLE X
CONTENTS OF CHEMISTRY-RELATED DATA BASES

AGRICOLA (Formerly CAIN) (U.S. Department of Agriculture)

Coverage:
  Agriculture, agricultural economics, agricultural engineering, food, nutrition, pesticides, plant science, fertilizers, consumer protection, animal science, veterinary medicine, entomology, forestry, natural resources
Content:
  Input: Over 1,500,000 items since January, 1970
    Additions: 12,000/month from journals, government reports, monographs
  Output: Citations, keywords

APILIT (American Petroleum Institute)

Coverage:
  Petroleum refining, petrochemicals
Content:
  Input: Over 280,000 items since January, 1964
    Additions: 1,500/month from journals, trade literature, government reports, and meeting papers
    Corresponds to *API Abstracts of Refining Literature*
  Output: Citations, keywords

TABLE X
CONTENTS OF CHEMISTRY-RELATED DATA BASES (Continued)

APIPAT (American Petroleum Institute)

Coverage:
  Petroleum refining, petrochemicals patents covering: United States, Belgium, Canada, France, Germany, Great Britain, Holland, Italy, Japan, South Africa
Content:
  Input: Over 1,500,000 items since January, 1964
    Additions: 8,400/month
    Corresponds to *API Patent Alert Abstracts*
  Output: Citations, keywords

APTIC (Air Pollution Technical Information Center, Environmental Protection Agency)

Coverage:
  Air pollution, its effects, prevention, and control; atmospheric interaction; economics; measurement methods; standards
Content:
  Input: Over 89,000 items since January, 1966
    Additions: Discontinued, from journals, books, proceedings, preprints, patents, reports, government documents
  Output: Citations, keywords, abstracts
  Retrospective search only

BIOSIS (*BIOSIS Previews*) (Biosciences Information Service)

Coverage:
  Aerospace biology, agriculture, bacteriology, behavioral sciences, biochemistry, bioinstrumentation, biophysics, cell biology, environmental biology, experimental medicine, genetics, immunology, microbiology, nutrition, parasitology, pathology, pharmacology, physiology, public health, radiation biology, systematic biology, toxicology, veterinary science, virology, zoology
Content:
  Input: Over 2,6000,000 items since January, 1969
    Additions: 20,000/month from journals, books, notes, research communications, symposia
    Corresponds to *Biological Abstracts* and *Bio-Research Index*
  Output: Citations, keywords, cross codes, biosystematic codes (for *Biological Abstracts* and *Bio-Research Index*)

CA-Files (*Chemical Abstracts*) (Chemical Abstracts Service) (CAS)

CAS generates a number of machine-searchable files. Listed below are the CA Master File and the currently available subfiles. Each individual data base is described in its place in alphabetical order. When CA Files is cited as a potential information source in Table X, Acronym Index to Chemistry-Related Data Bases, the appropriate data base to be searched will depend on the availability of the CA Master File of its subfiles.

CA (*Chemical Abstracts*) (Chemical Abstracts Service)

Coverage (Master File):
  Organic, bio-, macromolecular, applied, physical, analytical, or other chemistry; chemical engineering; selected biology; toxicology

TABLE X
CONTENTS OF CHEMISTRY-RELATED DATA BASES (Continued)

CA *(Chemical Abstracts)* **(Chemical Abstracts Service)** (Continued)

Content:
Input:
Additions: 26,000/month from 14,000 journals, proceedings, reports, patents, symposia, books, reviews, dissertations
Output: Citations, abstracts

CA CASIA (Chemical Abstracts Subject Index Alert) (Chemical Abstracts Service)

Coverage:
Subject volume index; classes of substances; incompletely defined substances; applications; uses; physicochemical concepts and phenomena; properties; reactions; engineering and industrial apparatus and processes; common and scientific names of chemicals, animals, and plants
Content:
Input: Over 2,300,000 items from January, 1972, to June, 1979
Output: CA Abstract Numbers, CA Registry Numbers, annual subject volume indexing terms, chemical names
Retrospective search only

CA CBAC (Chemical-Biological Activities) (Chemical Abstracts Service)

Coverage:
Pharmacodynamics, agrochemicals, hormone pharmacology, essential oils and cosmetics, biochemical interactions, pharmaceuticals, toxicology, pharmaceutical analysis
Content:
Input: Over 1 million items since January, 1965
  Additions: 4,800/month from normal CA coverage
  Corresponds to CA content in CA section numbers 1–5, 63–64
Output: Abstracts, citations, keywords
Tapes are available for lease for retrospective searching and SDI

CA CONDENSATES (also called CAS-CON, CHEM-CON, CAC) (Chemical Abstracts Service)

Coverage:
Contents same as CA except that no abstracts provided
Content:
Input: Over 4,000 items from June, 1967, to June, 1979
Output: Citations, keywords, CA Abstract Numbers
Retrospective search only

CA CT (Chemical Titles) (Chemical Abstracts Service)

Coverage:
Titles of pages containing new and significant chemical information
Content:
Input: Over 2 million since January, 1962
  Additions: Over 12,000/month from normal *Chemical Abstracts* coverage
Output: Citations
Tapes are available for lease for retrospective searching and SDI

TABLE X
CONTENTS OF CHEMISTRY-RELATED DATA BASES (Continued)

CA ECOENV (Ecology and Environment) (Chemical Abstracts Service)

Coverage:
  Chemical and chemical engineering aspects of ecology and environment including: toxicology, foods, fertilizers, soils, plant nutrition, mineralogical and geological chemistry, air pollution and industries hygiene, sewage and wastes, water treatment
Content:
  Input: Over 300,000 items from January, 1975
    Additions: 5,000/month from normal *Chemical Abstracts* coverage
    Corresponds to content in CA section numbers 4, 17, 19, 53, 59, 60, 61
  Output: Citations, keywords, abstracts
  Tapes available for lease for retrospective searching and SDI

CA ENERGY (Chemical Abstracts Service)

Coverage:
  Chemical, chemical engineering aspects of energy sources; production and technology, including propellants, explosives; fossil fuels, derivatives, and related products; electrochemical, radiational, and thermal energy technology; thermodynamics, thermochemistry and thermal properties; nuclear phenomena; nuclear technology; electrochemistry
Content:
  Input: Over 200,000 items from January, 1975
    Additions: 3,600/month from normal *Chemical Abstracts* coverage
    Corresponds to CA section numbers 50–52, 69–72
  Output: Citations, keywords, abstracts
  Tapes available for lease for retrospective searching and SDI

CA FAC (*Food and Agricultural Chemistry*) (Chemical Abstracts Service)

Coverage:
  Chemical aspects of food production, preparation, consumption; toxicology, agrochemicals, fermentations, foods, animal nutrition, fertilizers, soils, plant nutrition
Content:
  Input: Over 200,000 items since January, 1975
    Additions: 3,500/month from normal *Chemical Abstracts* coverage
    Corresponds to content in CA section numbers 4, 5, 16–19
  Output: Citations, keywords
  Tapes available for lease for retrospective searching and SDI

CA MATERIALS (Chemical Abstracts Service)

Coverage:
  Chemical and chemical engineering aspects of production, properties, applications of industrially important materials; natural resources, derived materials used by industry for the production of goods, including synthetic high polymers; plastics manufacture and processing; plastics fabrication and uses; elastomers, including natural rubber; textiles; leather and related materials; coatings, inks, and related products; cellulose, lignin, paper; mineralogical and geological chemistry; extractive metallurgy; ferrous metals and alloys; nonferrous metals and alloys; ceramics; cement and concrete products

TABLE X
CONTENTS OF CHEMISTRY-RELATED DATA BASES (Continued)

CA MATERIALS (Chemical Abstracts Service)   (Continued)

Content:
Input: Over 400,000 items since January, 1975
    Additions: 6,800/month from normal *Chemical Abstracts* coverage
    Corresponds to content in CA section numbers 35–39, 41–43, 53–58
Output: Citations, keywords, abstracts
Tapes available for lease for retrospective searching and SDI

CANCERLIT (International Cancer Research Data Base) (National Cancer Institute)

Coverage:
Cancer in humans; treatment by chemotherapy, radio therapy, immunotherapy, surgery,
clinical studies, epidemiology, pathogenesis, immunology
Content:
Input: Over 200,000 items since 1963
    Additions: 3,000/month from journals, proceedings, government reports, symposia,
    monographs, books, theses
    Corresponds to *Cancer Therapy Abstracts* (1967–1973), *Carcinogenesis Abstracts*
    (1963–1973), *Cancerlit* (1973 to date)
Output: Citations, keywords, abstracts

CANCERPROJ (International Cancer Research Data Bank) (National Cancer Inst.)

Coverage:
Ongoing cancer research projects (Federal and privately funded)
Content:
Input: Over 20,000 items for the most recent 3 years
    Additions. as received
Output: Citations, keywords, abstracts

CA PATCON (CA Patent Concordance) (Chemical Abstracts Service)

Coverage:
Chemistry-related patents in CA
Content:
Input: Over 300,000 items from January, 1972, to January, 1981
    Additions: 40,000/year from patents
    Corresponds to CA Patent Concordance
Output: Citation, keywords, patent numbers

CA PS&T (*Polymer Science and Technology*) (Chemical Abstracts Service)

Caverage:
Synthetic high polymers; research, development, production, uses, equipment, plastics
manufacture and processing; plastics fabrication and uses; elastomers, including natural
rubber; textiles; dyes, flourescent whitening agents: photosensitizers; leather and related
materials; coatings, inks and related products; cellulose, lignin, paper, industrial
carbohydrates; fats and waxes; surface-active agents, detergents

TABLE X
CONTENTS OF CHEMISTRY-RELATED DATA BASES (Continued)

CA PS&T *(Polymer Science and Technology)* **(Chemical Abstracts Service)**   **(Continued)**

Content:
    Input: Over 270,000 items since May, 1971
        Additions: Over 3,300/month from normal *Chemical Abstracts* coverage
        Corresponds to CA section numbers 35–46
    Output: Citations, keywords, abstracts
    Tapes available for lease for retrospective searching and SDI

CA SEARCH (Chemical Abstracts Service)

Coverage:
    All items in CA Condensates and CASIA
Contnet:
    Input: Over 320,000 items since June, 1978
        Additions: 24,000/month from CA
    Output: Citations, keywords, CA Abstract Numbers, CA Registry Numbers, annual subject
    volume indexing terms

CAS ON-LINE (Chemical Abstracts Service)

Coverage:
    All chemical substances in CA Registry System
Content:
    Input: Over 5 million items since 1965
        Additions: As unique chemicals are registered
    Output: CA Registry Numbers, molecular formula, structure diagrams

---

CAB (Commonwealth Agricultural Bureau Abstracts) (U.K.)

Coverage:
    Agricultural engineering, economics, policy, entomology; animal breeding, genetics, feeds,
    feeding, health, production; crop ecology, physiology, husbandry; dairy science and
    technology; fertilizers; forestry and forest products; horticulture, human nutrition, entomology,
    helminthology, mycology, protozoology; pastures, pests, and disease control; plant breeding,
    genetics, growth regulators, nematology, pathology; plantation crops; pollination; reproductive
    physiology; soil science; soil management; storage of grain and foodstuffs; taxonomy;
    theoretical genetics; weed biology and control
Content:
    Input: Over 900,000 itmes since January, 1972
        Additions: 10,000/month from over 8,500 journals, books, reports, other publications
        (international coverage)
    Output: Citations, keywords, abstracts

CASSI (CAS Source Index) (Chemical Abstracts Service)

Coverage:
    Serials and nonserials abstracted in *Chemical Abstracts,* and specific libraries holding these
    publications
Content:
    Input: Over 50,000 items through 1978
        Updates: quarterly
        Corresponds to CAS *Source Index*
    Output: Same as content

TABLE X
CONTENTS OF CHEMISTRY-RELATED DATA BASES (Continued)

CDI (*Comprehensive Dissertation Index*) (University Microfilms International)

Coverage:
  Dissertations and theses accepted for academic doctoral and masters degrees granted by
  many United States educational institutions and some non–U.S. universities
Content:
  Input: Over 700,000 items from 1861 (M.S. theses since 1962)
    Additions: 3,500/items/month, PH.D. theses from *Dissertation Abstracts International,*
    American Doctoral Dissertations; Comprehensive Dissertation Index; also M.S. theses from
    Masters Abstracts
  Output: Citations

CHEMDEX (System Development Corp. Search Service)

Coverage:
  A chemical name dictionary: CA Registry Numbers, preferred CA nomenclature, molecular
  formulas, molecule formula fragments, ring descriptions, number of rings, rings present
Content:
  Input: Over 2,900,000 items since January, 1972
    Additions: 3,300/month from master nomenclature tapes of *Chemical Abstracts*
  Output: CA Registry Numbers, preferred CA nomenclature, molecular formulas, tradenames,
  synonyms, ring system descriptions
  Retrospective search only

CHEMLINE (National Library of Medicine)

Coverage:
  A chemical name dictionary for chemical compounds in TOXLINE file: chemical nomenclature,
  synonyms, trade names, molecular formulas, CA Registry Numbers, Wiswesser Line
  Notations, numbers of rings, ring element analyses, component line formulas, TSCA* numbers
  (no bibliographical references)
Content:
  Input: Over 362,000 compounds
    Additions: as available
  Output: Same as subject coverage
  Retrospective search only

CHEMNAME (Lockheed Information Services)

Coverage:
  A chemical name dictionary: CA Registry Numbers, molecular formulas, ring element
  analyses, ring sizes, CA preferred nomenclature for the ninth collective period, synonyms,
  tradenames
Content:
  Input: Over 400,000 compounds
    Additions: as available
  Output: Same as subject coverage
  Retrospective search only

---

*Toxic Substances Control Act.

TABLE X
CONTENTS OF CHEMISTRY-RELATED DATA BASES (Continued)

CIN (*Chemical Industry Notes*) (Chemical Abstracts Service)

Coverage:
  Chemical business literature: chemicals production, pricing, sales, facilities, products,
  and processes; corporate activities; government activities; personnel in the pharmaceutical,
  petroleum, paper and pulp, agriculture, and food industries
Content:
  Input: Over 240,000 items since December, 1974
    Additions: 100/week from over 80 worldwide industry periodicals
    Corresponds to *Chemical Industry notes*
  Output: Citations, keywords/descriptors, abstracts

CLAIMS/CHEM (IFI/Plenum Data Co.)

Coverage:
  U.S. chemical and chemistry-related patents
Content:
  Input: Over 265,000 U.S. patents (1950–1970)
  Output: U.S. patent numbers, titles, assignees, supplemental terms, U.S. Patent Office
  main/cross references, *Chemical Abstracts* equivalents and CA abstract numbers
  Retrospective search only

CLAIMS/U.S. PATENTS (IFI/Plenum Data Co.)

Coverage:
  U.S. chemical and chemistry-related patents; foreign equivalents for chemical patents from
  Belgium, France, Great Britian, Germany, The Netherlands
Content:
  Input: Over 700,000 U.S. patents since 1971
    Additions: 5,500 items/month
  Output: U.S. patent numbers, titles, assignees, inventors, classification codes and groups,
  cross-reference numbers, foreign patent numbers, publication dates

---

CLINPROT (International Cancer Data Bank) (National Cancer Institute)

Coverage:
  Anticancer agents, clinical trials, patent entry criteria, therapy regimen, study parameters
Content:
  Input: Over 1,000 items
    Additions: as received
  Output: Citations, keywords, abstracts

CNMR (Carbon-13 Nuclear Magnetic Resonance Spectral Search System) (available via National Institutes of Health/Environmental Protection Agency Chemical Information System) (NIH/EPA CIS)

Coverage:
  Carbon-13 nuclear magnetic resonance (NMR) spectral data
Content:
  Input: Over 8,500 spectra
  Additions: as available
  Output: CA collective index names, CA Registry Numbers, internal ID numbers, NMR spectra

TABLE X
CONTENTS OF CHEMISTRY-RELATED DATA BASES (Continued)

CONFPAP (*Conference Papers Index*) (Data Courier, Inc.)

Coverage:
Life sciences: clinical medicine, experimental biology and medicine, animal and plant science, biochemistry, pharmacology. Engineering: aerospace, mechanical, civil, electronics, chemical, nuclear, power. Physical sciences: chemistry, geosciences, physics, astronomy, mathematics, operational research, materials science and technology
Content:
Input: Over 800,000 items since January, 1973
  Additions: 10,000/month, references to scientific and technical papers presented annually at 1000 regional, national, and international meetings
  Corresponds to *Conference Papers Index*
Output: Citations, keywords

CRDS (Chemical Reactions Documentation Service) (Derwent Publications, Ltd.)

Coverage:
Synthetic organic chemistry
Content:
Input: Over 40,000 items since January, 1942
  Additions: 400 items/year from "Theilheimer's Synthetic Methods of Organic Chemistry," vols. 1–30; and *Journal of Synthetic Methods,* vol. 1 (1975 to date)
Output: Citations, keywords

CRYST (Crystallographic Search System) (Crystallographic Data Centre, U.K.) (available via National Institures of Health/Environmental Protection Agency Chemical Information System) (NIH/EPA CIS)

Coverage:
Organic and organometallic structures determined by x-ray and neutron diffraction— except proteins and high polymers
Content:
Input: Over 11,000 compounds, over 25,000 citations
  Additions: as available
Output: Citations, structures, reference numbers

DDC (Defense Documentation Center) (Department of Defense)

Coverage:
Chemistry, biology, engineering, general science, and technology
Content:
Input: Over 1,500,000 items since early 1950s
  Additions: 2,500/month
  Corresponds to documents published in the DDC's *Technical Abstract Bulletin* (TAB) and U.S. Department of Commerce *Government Reports Abstracts* (GRA)
Output: Citations, keywords, abstracts, government report numbers

---

DRI-CHEM (DRI Chemical Services) (Data Resources, Inc.)

Coverage:
Data and analytical techniques to serve as links between industry and the economy; relationships between products demand, market growth, and product prices; estimates of price; availability of raw materials; operating costs, yields, and capacities—U.S. only

TABLE X
CONTENTS OF CHEMISTRY-RELATED DATA BASES (Continued)

DRI-CHEM (DRI Chemical Services) (Data Resources, Inc.)   (Continued)

Content:
  Input: Since 1975, for over 130 chemical processes including forecasts to 1984
    Updates: as data are received
    Corresponds to semiannual *DRI Chemical Review*
  Output: Data relating to items in subject

DRUGDOC (Excerpta Medica Foundation)

Coverage:
  Biological effects of chemical compounds, especially drugs and potential drugs; chemistry;
  chemical engineering; life sciences; medicine
Content:
  Input: Over 550,000 items since January, 1968
    Additions: 50,000 items/year from journals, government reports, proceedings, reprints,
    theses, monographs
    Corresponds to *Drug Literature Index,* and *Adverse Reaction Titles*
  Output: Citations, keywords

EDB (Energy Data Base) (Energy Research and Development Administration)

Coverage:
  Energy: coal, oil, natural gas, electric, nuclear, solar, conservation, geothermal; advanced
  energy systems
Content:
  Input: Over 750,000 items since January, 1976
    Additions: 15,000/month from 2,000 journals, reports, patents, proceedings, monographs,
    books
    Corresponds to *Energy Research Abstracts*
  Output: Citations, keywords

EI COMP (*EI Compendex*) (Engineering Index, Inc.)

Coverage:
  Engineering: civil, environmental, geological, biological, electrical, electronic, control, chemical,
  agricultural, food, mining, metals, fuel, mechanical, automotive, nuclear, aerospace, industrial
  and management; technology applications
Content:
  Input: Over 550,000 items since January, 1970
    Additions: 6000/month from 1,500 journals, books, proceedings, reports
    Corresponds to *Engineering Index Monthly*
  Output: Citations, keywords, abstracts, Card-Alert codes

ENERGY ABS (*Energy Abstracts*) (Engineering Index, Inc.)

Coverage:
  Energy: Chemistry, chemical engineering, mining, metallurgy
Content:
  Input: Over 50,000 items since December, 1975
    Additions: 15,000/year from 1,500 journals, books, proceedings, reports
    Corresponds to *Energy Abstracts* subset of *Engineering Index*
  Output: Citations, keywords, abstracts

TABLE X
CONTENTS OF CHEMISTRY-RELATED DATA BASES (Continued)

ENERGYLINE (Environmental Information Center, Inc.)

Coverage:
Economics, U.S. policy and planning, international political and economic issues, research and development, petroleum and natural gas reserves, coal resources, solar energy, fuel production, fuel transport and storage, electric power, nuclear power, residential consumption, environmental impact
Content:
Input: Over 50,000 items since January, 1971
  Additions: 800 items bimonthly from over 3000 journals, symposia, reports
  Corresponds to energy items in *Environment Abstracts* (1971–1975), and *Energy Information Abstracts* (from 1976)
Output: Citations, keywords, abstracts

ENVIROLINE (Environmental Information Center, Inc.)

Coverage:
Air environment: air pollution, noise pollution, and weather modification; environmental health: food and drugs, population planning and control, chemical, biological, and radiological contamination; land environment: land use and misuse, terrestrial resources, solid wastes, transportation, wildlife management; resource management: renewable and nonrenewable resources; water environment: water pollution, water resources, oceans, estuaries
Content:
Input: Over 85,000 items since January, 1971
  Additions: 1,000 items/month from 5,000 journals, symposia, reports, Congressional hearing transcripts, newspapers
  Corresponds to *Environmental Abstracts*
Output: Citations, keywords, abstracts

EPIC (Electronic Properties Information Center) (Center for Information and Numerical Data Analysis and Synthesis) (CINDAS)

Coverage:
Chemistry, chemical engineering, electronic materials, electronics, electrical engineering
Content:
Input: Over 70,000 items since January, 1965
  Additions: 10,000 items/year from 6,500 journals, government reports, monographs, proceedings, theses
Output: Citations

EX MEDICA (*Excerpta Medica*) (Excerpta Medica Foundation)

Coverage:
Anatomy, anthropology, anesthesiology, bioengineering, dermatology, drug dependence, environmental health and pollution control, forensic science, pediatrics, pharmacology, toxicology, physiology, psychiatry, public health, radiology, psychiatry, public health, radiology, surgery, urology, genetics, hematology, immunology, microbiology, neurology, nuclear medicine, occupational health and industrial medicine, pathology, nursing, dentistry, psychology (podiatry, optometry, chiropractic excluded)
Content:
Input: Over 1 million items since April, 1975
  Additions: 5,000/week from over 3,500 journals, books, monographs, dissertations, proceedings
  Corresponds to *Excerpta Medica*
Output: Citations, keywords, abstracts

TABLE X
CONTENTS OF CHEMISTRY-RELATED DATA BASES (Continued)

FIRST—1 (Fast Infrared Search Technique) (American Society for Testing and Materials) (ASTM)

Coverage:
Infrared Spectra
Content:
Input: Over 92,000 spectra, from open literature and from American Petroleum Institute, Sadtler Research Laboratories, National Bureau of Standards, Butterworths Scientific Publications Ltd., Coblenz Society, Chemical Manufacturers Association, Infrared Data Committee of Japan
Output: Accession numbers to reference collection, absorption peaks band numbers, coded functional groups

---

FOODS ADLIB (*Foods Adlibra*) (K&M Publications, Inc.)

Coverage:
Food marketing: agribusiness, meat packaging, millers, dairies, bakers, retailers, equipment suppliers; new products; ingredient developments; food economics; market research, nutrition; toxicology; processing; packaging
Content:
Input: Over 70,000 items since January, 1974
  Additions: 2,000/month from journals
  Corresponds to *Food Adlibra*
Output: Citations, keywords, abstracts

FSTA (*Food Science Technology Abstracts*) (International Food Information Service)

Coverage:
Food composition and properties, handling and transport, engineering, processing and preprocessing, analysis, quality control, legislation, hygiene, storage, packaging, management of processes and plants, microbiology, toxicology, economics, statistics, food engineering, food additives, fish and marine products, prices, food laws, regulations, standards
Content:
Input: Over 210,000 items since 1969
  Additions: 1,500 items/month from 1,200 journals, books, standards, patents
  Corresponds to *Food Science and Technology Abstracts*
Output: Citations, keywords, abstracts

GEOARCHIVE (Geosystems, U.K.)

Coverage:
Energy sources, engineering geology, exploration, geochemistry, geochronology, geology, geomathematics, geomorphology, petrology, physical geology, sedimentology, stratigraphy, tectonics, water, geophysics, historical geology, mineral deposits, mineralogy, oceanology, ore analysis, paleogeology, paleontology
Content:
Input: Over 350,000 items since January, 1974
  Additions: 5,000 items/month from 5,000 serials, books, dissertations, maps, proceedings, reports
Output: Citations, keywords

TABLE X
CONTENTS OF CHEMISTRY-RELATED DATA BASES (Continued)

GEO-REF (American Geological Institute)

Coverage:
Fields of geology: economic, engineering, environment, marine, extraterrestrial, aerial, geochemistry, geochronology, geomorphology, igneous and metamorphic petrology, solid earth geophysics, stratigraphy, mineralogy
Content:
Input: Over 950,000 items since January, 1961
  Additions: 4,100 items/month from 3,000 journals, proceedings, monographs, government documents
  Corresponds to publications of American Geological Institute
Output: Citations, keywords

HEEP (*Health Effects of Environmental Pollutants*) (National Library of Medicine and Biosciences Information Service)

Coverage:
Occupational health and industrial medicine; effects of chemicals/substances in the environment on human health; potentially harmful effects of pollutants on humans; analytical methods for examining biological tissues/fluids; studies on lower vertebrates as indicators of substances toxic to humans, vertebrates, and invertebrates as vectors in the food chain
Content:
Input: Over 90,000 items since January, 1972
  Additions: 1,000 items/month from *Biological Abstracts, MEDLARS*
  Corresponds to printed *HEEP Journal*
Output: Citations, keywords, abstracts
Tapes available for lease or retrospective search and SDI

ICRS (*Index Chemicus Registry System*) (Institute for Scientific Information)

Coverage:
Synthesis, application of organic chemicals, chemistry, chemical engineering
Content:
Input: Over 210,000 items since 1966
  Additions: 15,000/year from journal literature
  Corresponds to *Current Abstracts of Chemistry,* and *Index Chemicus; Chemical Structure Index,* 1970; and *ICRA* 1966–1970
Output: Citations

IFI (IFI Comprehensive Data Base) (IFI/Plenum Data Co.)

Coverage:
Chemical and chemistry-related patents: specific chemicals, chemical fragments, role designations
Content:
Input: Over 470,000 items since 1950
  Additions: quarterly
Output: Patent numbers, U.S. Patent and Trademark Office original references, *Chemical Abstracts* references (if available), assignees, titles, expanded to foreign equivalents
Tapes available for lease for retrospective search and SDI

TABLE X
CONTENTS OF CHEMISTRY-RELATED DATA BASES (Continued)

INSPE (Institution of Electrical Engineers, London)

Coverage:
Physics: general, atomic and molecular, elementary particle, nuclear
Engineering: electronic, electrical, nuclear, computer, control
Properties: magnetic, optical, structure, thermal mechanical
Condensed matter: gases, fluid dynamics, plasmas
Content:
Input: Over 1,250,000 items since 1969
   Additions: 12,500/month from 2,000 journals, dissertations, proceedings, reports, books, patents
   Corresponds to *Physics Abstracts, Electrical & Electronics Abstracts, Computer & Control Abstracts*
Output: Citations, keywords, abstracts

IPA (*International Pharmaceutical Abstracts*) (American Society of Hospital Pharmacists International)

Coverage:
Pharmaceutical technology, institutional pharmacy practice, adverse drug reactions, toxicity, investigational drugs, drug evaluations, drug interactions, pharmaceutics, drug metabolism and body distribution, pharmacognosy, legislation and regulations, history, sociology, economics, ethics, pharmacy practice, information processing and literature
Content:
Input: Over 60,000 items since January, 1970
   Additions: 1,200 bimonthly from over 500 journals
   Corresponds to *Pharmaceutical Abstracts*
Output: Citations, keywords

IRGO (Singer Technical Services, Inc.)

Coverage:
Infrared spectra
Content:
Input: Over 90,000 items
   Additions: as received from open literature, and from American Petroleum Institute, Sadtler Research Laboratories, National Bureau of Standards, Butterworth Scientific Publications Ltd., Coblentz Society, Chemical Manufacturers Association, Infrared Data Committee of Japan
Output: Accession numbers to reference collection, absorption peaks, band numbers, coded functional groups

---

IRIS (Infrared Information System) (Sadtler Research Laboratories)

Coverage:
Infrared spectra
Content:
Input: Over 100,000 items
   Additions: as received from open literature, and from American Petroleum Institute, Sadtler Research Laboratories, National Bureau of Standards, Butterworth Scientific Publications Ltd., Coblentz Society, Chemical Manufacturers Association, Infrared Data Committee of Japan
Output: Accession numbers to reference collection, absorption peaks, band numbers, coded functional groups

TABLE X
CONTENTS OF CHEMISTRY-RELATED DATA BASES (Continued)

ISMEC (Information Service in Mechanical Engineering) (Data Courier, Inc.)

Coverage:
   Mechanical engineering: mechanics, materials, devices, production processes, tools,
   equipment, energy and power, transportation and handling, measurement and control,
   management and production, applications in science and industry
Content:
   Input: Over 100,000 items since 1973
      Additions: 15,000/year, worldwide from journals, periodicals, proceedings, reports, books
   Output: Citations, keywords

MANLAB-NPL (Materials Data Bank) (Manlabs, Inc.)

Coverage:
   Numerical thermochemical data
Content:
   Input: Over 1,800 inorganic chemicals, metal alloys
      Additions: annual updates
   Output: Thermochemical data tables; binary, ternary phase diagrams

---

MCIC (Metals and Ceramics Information Center) (Battelle Memorial Institute, Columbus, Ohio)

Coverage:
   Metals, composites, ceramics, corrosion/compatibility, super alloys, refractory metals,
   metalworking, joining, testing
Content:
   Input: Over 110,000 items since 1955
      Additions: 4,000/year from journals, symposia and conference proceedings, theses,
      government reports, books, meeting reprints
   Output: Citations, keywords, abstracts

MEDLARS (Medical Literature Analysis and Retrieval System) (National Library of Medicine)

Coverage:
   Medicine and medical research, drug administration and effects, anatomy and histology,
   biosynthesis, cytology, taxonomy, congenital diseases, medical diagnosis, enzymology,
   etiology, immunology, injuries, instrumentation, surgery, toxicology
Content:
   Input: Over 2,700,000 items since January, 1966
      Additions: 20,000/month from over 3,000 journals, monographs, and other recurring
      bibliographies
      Corresponds to *Index Medicus*
   Output: Citations, keywords, abstracts

MEDLINE (MEDLARS Online)

Coverage:
   See MEDLARS

METADEX (American Society for Metals, and Metals Society, London)

Coverage:
   Materials, processes, properties, products, forms, influencing factors, metallics systems,
   intermetallic compounds

TABLE X
CONTENTS OF CHEMISTRY-RELATED DATA BASES (Continued)

METADEX (American Society for Metals, and Metals Society, London) **(Continued)**

Content:
  Input: Over 350,000 items since january, 1966
    Additions: 2,500/month from journals, proceedings, reports, books
    Corresponds to *Metal Abstracts* and *Alloys Index*
  Output: Citations, keywords, names

MSSS (Mass Spectral Search System) (Mass Spectrometry Data Centre, U.K.) (available via National Institutes of Health/Environmental Protection Agency Chemical Information System) (NIH/EPA CIS)

Coverage:
  Mass spectral data, references to mass spectral data
Content:
  Input: Over 32,000 compounds, over 60,000 spectra
    Additions: as available
    Corresponds to *Mass Spectrometry Bulletin*
  Output: Mass spectral data, citations, keywords

NASA (National Aeronautics and Space Administration)

Coverage:
  Physical (chemistry, biology, physics) sciences, social sciences, management, materials, electronics, instrumentation, controls, mechanisms, aeronautics, space sciences, and engineering
Content:
  Input: Over 1 million items since 1961
    Additions: 5,000/month from journals, NASA reports
    Corresponds to *International Aerospace Abstracts (IAA)* and *Scientific and Technical Aerospace Reports* (STAR)
  Output: Citations, keywords, abstracts

NIH/EPA CIS (National Institutes of Health/Environmental Protection Agency Chemical Information System)

Coverage:
  NIH/EPA CIS contains many files. Each is listed separately by its own acronym. These include:

SANSS	Structure and Nomenclature Search System
CNMR	C13 Nuclear Magnetic Resonance Search System
CRYST	X-Ray Crystallography Search System
MSSS	Spectral Search System
OHM-TADS	Oil and Hazardous Materials Technical Assistance Data System
PDSM	Powder Diffraction Search Match
RTECS	Registry of Toxic Effects of Chemical Substances

NSA (*Nuclear Science Abstracts*)

Coverage:
  Nuclear science, technology; energy, medicine

TABLE X
CONTENTS OF CHEMISTRY-RELATED DATA BASES (Continued)

NSA *(Nuclear Science Abstracts)* (Continued)

Content:
Input: Over 550,000 items from January, 1966, to June, 1976
    Additions: from journals, government reports, monographs, proceedings, theses, patents
    NSA incorporated into ERDA Energy Information Data Base, now U.S. Department of
    Energy's *Energy Research Abstracts*
Output: Citations, keywords
For retrospective search only

NTIS (National Technical Information Service) (U.S. Department of Commerce)

Coverage:
Aeronautics, agriculture, astronomy, astrophysics, atmospheric sciences, behavioral and social
sciences, biological and medical sciences, chemistry, earth science, oceanography,
electronics, energy, materials, methods, equipment, communications, ordnance, physics,
space technology. Engineering: civil, electronic, electrical, chemical, industrial, mechanical,
mining, aeronautical, marine
Content:
Input: Over 750,000 items since January, 1964
    Additions: 5,000/month from reports from government agencies, contractors, grantees on
    federally sponsored research
    Corresponds to *Weekly Government Abstracts and Government Report Announcements*
Output: Citations, keywords, abstracts

OA *(Oceanic Abstracts)* (Data Courier, Inc.)

Coverage:
Marine biology, biological oceanography, marine pollution, living marine resources (fisheries,
mariculture), nonliving marine resources (including oil, gas, minerals), physical and chemical
oceanography, desalination, meteorology, geology and geophysics, ships, navigation,
shipping, diving, acoustics, optics, remote sensing, government and law, coastal and deep-sea
environments, engineering instrumentation and techniques
Content:
Input: Over 100,000 items since January, 1964
    Additions: 1,200 bimonthly from 2,000 primary sources, including journals, proceedings,
    documents of limited circulation, reports, books, trade publications
    Corresponds to *Oceanic Abstracts*
Output: Citations, keywords

**OHM-TADS (Oil and Hazardous Materials—Technical Assistance Data System) (Office of
Hazardous Materials, Environmental Protection Agency) (available via National Institutes of
Health/Environmental Protection Agency Chemical Information System) (NIH/EPA CIS)**

Coverage:
Data about materials shipped by rail, truck, ship, barge. Physical, chemical, biological,
toxicological properties; selected commercial data; deleterious effects on water quality;
handling procedures
Content:
Input: Over 1,200 materials
    Additions: as available
Output: Up to 126 different individual categories/material describing items under subject
coverage
Retrospective search only

TABLE X
CONTENTS OF CHEMISTRY-RELATED DATA BASES (Continued)

PAPERCHEM (Institute of Paper Chemistry)

Coverage:
Pulp, paper, and board manufacturing and utilizing industries: raw materials, pulp products, packaging, graphic arts, forestry, silviculture, chemistry of carbohydrates, cellulose, hemicelluloses, lignins and wood extractives; technology, such as engineering, process control, machinery, equipment, maintenance, pollution control, water, power
Content:
Input: Over 130,000 items since July, 1969
Additions: 1,000/month from over 800 periodicals, reviews, symposia and conference proceedings, books, patents, dissertations, translations
Corresponds to *Abstract Bulletin of the Institute of Paper Chemistry*
Output: Citations, keywords, abstracts

PATS (Patsearch) (Pergamon International Information Corporation)

Coverage:
U.S. patents
Content:
Input: Over 500,000 U.S. patents since 1971
Additions: 1,200 items/week
Output: Patent numbers, subject matter, inventors, assignees, citations to U.S./foreign patents, U.S. and International Patent Classification System codes

---

PDSM (Powder Diffraction Search-Match) (Joint Committee on Powder Diffraction Standards —International Centre for Diffraction Data) (available via National Institutes of Health/ Environmental Protection Agency Chemical Information System) (NIH/EPA CIS)

Coverage:
X-ray diffraction patterns for organics, inorganics, minerals, and alloys
Content:
Input: Over 33,000 powder diffraction patterns
Additions: as available
Output: Same as input

P/E NEWS (American Petroleum Institute)

Coverage:
News items from seven major suppliers of petroleum and energy business news: *Middle East Economic News, Oil Daily, Oil and Gas Journal, Petroleum Economist, Petroleum Intelligence Weekly, Platts Oilgram News Service,* and *National Petroleum News*
Content:
Input: Over 125,000 items since January, 1975
Additions: 26,000 items/year from the periodicals listed above
Output: Citations, keywords

PESTDOC (Pesticidal Literature Documentation) (Derwent Publications Ltd.)

Coverage:
Pesticides, agriculture, agricultural engineering, chemistry, chemical engineering, environment
Content:
Input: Over 100,000 items since July, 1968
Additions: 9,000 items/year from over 180 journals, government documents, monographs, preprints, conference papers, proceedings, theses
Output: Citations, keywords

TABLE X
CONTENTS OF CHEMISTRY-RELATED DATA BASES (Continued)

PIRA (Research Association for Paper and Board, Printing and Packaging Industries)

Coverage:
  Adhesives, binding, composition, education and training, finishing, food packaging,
  forecasting, graphic arts, industrial relations, inks, materials, machinery, occupational safety,
  plastics packaging, pollution, production, pulps, recycling, reprography, retailing, testing
Content:
  Input: Over 50,000 items since January, 1975
    Additions: 800/month from over 600 journals, books, pamphlets, standards, specifications,
    legislation, regulations, translations, conference papers, reports, trade literature
    Corresponds to *Paper and Board Abstracts, Printing Abstracts, Packaging Abstracts,
    Managing and Marketing Abstracts*
  Output: Citations, keywords, abstracts

PNI (*Pharmaceutical News Index*) (Data Courier Inc.)

Coverage:
  Drugs and cosmetics; Food and Drug Administration approvals and recalls, health legislation,
  government regulations, pharmaceutical research
Content:
  Input: Over 40,000 items since December, 1975
    Additions: 100 items/month from *Drug Research Reports* ("The Blue Sheet"), *FDA Reports*
    ("The Pink Sheet"); *Quality Control Reports* ("The Gold Sheet"); others
  Output: Citations, descriptions

POLLN ABS *(Pollution Abstracts)* (Data Courier, Inc.)

Coverage:
  Air, water, marine, land, thermal, and noise pollution; waste treatment; pesticides; radiation;
  general environmental quality; solid wastes
Content:
  Input: Over 65,000 items since January, 1970
    Additions: 1,000 bimonthly from technical and nontechnical reports; 2,500 journals,
    worldwide; research papers, contracts, proceedings, patents, government documents
    (executive actions, treaties, legislation, regulations, court decisions)
    Corresponds to *Pollution Abstracts*
  Output: Citations, keywords, abstracts

PPDS (Physical Property Data Service) (Institution of Chemical Engineers, London)

Coverage:
  Stream constant properties: formula weight; critical temperature, pressure, and volume;
  melting and boiling points; parachor; heat of formation—vapor and liquid; flash points; upper
  and lower flammability limits, autoignition temperature; solubility parameter, acentric factor;
  acentric factor of homomorph; dipole moment
  Stream variable properties: vapor heat capacity, viscosity, thermal conductivity, enthalpy;
  liquid heat capacity, density, coefficient of cubical expansion; latent heat of vaporization;
  surface tension; vapor pressure; liquid viscosity; vapor density; total heat of formation

TABLE X
CONTENTS OF CHEMISTRY-RELATED DATA BASES (Continued)

PPDS (Physical Property Data Service) (Institution of Chemical Engineers, London)

(Continued)

Content:
  Input:
    Additions as available, from critical evaluations of published and unpublished sponsored
    and unsponsored research (some estimated data)
  Output: Tabular data relating to requested properties of about 500 compounds

PTS (Predicasts Terminal System) (Predicasts, Inc.)

Coverage:
  Company acquisitions, plant capacities, market data, new products, technology, domestic and
  international company forecasts, statistical reports, reviews of domestic and foreign industry
Content:
  Input: Over 1,700,000 items since January, 1972
    Additions: 2,500/week from newspapers, business magazines and trade journals, bank
    letters, special reports, including PROMPT (formerly CMA—*Chemical Market Abstracts* and
    Equipment Market Abstracts)
  Output: Citations, keywords, abstracts

RAPRA (Rubber and Plastics Research Association of Great Britain)

Coverage:
  Synthesis, polymerization, economics, natural rubber cultivation, polymer applications, raw
  materials, monomers, properties, testing, compounding ingredients, toxicity, processing
  technology, industrial organization/administration, machinery, test equipment, environmental
  and industrial hazards
Content:
  Input: Over 110,000 items since January, 1972
    Additions: 1,800/month from journals, proceedings, preprints, books, patents, trade
    literature, technical reports, government publications
  Output: Citations, keywords

RINDOC (Pharmaceutical Literature Documentation) (Derwent Publications Ltd.)

Coverage:
  Pharmaceuticals, chemistry, chemical engineering, life sciences, medicine
Content:
  Input: Over 270,000 items since July, 1964
    Additions: 50,000/year from over 400 journals
  Output: Citations, keywords

RTECS (Registry of Toxic Effects of Chemical Substances) National Institute of Occupational Safety and Health) (NIOSH) (available via NIH/EPA CIS)

Coverage:
  Toxicology of chemical substances

TABLE X
CONTENTS OF CHEMISTRY-RELATED DATA BASES (Continued)

RTECS (Registry of Toxic Effects of Chemical Substances) National Institute of Occupational Safety and Health) (NIOSH) (available via NIH/EPA CIS) (Continued)

Content:
Input: Over 39,000 compounds
    Additions: as available
Output: Preferred CA chemical names, molecular weights, molecular formulas, synonyms, CA Registry Numbers (if available), toxic doses, test animals used, references if available

---

SAE (*SAE Abstracts*) (Society of Automotive Engineers, Inc.)

Coverage:
Aircraft, missiles and spacecraft, ground support equipment, passenger cars, military equipment, aircraft propulsion, electric vehicles, energy conversion, fuels and lubricants, manufacturing and production, transportation systems, emissions, safety, noise, management, testing and instrumentation
Content:
Input: Over 11,000 items since January, 1965
    Additions: 800/year from meeting or conference papers and proceedings, books, special publications, handbooks from SAE
Output: Citations, keywords, abstracts

SANSS (Structure and Nomenclature Search System) (Available via NIH/EPA CIS)

Coverage:
Structures, chemical names, CA Registry Numbers, structural diagrams, systematic names, synonyms
Content:
Input: Over 116,000 compounds, 330,000 names
    Additions: as available from over 40 different sources
Output: Same as subject coverage; refers to sources

SCA (*Surface Coatings Abstracts*) (Paint Research Association)

Coverage:
Paints, surface coatings, resins, solvents, plasticizers, printing inks, insulations, fire retardants, deterioration, testing, industrial hazards, pollution marketing
Content:
Input: Over 30,000 items since January, 1976
    Additions: Over 8,000 items/year from journals, patents, proceedings and books
    Corresponds to *Surface Coatings Abstracts*
Output: Citations, keywords, abstracts

SCISEARCH (Institute for Scientific Information)

Coverage:
Chemistry, physics, and biological, medical sciences
Content:
Input: Over 3 million items since January, 1974
    Additions: 42,000/month from journals, reviews, editorials
    Corresponds to *Science Citation Index*
Output: Citations

TABLE X
CONTENTS OF CHEMISTRY-RELATED DATA BASES (Continued)

SDILINE (National Library of Medicine)

Coverage:
  Medical care, endocrinology, genetics, veterinary medicine, autonomic drugs, pharmacy, organic chemistry, behavioral sciences
Content:
  Input: From 3,000 worldwide biomedical periodicals, monographs, recurring bibliographies
    Additions: 20,000 items/month
    Corresponds to *Index Medicus*
  Output: Citations, keywords, some abstracts
  For current awareness (SDI) only

SPIN (*Searchable Physics Information Notices*) (American Institute of Physics) (AIP)

Coverage:
  Physics: general, elementary particle, nuclear, atomic and molecular classical, medical, bio- and medical, geo-, astro-condensed matter; fluid plasmas, electrical discharges, physical chemistry, biomedical engineering, astronomy
Content:
  Input: Over 120,000 items since January, 1975
    Additions: 2,000/month from journals published by AIP and member societies, Russian journals, proceedings, other selected American physics journals
  Output: Citations, keywords, abstracts

SSIE (*SSIE Current Research*) (Smithsonian Science Information Exchange)

Coverage:
  Ongoing and recently completed basic and applied research: agricultural sciences, behavorial sciences, biological sciences, earth sciences, chemistry and chemical engineering, electronics, engineering materials, mathematics, medical sciences, physics, social sciences
Content:
  Input: Over 215,000 items at any one time
    Changes: 9,000/month, about research in progress from over 1,300 funding organizations, such as Federal, state, and local government; nonprofit associations; colleges and universities; nonaffiliated investigators; and some non–U.S. organizations and private industry
  Output: Citations, keywords, abstracts, contractor, agency

TDB (Toxicology Data Bank) (National Library of Medicine)

Coverage:
  Chemical, physical, biological, pharmacological, toxicological, environmental facts/data
Content:
  Input: Over 1,000 items from 80 standard reference textbooks, handbooks, monographs, and criteria documents
    Additions: as appropriate
  Output: 60 different data categories, including CA Registry Numbers, preferred chemical names, synonyms, molecular formulas, molecular weight, human and animal toxicology, antidotes, treatments

TABLE X
CONTENTS OF CHEMISTRY-RELATED DATA BASES (Continued)

TEPIAC (Thermophysical and Electronic Properties Information and Analysis Center) (Purdue University)

Coverage:
Thermophysical properties, chemical engineering, electronics, electrical engineering, mechanical engineering, physics.
Content:
Input: Over 100,000 items since 1900
Additions: 5,000/year from over 5,000 journals, government documents
Corresponds to *Thermophysical Properties Research Literature Retrieval Guide* (editions 1900–1964 and 1964–1970)
Output: Citations

TEX TECH (*Textile Technology*) (Institute of Textile Technology)

Coverage:
Natural fibers, fiber sciences, yarns, fiber production, knitting, bleaching, printing, chemical and mechanical finishing, cutting, sewing, pressing, apparel and fabric products, fiber and yarn testing and measurement, plant and equipment, man-made fibers, fiber treatment, yarn apparel modification, weaving, nonwoven materials, dyeing, mill management, production, laundering, dry cleaning.
Content:
Input: Over 200,000 items since 1966
Additions: 1,500/month from worldwide periodicals, reports, patents
Corresponds to *Textile Technology Digest*
Output: Citations, keywords

TITUS (Textile Information Treatment Users' Service) (Institut Textile de France)

Coverage:
Processes and machines for the production of fibrous materials, fibers, yarns, woven fabrics, knitted fabrics and nonwoven fabrics; processes, products, and machines for textile finishing, including bleaching, dyeing, and printing; processes and machines for washing, laundering, and dry cleaning; processes and machines for industrial fabrics, including tire cord, belts, filters, carpets, wall covering, analysis and testing of textile products and equipment; textile engineering; textile industry management; environmental protection and pollution control related to the textile industry; biological effects of textiles related to health protection
Content:
Input: Over 120,000 items since January, 1967
Additions: 7,500/year from 8,000 journals, reports, patents, standards, technical data sheets
Output: Citations, keywords, abstracts

TOXLINE (National Library of Medicine)

Coverage:
Human and animal toxicity, adverse drug reactions, poisoning cases, environmental pollutant effects, chemicals toxicity, analytical methodology
Content:
Input: Over 950,000 items since July, 1950
Additions: 7,000/month, worldwide from 12,000 journals, and from major secondary sources and special collections
Output: Citations, keywords, abstracts

TABLE X
CONTENTS OF CHEMISTRY-RELATED DATA BASES (Continued)

TULSA (University of Tulsa)

Coverage:
  Exploration: development and production of oil and natural gas; geology; geophysics; geochemistry; drilling, well logging; petroleum; transportation and storage; ecology; pollution; alternative fuels; energy sources
Content:
  Input: Over 250,000 items since 1965
    Additions: 1,500/month from journals, patents
    Corresponds to *Petroleum Abstracts*
  Output: Citations, keywords

VETDOC (Veterinary Literature Documentation) (Derwent Publications Ltd.)

Coverage:
  Veterinary science, agricultural engineering chemistry, chemical engineering, life sciences
Content:
  Input: Over 65,000 items since January, 1968
    Additions: 5,000/year, from over 130 journals, government reports
  Output: Citations

WAA (*World Aluminum Abstracts*) (American Society for Metals)

Coverage:
  Aluminum industry: general, ores (exclusive of mining), alumina production and extraction, melting, casting, foundry, metalworking, fabrication, finishing, physical and mechanical metallurgy, engineering properties and tests, quality control and tests, end uses
Content:
  Input: Over 70,000 items since January, 1968
    Additions: 550/month, worldwide from 1,600 journals, proceedings, reports, books, patents
    Corresponds to *World Aluminum Abstracts*
  Output: Citations, keywords, abstracts

WORLD PET (*World Petrochemicals*) (SRI International, Inc.)

Coverage:
  Worldwide coverage of products, countries, region locations, plant capacities, company ownerships, production imports, exports, consumption
Content:
  Input: Over 6,000 plant/product listings from 1974 to date, including forecasts to 1984, with updates as needed
    Corresponds to the annually published *World Petrochemicals*
  Output: Same as coverage

WORLDTEX (*World Textiles*) (Shirley Institute, U.K.)

Coverage:
  Fabrics, fibers, yarns: chemical, finishing production processes; properties; clothing; makeup goods, laundering, and dry cleaning; mill engineering; management, economics, and statistics; analysis, testing, quality control; polymer science; production, consumption, and international trade data; pollution; safety and health hazards

TABLE X
CONTENTS OF CHEMISTRY-RELATED DATA BASES (Continued)

WORLDTEX (*World Textiles*) (Shirley Institute, U.K.) (Continued)

Content:
Input: Over 85,000 items since January, 1970
Additions: 700 items/month, from journals, books, patents, reports, pamphlets, monographs, standards
Corresponds to *World Textile Abstracts*
Output: Citations, keywords

WPI (*World Patent Index*) (Derwent Publications, Ltd.)

Coverage:
Pharmaceuticals (from 1963), agricultural chemicals (from 1965), plastics (from 1966), chemistry (from 1970), mechanical, electrical (from 1974)
Content:
Input: Over 1,500,000 items since 1963
Additions: 19,500/month
Output: Citations, keywords, patent numbers, countries, International Patent Classification

---

ACRONYM INDEX TO CHEMISTRY-RELATED DATA BASES

More often than not, references to a particular subject or subject area are found in more than one data base. To be sure of retrieval of all available references, several data bases may have to be searched in sequence, and sometimes even in parallel. Quite often, the same references are in more than one data base.

Table XI is an index to computer-readable data bases described in Table X, Contents of Chemistry-Related Data Bases. In Table XI one or more potentially useful data bases are listed for each subject area. Each data base cited probably has material pertinent to that subject area. In many subject areas, more than five data bases contain related material. To use the index, identify the subject area, note the data bases listed by acronym for that subject area, and refer to Table X to obtain an indication of the extent of coverage by the indicated data base(s).

Note that the data bases in Table X are not referenced to every subject area covered. The index would have had to be quite massive to do this. The intent is to suggest searching possibilities.

Vendor use manuals and thesauri or word lists for each subject area are essential to selection of the data bases more likely to be useful.

This index was developed in part from 10 years of computer searching and SDI experience, and in part from vendors' manuals, data-base generators' descriptive literature and indexes, user group discussions, thesauri, and articles from the open literature.

TABLE XI

ACRONYM INDEX TO CHEMISTRY-RELATED DATA BASES WHICH CONTAIN REFERENCES/DATA ABOUT THE SUBJECT*

Subject	Data-base acronym				
Acentric factors	PPDS	CA-FILES			
Acoustics	CA-FILES	CDI	CONF PAP	NASA	SCISEARCH
Adhesives	CA-FILES	EI COMP	ISMEC	NASA	NTIS
Aerospace engineering	EI COMP	ISMEC	MCIC	NASA	NTIS
Agricultural engineering	AGRICOLA	BIOSIS	EI COMP	CAB	NTIS
Agricultural industry	CIN	PTS			
Agriculture	AGRICOLA	BIOSIS	CA-FILES	CAB	NTIS
Agrochemicals	AGRICOLA	BIOSIS	CA-FILES	CAB	
Air measurements	APTIC	CA-FILES	ENVIROLINE	POLLN ABS	
Air pollution	APTIC	CA-FILES	EI COMP	POLLN ABS	NTIS
Alumina	CA-FILES	EI COMP	NASA	NTIS	WAA
Aluminum	CA-FILES	EI COMP	METADEX	NTIS	WAA
Analytical chemistry	CA-FILES	EI COMP	INSPEC	NASA	NTIS
Anesthetics	BIOSIS	CA-FILES	EX MEDICA	MEDLARS	TOXLINE
Animal science	AGRICOLA	BIOSIS	CAB	VETDOC	
Anticancer agents	BIOSIS	CANCERLIT	CLINPROT	EX MEDICA	MEDLINE
Applied chemistry	APILIT	CA-FILES	EI COMP	NASA	NTIS
Atomic energy	EDB	EI COMP	ENERGY	NSA	NTIS
Autoignition temperatures	CA-FILES	OHM-TADS	PPDS		
Automobiles	EI COMP	ISMEC	METADEX	NTIS	SAE
Bacteriology	CA-FILES	EX MEDICA	MEDLARS		
Baking	AGRICOLA	BIOSIS	FOODS ALIBRA	FSTA	
Binders	See Adhesives.				
Biochemistry	BIOSIS	CA-FILES	EX MEDICA	MEDLARS	TOXLINE
Bioinstrumentation	BIOSIS	CA-FILES	INSPEC	NASA	NTIS
Biology	BIOSIS	CA-FILES	CAB	EX MEDICA	FSTA
Boiling points	CA-FILES	PPDS	RAPRA	TEPIAC	
CA nomenclature	CHEMDEX	CHEMLINE	CHEMNAME	RTECS	
CA Registry Numbers	CA-FILES	CAS ONLINE	CHEMDEX	CHEMLINE	CHEMNAME
Cancer projects	CANCERPROJ	SSIE			
Carbohydrates	BIOSIS	CAB	CA-FILES	FSTA	PAPERCHEM

Subject					
Carbon-13 spectra	CA-FILES	CNMR			
CAS journals	CASSI				
Cellulose	BIOSIS	CAB	CA-FILES	FSTA	PAPERCHEM
Cellulose fibers	CA-FILES	PAPERCHEM	TEXTECH	TITUS	WORLDTEX
Cement	CA-FILES	EI COMP	MATERIALS	MCIC	NTIS
Ceramics	CA-FILES	EI COMP	MCIC	NASA	NTIS
Chemical business	CIN	P/E NEWS	PNI	PTS	WORLD PET
Chemical dictionary	CHEMDEX	CHEMLINE	CHEMNAME		
Chemical equipment	CA-FILES	EI COMP	NASA	NTIS	PAPERCHEM
Chemical engineering	CA-FILES	EI COMP	ICRS	NASA	NTIS
Chemical manufacture	CA-FILES	EI COMP			
Chemical markets	CIN	DRI-CHEM	PNI	PTS	WORLD PET
Chemical products	CIN	DRI-CHEM	PNI	PTS	WORLD PET
Chemical rings	CA-FILES	CAS ONLINE	CHEMDEX	CHEMLINE	CHEMNAME
Chemical structures	CAS ONLINE	CHEMDEX	CHEMLINE	CHEMNAME	SANSS
Chemical synonyms	CHEMDEX	CHEMLINE	CHEMNAME	OHM-TADS	RTECS
Chemotherapy	CA-FILES	CANCERLIT	EX MEDICA	TDB	TOXLINE
Civil engineering	AGRICCLA	EI COMP	NTIS		
Clinical medicine	EX MEDICA	MEDLARS			
Coal energy	CA-FILES	EDB	EI COMP	ENERGYLINE	NTIS
Coatings	CA-FILES	EI COMP	NTIS	RAPRA	SCA
Coefficients of expansion	EI COMP	NASA	NTIS	PPDS	TEPIAC
Concrete products	See Cement.				
Consumerism	AGRICOLA				
Control engineering	CA-FILES	EI COMP	INSFEC	ISMEC	NASA
Corporation information	CIN	PTS	PNI		
Cosmetics	BIOSIS	CA-FILES			
Critical points	CA-FILES	PPDS	TEPIAC		
Crop husbandry	AGRICOLA	BIOSIS	CAB		
Crystallography	CA-FILES	CRYST	INSFEC	SPIN	
Current research	CANCERPROJ	SSIE			
Dairy science	AGRICOLA	CAB			
Dentistry	EX MEDICA	MEDLARS	NTIS		
Dermatology	EX MEDICA	MEDLARS	TOXLINE		
Desalination	CA-FILES	EI COMP	NTIS	OA	POLLN ABS

*For CA-FILES, see individual data bases produced by Chemical Abstract Service.

TABLE XI
ACRONYM INDEX TO CHEMISTRY-RELATED DATA BASES WHICH CONTAIN REFERENCES/DATA ABOUT THE SUBJECT* (Continued)

Subject	Data-base acronym				
Detergents	CA-FILES	EI COMP	INSPEC	PDSM	SPIN
Diffraction	CA-FILES	CRYST			
Dissertations	CA-FILES	CDI			
Drugs	DRUG DOC	IPA	MEDLARS	PNI	TOXLINE
Drycleaning	CA-FILES	TEXTECH	TITUS	WORLDTEX	
Dyes	CA-FILES	TEXTECH	TITUS	WORLDTEX	
Earth sciences	GEOARCHIVE	GEO-REF	NTIS	TULSA	
Ecology	BIOSIS	CA-FILES	CAB	ENVIROLINE	NTIS
Economics	AGRICOLA	CIN	ENERGYLINE	PTS	
Elastomers	CA-FILES	EI COMP	NASA	NTIS	RAPRA
Electrical engineering	DDC	EI COMP	INSPEC	NASA	NTIS
Electrochemistry	CA-FILES	DDC	ENERGY	INSPEC	NTIS
Electronics engineering	DDC	EI COMPEND	INSPEC	NASA	NTIS
Electronic materials	EI COMPEND	EPIC	INSPEC	METADEX	NASA
Emissions	APTIC	CA-FILES	NTIS	POLLN ABS	TOXLINE
Energy	CA-FILES	EDB	ENERGYLINE	NTIS	NASA
Energy conservation	EDB	EI COMP	ENERGYLINE	NASA	NTIS
Engineering	DDC	EI COMP	INSPEC	ISMEC	NTIS
Engineering	See engineering discipline of interest.				
Engineering, aerospace	See Aerospace engineering.				
Engineering, agricultural	AGRICOLA	BIOSIS	CAB	EI COMP	NTIS
Engineering, chemical	See Chemical engineering.				
Engineering, civil	See Civil engineering.				
Engineering, electronics	See Electrical engineering.				
Engineering, environmental	See Environmental engineering.				
Engineering, food	See Food.				
Engineering, fuel	See Fuels.				
Engineering, geological	See Geology.				
Engineering, industrial	See Industrial engineering.				
Engineering, mechanical	See Mechanical engineering.				
Engineering, metals	See Metals engineering.				

Subject					
Engineering, nuclear	See Nuclear engineering.				
Enthalpies	CA-FILES	E COMP	PPDS	TEPIAC	
Environmental engineering	CA-FILES	E COMP	ENVIROLINE	NTIS	POLLN ABS
Environmental health	CA-FILES	ENVIROLINE	EX MEDICA	NTIS	TOXLINE
Enzymology	BIOSIS	CA-FILES	EX MEDICA	IPA	MEDLINE
Essential oils	CA-FILES				
Explosives	CA-FILES	DDC	EI COMP	NASA	NTIS
Fabrics	NTIS	TEXTECH	TITUS	WORLDTEX	
Fats	CA-FILES	FSTA	NTIS		
Fermentations	BIOSIS	CA-FILES	FSTA	NTIS	
Fertilizers	AGRICOLA	CA-FILES	CAB	NTIS	
Fibers	CA-FILES	PAPERCHEM	TEXTECH	TITUS	WORLDTEX
Fibers trade data	CIN	PTS	WORLDTEX		
Finishes	CA-FILES	NTIS	SCA		
Fire retardants	CA-FILES	PAPERCHEM	RPRA	SCA	TITUS
Flammabilities	CA-FILES	NTIS	O-M-TADS	PPDS	TEPIAC
Flash points	CA-FILES	E COMP	O-M-TADS	PPDS	TEPIAC
Fluorescent agents	CA-FILES	PAPERCHEM			
Fluid dynamics	EI COMP	IN-SPEC	ISMEC	PPDS	
Food	AGRICOLA	CA-FILES	CAB	FOODS ADLIB	FSTA
Food additives	See Food.				
Food composition	See Food.				
Food engineering	See Food.				
Food industry	ALIB	CIN	FOODS ADLIB	PTS	
Food properties	See Food.				
Forensic science	CA-FILES	EX MEDICA	MEDLARS		
Forest products	See Forestry.				
Forestry	AGRICOLA	CAB	NTIS	PAPERCHEM	
Fossil fuels	See Fuels.				
Fragment searching	CAS ONLINE	CHEMLINE	CHEMNAME	SANSS	
Fuel engineering	See Fuels.				
Fuels	APILIT	EDB	EI COMP	NASA	NTIS
General science	DDC	NASA	NTIS		
Genetics	AGRICOLA	BIOSIS	CAB	EX MEDICA	MEDLARS

*For CA-FILES, see individual data bases produced by Chemical Abstract Service.

Subject	Data-base acronym				
Geological chemistry	See Geology.				
Geological engineering	See Geology				
Geology	CONFPAP	GEOARCHIVE	GEO-REF	NTIS	TULSA
Geosciences	See Geology.				
Geothermal energy	CA-FILES	EDB	ENERGY	NTIS	
Glues	See Adhesives.				
Graphic arts	PAPERCHEM	PIRA			
Hazardous materials	CA-FILES	HEEP	OHM-TADS	RTECS	TOXLINE
Health and safety	CA-FILES	EI COMP	NTIS	WORLDTEX	
Heat capacities	CA-FILES	PPDS	TEPIAC		
Heats of formation	CA-FILES	PPDS			
Helminthology	AGRICOLA	BIOSIS	CAB		
Hematology	CA-FILES	BIOSIS	EX MEDICA	MEDLARS	
Hormone pharmacology	CA-FILES	CBAC	EX MEDICA	MEDLINE	TOXLINE
Horticulture	AGRICOLA	CAB	CA-FILES		
Immunology	BIOSIS	CA-FILES	EX MEDICA	MEDLARS	NTIS
Immunotherapy	BIOSIS	CANCERLIT	EX MEDICA	MEDLARS	NTIS
Industrial engineering	DDC	EI COMP	ISMEC	NASA	
Industrial hazards	See Hazardous materials.				
Industrial hygiene	CA-FILES	EX MEDICA	HEEP	TOXLINE	
Infrared spectra	FIRST-1	IRGO	IRIS		
Inks	CA-FILES	PAPERCHEM	PIRA	SCA	
Instrumentation	CA-FILES	EI COMP	ISMEC	INSPEC	NTIS
Insulations	CA-FILES	EI COMP	SCA		
Intermetallics	CA-FILES	MCIC	METADEX	NASA	NTIS
Journals—CAS	CASSI				
Leather	CA-FILES				
Legislation	CIN				
Life sciences	AGRICOLA	BIOSIS	CA-FILES	CAB	
Lignin	CA-FILES	PAPERCHEM			
Liquid wastes	See Pollution.				

Subject					
Lubricants	CA-FILES	EI COMP	METADEX	NASA	SAE
Macromolecular chemistry	CA-FILES	CDC	NTIS	NASA	
Magnetic properties	INSPEC	ISMEC	SPIN	NTIS	OA
Marine geology	EI COMP	GEOARCHIVE	GEOREF	NTIS	
Mass spectral data	MSSS				
Materials	CA-FILES	METADEX	NASA	NTIS	PIRA
Materials engineering	EI COMP	ISMEC	METADEX	NASA	NTIS
Mechanical engineering	EI COMP	ISMEC	NASA	NTIS	
Mechanical properties	EI COMP	NASA	NTIS	PPDS	TEPIAC
Medicine	CANCERLINE	CANCERPROJ	DRUGDOC	EX MEDICA	MEDLARS
Melting points	CA-FILES	PPDS	TEPIAC		
Metallic systems	See Metallurgy.				
Metallurgy	CA-FILES	MCIC	METADEX	NTIS	WAA
Metals and alloys	CA-FILES	MCIC	METADEX	NASA	NTIS
Metals composites	EI COMP	MCIC	METADEX	NTIS	WAA
Metals engineering	EI COMP	MCIC	METADEX	NASA	NTIS
Metals processes	EI COMP	METADEX	MCIC	NASA	NTIS
Metals products	EI COMP	MCIC	METADEX	NASA	NTIS
Metals properties	MCIC	METADEX	NASA	NTIS	TEPIAC
Microbiology	BIOSIS	CA-FILES	EX MEDICA	MEDLARS	
Mineral chemistry	See Mineralogy.				
Mineralogy	EDB	ENERGY ABS	GEOARCHIVE	GEO-REF	NTIS
Mining engineering	EDB	EI COMP	ENERGY ABS	GEO-REF	NTIS
Molecular formulas	CA-FILES	CHEMDEX	CHEMLINE	CHEMNAME	RTECS
Monomers	CA-FILES	EI COMP	PAPERCHEM	RAPRA	
Mycology	AGRICOLA	BIOSIS	CAB	NTIS	
Natural gas	CA-FILES	EDB	ENERGYLINE	ENERGY	TULSA
Natural rubber	CA-FILES	NTIS	RAPRA	RTECS	
Nematology	AGRICOLA	BIOSIS	CAB		
Neurology	EX MEDICA	MEDLARS			
Noise pollution	EI COMP	NTIS	POLLN ABS	SAE	
Nomenclature	CHEMDEX	CHEMLINE	CHEMNAME	RTECS	
Nuclear energy	See Atomic energy.				
Nuclear engineering	EDB	EI COMP	NSA	NTIS	

*For CA-FILES, see individual data bases produced by Chemical Abstract Service.

TABLE XI
ACRONYM INDEX TO CHEMISTRY-RELATED DATA BASES WHICH CONTAIN REFERENCES/DATA ABOUT THE SUBJECT* (Continued)

Subject	Data-base acronym				
Nuclear magnetic resonance	CA-FILES	CNMR			
Nuclear medicine	See Medicine.				
Nuclear power	See Atomic energy; See Nuclear engineering.				
Nuclear properties	CNMR				
Nursing	See Medicine.				
Nutrition	AGRICOLA	BIOSIS	CA-FILES	CAB	FSTA
Occupational health	CA-FILES	EX MEDICA	HEEP	TOXLINE	
Occupational safety	CA-FILES	HEEP	NTIS	PIRA	TOXLINE
Oceanography	GEO-REF	OA	NTIS		
Operations research	EI-COMP	NASA	NTIS		
Optical properties	See Optics.				
Optics	EI COMP	INSPEC	NASA	NTIS	SPIN
Ore analysis	CA-FILES	GEOARCHIVE	NTIS		
Organic chemistry	CA-FILES	CRDS	ICRS	NTIS	SANSS
Packaging	CA-FILES	EI COMP	FOODS ADLIB	PAPERCHEM	PIRA
Parasitology	AGRICOLA	BIOSIS	CAB	EX MEDICA	MEDLARS
Paper	CA-FILES	PAPERCHEM	PIRA		
Paper industry	CIN	PTS			
Parachors	CA-FILES	PPDS			
Patent equivalents	CA-FILES	CLAIMS			
Patents	APIPAT	CLAIMS	IFI	PATS	WPI
Pathology	BIOSIS	EX MEDICA	MEDLARS		
Pest control	See Pesticides.				
Pesticides	AGRICOLA	CAB	CA-FILES	PESTDOC	TOXLINE
Petrochemicals	APILIT	CA-FILES	NTIS	P/E NEWS	WORLD PET
Petroleum energy	APILIT	APIPAT	EDB	EI COMP	P/E NEWS
Petroleum refining	APILIT	APIPAT	EI COMP		
Petrology	GEOARCHIVE	GEO-REF			
Pharmaceuticals	DRUGDOC	IPA	MEDLARS	PNI	TOXLINE
Pharmacognosy	DRUGDOC	IPA	MEDLARS	PNI	RINGDOC

Pharmacology	DRUGDOC	IPA	MEDLARS	PNI	RINGDOC
Photosensitizers	CA-FILES	EX MEDICA	MEDLARS	TITUS	WORLDTEX
Physical chemistry	CA-FILES	EI COMP	INSPEC	NTIS	SPIN
Physics	INSPEC	ISMEC	NASA	NTIS	
Physiology	BIOSIS	CAB	EX MEDICA	MEDLARS	NTIS
Plant breeding	See Horticulture.				
Plant growth	AGRICOLA	EIOSIS	CA-FILES	CAB	TOXLINE
Plant manufacturing	CIN				
Plant science	AGRICOLA	EIOSIS	CAB	NTIS	
Plasmas	CA-FILES	EI COMP	INSPEC	NASA	NTIS
Plasticizers	CA-FILES	EI COMP	PAPERCHEM	RAPRA	SCA
Plastics	CA-FILES	EI COMP	NASA	NTIS	RAPRA
Plastics fabrication	See Plastics.				
Poisons	EX MEDICA	MEDLARS	RTECS	TOXLINE	
Pollination	AGRICOLA	EIOSIS	CAB		
Pollution	APTIC	CA-FILES	ENVIROLINE	NTIS	POLLN ABS
Polymers	CA-FILES	EI COMP	RAPRA	TITUS	WORLDTEX
Powder diffraction	PDSM				
Pricing of chemicals	CIN				
Propellants	CA-FILES	DDC	EI COMP	NASA	NTIS
Properties	CA-FILES	EPIC	CHM-TADS	PPDS	TEPIAC
Property estimation	PPDS				
Public health	BIOSIS	EX MEDICA	MEDLARS	NTIS	
Pulp	CA-FILES	NTIS	PAPERCHEM	PIRA	
Pulp industry	CIN	FIRA	PTS		
Quality control	FSTS	INSPEC	PAPERCHEM	WAA	WORLDTEX
Radiation	BIOSIS	EDB	INSPEC	NSA	NTIS
Radiology	BIOSIS	CANCERLIT	EX MEDICA	MEDLARS	
Raw materials	CA-FILES	EDB.	EI COMP	NTIS	PAPERCHEM
Refractories	CA-FILES	NCIC	NASA	NTIS	
Regulations	FSTA	IPA	PNI		
Registry Numbers	See CA Registry Numbers				
Remote sensing	CA-FILES	EI COMP	ISMEC	NASA	NTIS
Reprography	PIRA				

*For CA-FILES, see individual data bases produced by Chemical Abstract Service.

TABLE XI

ACRONYM INDEX TO CHEMISTRY-RELATED DATA BASES WHICH CONTAIN REFERENCES/DATA ABOUT THE SUBJECT* (Continued)

Subject		Data-base acronym				
Resins	CA-FILES	PAPERCHEM	RPRA	SCA		
Sewage wastes	See Pollution.					
Soil science	AGRICOLA	CAB	NTIS			
Solar energy	CA-FILES	EDB	EI COMP	INSPEC	NASA	
Solid wastes	See Pollution.					
Solubilities	CA-FILES	PPDS	TEPIAC			
Solvents	CA-FILES	TITUS				
Spectral data	CNMR	FIRST-1	IRGO	IRIS	MSSS	
Standards	APTIC	FSTA	NTIS			
Surface-active agents	CA-FILES	EI COMP	NTIS			
Surface tensions	CA-FILES	PAPERCHEM	PPDS	TEPIAC		
Surgery	EX MEDICA	MEDLARS				
Synonyms	CHEMDEX	CHEMLINE	CHEMNAME	OHM-TADS	RTECS	
Systematic names	See Nomenclature.					
Taxonomy	AGRICOLA	BIOSIS	CAB	SCA	WAA	
Testing	EI COMP	ISMEC	NTIS	NTIS		
Textile printing	See Textiles.					
Textiles	CA-FILES	NTIS	TEX TECH	TITUS	WORLDTEX	
Thermal conductivity	CA-FILES	PPDS	TEPIAC			
Thermal energy	See Energy.					
Thermal properties	See Properties.					
Thermochemical data	CA-FILES	MANLAB-NPL	PPDS	NTIS	TEPIAC	
Thermochemistry	CA-FILES	EI COMP	NASA	NTIS	TEPIAC	
Thermodynamics	CA-FILES	EI COMP	NASA	NTIS	PPDS	
Tires	CA-FILES	NTIS	RAPRA	SAE		
Toxicology	BIOSIS	CA-FILES	EX MEDICA	TDB	TOXLINE	
Trade names	CHEMDEX	CHEMLINE	CHEMNAME	OHM-TADS	RTECS	
Transportation	EI COMP	ENVIROLINE	ISMEC	NTIS	SAE	
TSCA Number	CHEMLINE					
Urology	EX MEDICA	MEDLARS				

360

Vapor densities	CA-FILES	PPDS	TEPIAC	
Vapor pressure	CA-FILES	PPDS	TEPIAC	
Veterinary medicine	AGRICOLA	BIOSIS	CAB	EX MEDICA	VETDOC
Virology	BIOSIS	EX MEDICA	MEDLARS	
Viscosities	CA-FILES	EI COMP	PPDS	TEPIAC
Wallboards	PAPERCHEM			
Water pollution	See Pollution.			
Waxes	CA-FILES	NTIS		
Weed control	AGRICOLA	CAB		
Whiteners	CA-FILES	PAPERCHEM	CA-FILES	NTIS
Wiswesser notation	CHEMLINE	ICRS	PIRA	TEXTECH	WORLDTEX
Wood extractives	PAPERCHEM		TOXLINE	
X-ray diffraction	CRYST	PDSM		
Zoology	AGRICOLA	BIOSIS	CAB	NTIS

*For CA-FILES, see individual data bases produced by Chemical Abstract Service.

BIBLIOGRAPHY

The items presented in the bibliography are intended as guides to literature, word lists, journals, newsletters, and training which will promote the understanding and use of computer-searchable data bases.

Books/Papers

Atherton, P., and R. W. Christian: "Librarians and Online Services," Knowledge Industry Publications, Inc., New York, 1977.

Christian, R. W.: "The Electronic Library Bibliographic Data Bases," 1978–1979, Knowledge Industry Publications, Inc., New York, 1978.

Cuadra, C. A.: *Ann. Rev. Inf. Sci. Techn.,* **9,** (1974); **10,** (1975); **11,** (1976); **12** (1977).

Dickman, J. T., and R. E. O'Dette: "The Structure and Chemical Engineering Content of the Chemical Abstracts Service Data Base," Paper No. 74B, Symposium on Sources of Chemical Engineering and Information; 68th National Meeting, AICHE, Houston, Texas, March, 1971.

Gilchrist, A.: "The Thesaurus in Retrieval," ASLIB, London, 1971.

Hartiner, E. P., "An Introduction to Automated Literature," Marcel Dekker, New York, 1981.

Heller, S. R., and G. W. Milne: "The NIH-EPA Chemical Information System," *Environ. Sci. Techn.,* **13,** 798–803 (1979).

Jaxel, R.: "Directory of Computerized Data Files and Related Software Available from Federal Agencies," NTIS-SR-74-01, National Technical Information Service, Springfield, VA, 1974.

Keenan, S.: "Key Papers on the Use of Computer-Based Bibliographic Services," American Society for Information Science, Washington, DC, and National Federation of Abstracting and Indexing Services, Philadelphia, PA, October, 1973.

Krentz, David M.: "On-Line Searching-Specialist Required," *J. Chem. Inf. Comput. Sci.,* **18,** 4–9, (1978).

Lancaster, F. W., and E. G. Fayen: "Information Retrieval On-Line," Melville Publishing Co., Los Angeles, CA, 1973.

Lynch, M. F., et al.: "Computer Handling of Chemical Structure Information," American Elsevier, Inc., New York, 1971.

Marron, B., and Dennis W. Fife: "On-Line Systems—Techniques and Services," PB-260 902, National Technical Information Service, Springfield, VA, 1976.

Mathies, M. L., and P. G. Watson: "Computer-Based Reference Service," American Library Association, Chicago, 1973.

Meadow, C. T., and Pauline Cochrane, "Basics of Online Searching," John Wiley & Sons, New York, 1981.

Michaels, C. J.: "Searching CA Condensates On-Line vs. the CA Keyword Indexes," *J. Chem. Inf. Comput. Sci.,* **15,** 113–125 (1975).

Milne, G. W. A., et al.: "The NIH-EPA Substructure and Nomenclature Search System," *J. Chem. Inf. Comput. Sci.,* **18,** 181–186 (1978).

Prewitt, B. G.: "Searching the Chemical Abstracts Condensates Data Base via Two On-Line Systems," *J. Chem. Inf. Comput. Sci.,* **15,** 178–183 (1975).

Sharp, D. E.: "Handbook of Interactive Computer Terminals," Reston Publishing Co., Inc., Reston, VA, 1977.

Sherrod, J.: "Information Systems and Networks," Eleventh Annual Symposium, March 27–29, 1974, Greenwood Press, Westport, CT, 1975.

Tate, Fred A.: "Access to the Biochemical Literature through Services Produced at Chemical Abstracts Service," *Fed. Proc., Fed. Am. Soc. Exp. Biol.,* **33**, 1712–1714 (1974).

Tomberg, A.: "EUSIDIC Database Guide," Learned Information, New York, 1978.

Wanger, J., C. A. Cuadra, and M. Fishburn: "Impact of On-Line Retrieval Services: A Survey of Users, 1974–75," System Development Corp., Santa Monica, CA, 1976.

Walker, D. E.: "Interactive Bibliographic Search: The User/Computer Inferface," AFIPS Press, Montvale, NJ, 1971.

Williams, M. E.: *Ann. Rev. Inf. Sci. Techn.,* **12,** (1977); **13,** (1978); **14,** (1979).

Williams, M. E., and S. H. Rouse: "Computer-Readable Bibliographic Data Bases—A Directory and Sourcebook," American Society for Information Sciences, Washington, DC, 1976, 1979.

Young, M. E.: "On-Line Information Retrieval Systems,": Vol. 1, "1964–1976, A Bibliography with Abstracts," NTIS/PS-78/0978, National Technical Information Service, Springfield, VA, 1978; vol. 2, NITS/PS-78-0979.

Thesauri/Word Lists

American Psychological Association: "Thesaurus of Psychological Index Terms," APA, Washington, DC, 1977.

Backer, S., and E. I. Valko: "Thesaurus of Textile Terms," 2d ed., MIT Press, Cambridge, MA, 1969.

Management Information Services: "Management Contents Data Base Thesaurus," Management Information Services, Skokie, IL, 1977.

BioSciences Information Service: "BIOSIS Search Guide," BIOSIS Previews Edition, Philadelphia, PA, 1979.

Chemical Abstracts Service: "Chemical Abstracts Index Guide," Chemical Abstracts Service, Columbus, OH, 1978.

Engineering Index, Inc.: "SHE" (Subject Headings for Engineering), Engineering Index, Inc., New York, 1972.

ERDA: "ERDA Subject Indexing and Retrieval Thesaurus," National Technical Information Service, Springfield, VA, 1975.

Heckman, C.: "GEOREF Thesaurus and Guide to Indexing," 2d ed., American Geological Inst., Falls Church, VA, 1978.

INSPEC: "INSPEC Thesaurus 1977," The Gresham Press, Surrey, England, 1977.

NASA: "NASA Thesaurus Alphabetical Update," NASA SP 7040, National Technical Information Service, Springfield, VA, 1971.

National Library of Medicine: "Medical Subject Headings Annotated Alphabetic List, 1979," PB-285 356, National Technical Information Service, Springfield, VA, 1978.

Office of Naval Research, and Engineers Joint Council: "Thesaurus of Engineering and Scientific Terms 1967," AD-672 000, National Technical Information Service, Springfield, VA, 1967.

National Technical Information Service: "Integrated Energy Vocabulary 1976," PB-259 000, NTIS, Springfield, VA, 1976.

U.S. Atomic Energy Commission[2]: "Subject Headings Used by the USAEC Division of Technical Information," 8th rev. ed., TID-5001, National Technical Information Service, Springfield, VA, 1969.

[2]Now the Department of Energy.

U.S. Dept of Agriculture: "Agricultural Terms," 2d ed., ORTX Press, Phoenix, AZ, 1978.

Journals
ASIS Bulletin
DATABASE
Journal of the American Society for Information Science
Journal of Chemical Information and Computer Sciences
ONLINE
On-Line Review

Newsletters
"BRS Bulletin," Bibliographic Retrieval Services, Inc., 1462 Erie Boulevard, Schenectady, NY, 12305.
"CHRONOLOG," (Lockheed) DIALOG Information Retrieval Service Organization, 52-08, Building 201, 3251 Hanover Street, Palo Alto, CA, 93403.
"CIS Newsletter," (NIH-EPA CIS SYSTEM) Information Sciences Corp., Suite 500, 918 16th Street, NW, Washington, DC, 20006.
"EURONET DIANE," (Direct Information Access Network for Europe) EURONET DIANE Information, Jean Monnet Building, B4 009 CEC, Luxembourg (Grand Duchy).
"ORBIT NEWS," SDC Search Service, 2500 Colorado Avenue, Santa Monica, CA, 90406.
"The NLM Technical Bulletin," National Library of Medicine, MEDLARS Management Section, 8600 Rockville Pike, Bethesda, MD, 20014.

Training
Bibliographic Retrieval Services, 1462 Erie Boulevard, Schenectady, NY, 12305.
Chemical Information System, Interactive Sciences Corp., Suite 400, 918 16th Street, N.W., Washington, DC, 20006.
Lockheed Information Systems, 3251 Hanover Street, Palo Alto, CA, 93403.
National Library of Medicine—MEDLARS Management Section, 8600 Rockville Pike, Bethesda, MD, 20014.
System Development Corp., 2500 Colorado Avenue, Santa Monica, CA 90406.
University of Pittsburgh—Training Center for Librarians and Information Specialists: School of Library and Information Science, Pittsburgh, PA.

LIBRARY PROBLEMS

A competent searcher should have at his command a knowledge of libraries, including where they are and what is in them, combined with an ability to find and use the information sought.

D. D. Berolzheimer

Lecturing to students on the contents of a library, and on the method of using the material contained therein, will not leave a very lasting impression, nor make the individuals proficient in finding desired information. Lectures and discussions do provide a general idea of the problems, but practical experience in handling the various sources of information is of prime importance in learning how and where to find desired data. What is true in this field applies to many others. One cannot expect, for example, to learn to make accurate chemical analyses by attending lectures on the subject. Actual experience in the laboratory is the sine qua non for acquiring proficiency in the work.

In general, it is desirable to give students individual assignments. When the course comes in the third year, judicious selection of individual topics aids in broadening searchers' general knowledge of chemistry, as well as having them gain some familiarity with the place and method of finding information in chemical publications.

Students having sufficient interest to do something beyond meeting minimum requirements should be encouraged to browse around the library. This should include looking at more than the specific publications used in connection with specific assignments.

To the Instructor In previous editions of this book the library problems consisted of lists of specific assignments for each part of every problem (24 assignments for each problem in the fourth edition). Accompanying each assignment was a suggested form or blank page which provided space to enter the various data requested. In using this arrangement a given student could be assigned a specific part number (such as 12) to be completed throughout the course. When mimeographed, the blank sheets provided for consistent presentation of the answers for grading. To illustrate these blank sheets, the form on pages 367 and 368 was used for Library Problem 1.

In this edition the suggested kinds of library assignments are much the same. However, only one illustrative blank form is included. Also, instead of the lists of assignments for each part of the problems, only two or three workable examples are suggested. The general nature of each illustrative entry is shown herewith.

Given: Subject of the assignment.
> *Examples:* Two or more illustrative assignments (for guidance of the instructor).

Find: A statement of the specific data desired.
> *Hint:* At least one reference source to consult (for guidance of the student).

Source: Citation of the publication containing the data. The bibliographical form should follow approved practice (see Chapter 8 for examples).

In order to save space, the *Source* entries are omitted in the problems. If an instructor uses blank mimeographed or photocopied sheets, a suitable form for each part of a given problem might be

Given: [Space to copy assigned subject]
Find: [Space to enter the specific data requested]
Source: [Space to enter bibliographic data for the citation]

Many years of teaching a course on the chemical literature convinced the author that each instructor should tailor the course to fit the situation. This includes especially the selection of library problems. Important considerations follow: (1) the educational status of the students (freshmen, juniors, graduates, etc.); (2) the curriculum in which they are enrolled (arts, science, agriculture, engineering, home economics, pharmacy, others); (3) the time available for the course; and (4) the local library resources and facilities.

Taking into account these aspects of the situation, it seems prudent for each instructor to adapt the problems to his or her students and institution. The morale of a class sinks rapidly if the assignments cannot be fulfilled with reasonable expenditure of time and effort. Also, to avoid personal embarrassment, an instructor should know that the specific assignments are accurate, and should know how to aid a student having difficulty locating specific data.

To heighten students' interest, topics of great current concern may be

Library Problem 1
Periodicals I

1. Supply, for the abbreviation indicated, the required information concerning the journal.

   a. Abbreviation . . . .

   b. Name . . . .

   c. Country . . . . .Language . . . .

   d. Frequency of appearance . . . . Present volume . . . .

   e. Volume. . . . . .appeared in the year. . . . .

   f. Nearest library containing journal. . . . .

2. . . . .

   is the official abbreviation for the journal. . . . .

   . . . .

   This periodical contains the following kinds of information:

3. Supply the information indicated for an important. . . . .

   journal publishing articles on. . . . .

   . . . .

Name . . . .

. . . .

Country . . . . .Place edited . . . .

Frequency of appearance . . . . Present volume. . . . .

4. The abbreviation . . . . .is for

. . . .

Source. . . . .

5. During the year. . . . . .the following article appeared on. . . . .

. . . .

Title . . . .

. . . .

Author(s) . . . .

Professional connection(s). . . . .

Abstract reference. . . . .

. . . .

Original reference . . . .

. . . .

assigned. Examples are pollution, chemical hazards, contraceptives, nuclear wastes, carcinogens, and atomic fusion.

The books by Bottle, Maizel, and Wilen (see Chapter 12) contain examples of various types of library problems. Solving the examples in the problem/solution workbook "CAS Printed Access Tools" is very helpful in gaining experience in using present-day *Chemical Abstracts*.

For a class meeting once a week for a semester the suggested kinds of problems will be more than sufficient for an average student. A maximum of 2 weeks may be allowed for completing each problem.

PROBLEMS

Library Problem 1

Periodicals I

1 *Given:* The standard abbreviation of a journal word title
   Examples: Tek. Tidskr.; Zavod. Lab.
   Find: The word title; country of origin; language used; frequency of appearance; present volume number; nearest library containing the periodical
   Hint: Consult "Source Index" of CAS. See this text, Chapter 7.
   Source[1]:

2 *Given:* A journal word title
   Examples: Chemical Engineering; Talanta; Journal of Organic Chemistry
   Find: The standard title abbreviation
   Hint: Consult "Source Index" (CAS). See this text, Chapter 7.

3 *Given:* The nature of articles and the name of a country
   Examples: Ceramics (Germany); pharmaceuticals (France); pesticides (U.K.)
   Find: The title of an important journal publishing such articles; the place where it is edited; the frequency of appearance; the present volume number[2]
   Hint: See the list of important journals in "Source Index" (CAS); also Table IV of this text (Chapter 7) for the appropriate section of *Chemical Abstracts* where note may be made of the frequently cited journals in several issues of this source.

4 *Given:* An abbreviation used in *Chemical Abstracts* (not a journal title)
   Examples: Alk; clin; ir; uv
   Find: The word(s) abbreviated
   Hint: See the Introduction to the first issue of a recent volume of *Chemical Abstracts*.

5 *Given:* The title of an article published in a given year
   Examples: Two Dimensional Plasma Equilibria (1979); Interdiffusion of Aluminum and Chromium Thin Films (1979)

[1]This entry space is omitted in subsequent parts.

[2]Give the full title. Underscore the standard abbreviation. Include series, volume, page, and year.

Find: The name(s) of the author(s); professional connections; abstract reference; original reference

> *Hint:* Consult "General Subject Index" of *Chemical Abstracts*; subject indexes of available journals from which the assignments might have been made.

Library Problem 2

Periodicals II

1 *Given:* The name of a product listed in current market reports

> *Examples:* Methanol; sulfuric acid

Find: The current price; the price range for a year (in the form of a graph)

> *Hint:* See market reports in *Chemical Marketing Reporter*; other journals carrying market reports.

2 *Given:* The name of a product being currently advertised

> *Examples:* Vacuum dryers; pH meters; catalysts

Find: Firms sponsoring the advertisements; journal(s) carrying the advertisements

> *Hint:* Consult journals carrying advertisements, such as *Chemical & Engineering News, Chemical Engineering, Analytical Chemistry, Journal of Chemical Education.*

3 *Given:* The title of a book and the name(s) of the author(s)

> *Examples:* "Transfer RNA" by Sidney Altman; "The Molten State of Matter" by A. R. Ubbelhode

Find: A review of the book, with the name of the reviewer and the place published

> *Hint:* Consult author indexes of periodicals likely to cover the kinds of books listed, such as *Techn. Book Rev. Index* and *New Techn. Books.*

4 *Given:* An author's name, the title of an article published by her or him, and its place of publication

> *Examples:* D. W. Margerum et al., Characterization of a Readily Accessible Copper (III)-Peptide Complex, *J. Am. Chem. Soc.,* **97,** 6895 (1975).

Find: Three citations to this article, with citing authors and journal references

> *Hint:* Consult *Science Citation Index.* See this text, Chapter 6A.

5 *Given:* An article by a given author, published in a specific journal

> *Examples:* Electrochemistry of Horse Heart Cytochrome C, by M. J. Eddowes and H. A. O. Hill; Free-Energy Theory of Inhomogeneous Fluids, by B. F. McCoy and H. T. Davis

Prepare: An abstract (follow the format of *Chemical Abstracts*)

> *Hint:* The article should be too recent to have been abstracted by *Chemical Abstracts.* Otherwise, the student may be allowed to select any article of interest.

Library Problem 3

Technical Reports

1 *Given:* The subject of a government investigation
 Examples: Automated nitrocellulose analysis; hydrogen by thermochemical reactions
 Find: The title of the publication; author(s); government agency involved; kind of publication (such as bulletin)
 Hint: See indexes of *Monthly Catalog of United States Government Publications.*

2a *Given:* The name of a chemical product
 Examples: Uranium; petroleum; helium
 Find: Production data for a 10-year period (construct a graph with production units as ordinates and years as abscissa)
 Hint: See *Minerals Yearbook.*

2b *Given:* The name of a state (country)
 Examples: Indiana; Maine; Nevada
 Find: The value of the two most important mineral products
 Hint: See *Minerals Yearbook.*

2c *Given:* The name (symbol) of a metal
 Examples: Gold; U; copper
 Find: A summary of mineral developments for a given year
 Hint: See *Minerals Yearbook.*

3 *Given:* The name of a chemical subject to U.S. tariff
 Examples: Niacin; oxasine; acrylonitrile
 Find: The amount produced; amount sold; unit value; name and address of one manufacturer
 Hint: See "Synthetic Organic Chemicals."

4 *Given:* The name of an imported (exported) chemical product
 Examples: Indigo (I); phenol (E); barytes (I); methanol (E)
 Find: The amount imported (exported) during . . . ; value
 Hint: Consult "Chemical Industry Facts Book."

5 *Given:* The name of a manufacturer of chemical products
 Examples: Hercules; Eli Lilly and Company
 Find: The following financial data: net income; net sales; net worth; dividend yield
 Hint: Consult "Chemical Industry Facts Book"; data in *Chemical & Engineering News.*

6 *Given:* The name (symbol) of a chemical product
 Examples: Methanol; Br; $CHCl_3$
 Find: U.S. census data: production; sales; value
 Hint: Consult *Annual Survey of Manufactures.*

7 *Given:* The name of a pesticide
 Examples: Pentiformic acid; sorbitan tritallate

Find: Government data available
  Hint: See "Toxic Substances Control Act Chemical Substance Inventory."

Library Problem 4

Patents

1 *Given:* A patent issued (abstracted) during a given year, and related to a given subject
  Examples: Amidinoureas (1978); diverter for use in fusion reactors (1979)
  Find: Patent title; abstract references; patentee(s); number; data; country
    Hint: Consult *Chemical Substance Index* and *General Substance Index* of *Chemical Abstracts*.
2 *Given:* A patent number for a specified country
  Examples: U.S. 4,152,117; Germany 2,701,395; Canada 1,050,176; U.S. Patent Application 914,098
  Find: The patent title; patentee(s); data
    Hint: Consult *Patent Number Index* of *Chemical Abstracts* (until 1981).
3 The classification system of United States patents has several aspects of searching interest. The reference hint is the same for all of them, namely, the "Manual of Classification."
  a *Given:* The title of an examination division
    Examples: Distillation; textiles
  Find: The classes included
  b *Given:* The title of a class
    Examples: Poisons; bleaching substances; adhesives
  Find: The examination division covering the class
  c *Given:* The name of a kind of patentable product
    Examples: Perfumes; coffee substitutes; cyclic terpenes
  Find: The appropriate classification number
  d *Given:* A classification number
    Examples: 62-22; 102-34; 260-291
  Find: The kind of product covered
4 *Given:* The name of a country and a patent number
  Examples: Canada 892,463; Romania 52,190
  Find: An equivalent patent in another country
    Hint: Consult Patent Concordance of *Chemical Abstracts* (until 1981).
5 *Given:* A U.S. patent (to be obtained from the U.S. Patent and Trademark Office)
  Examples: The instructor may maintain a collection for lending.
  Find: The Patent title; reference in *Official Gazette* (Patents); reference in *Chemical Abstracts*; patentee(s); number; date; international classification number; U.S. classification number; digest of description and claims
6 *Given:* A typical entry in the Patent Index of *Chemical Abstracts* (1981 or later)

Examples: Any typical entry containing most data elements used

Find: The meaning of each data element included in the entry

   Hint: See the Introduction to the 1981 CA Patent Index, and the example of a patent heading in the Introduction to issue 1 of volume 94.

Library Problem 5

Physical Constants I   The assignments for this problem give experience in using the smaller collections of physical constants, such as Dean's "Lange's Handbook of Chemistry," Weast's "Handbook of Chemistry and Physics," and Perry, Chilton, and Kirkpatrick's "Chemical Engineers' Handbook." In Parts 3 and 5 a check mark may be placed at the value to be found, in case not all of those indicated apply to the assignment made.

   Although only names of substances are given, formulas may be specified if the students have had sufficient training to enable them to determine the name for a given formula

1 *Given:* A specific kind of data

   Examples: Hardness tables; list of isotopes; x-ray data

   Find: Source of such data

   Hint: Consult small handbooks of physical constants.

2 *Given:* The name (formula) of a specific substance

   Examples: Geraniol; $COCl_2$; anisic acid; $CSeS$

   Find: Molecular weight; melting point; boiling point; specific gravity; other data included

   Hint: Consult small handbooks of physical constants.

3 *Given:* The name of an oil, fat, or wax

   Examples: Seal oil; Japan wax; horse fat

   Find: Specific gravity; melting point; acetyl value; iodine value; Hehner value; Reichert-Meisel number; saponification value; Maumene number; refractive index

   Hint: Consult small handbooks of physical constants.

4 *Given:* The formula or name of a substance

   Examples: Li_2Se (D); bibenzene or diphenyl (C); $SiHCl_3$ (V)

   Find: Appropriate thermal data (D = dissolution; C = combustion: F = fusion; V = vaporization; F = formation)

   Hint: Consult small handbooks of physical constants.

5 *Given:* A specific physical constant, such as melting point, boiling point, density, refractive index

   Examples: Melting point of $-30°C$; boiling point of $50°C$; refractive index of 1.80; density of 5.0

   Find: A substance having this value

   Hint: Consult property-substance tables in "International Critical Tables"; Weast, "Handbook of Chemistry and Physics."

6 *Given:* A solution of a given compound (for a specified solvent, concentration, and temperature)

   Examples: $CaCl_2$ at 20/4 (in water); $HClO_4$ at 15/4

   Find: The specific gravity; Baumé reading

   Hint: Consult small handbooks of physical constants.

7 *Given:* A subject in chemical engineering

   Examples: Filtration; dialysis

   Find: Data for this field

   Hint: Consult "Chemical Engineers' Handbook."

Library Problem 6

Physical Constants II   This problem is more difficult than Problem 5. The comprehensive compilations contain original references, i.e., answers to the questions "Who stated this?" and "Where did he or she state it?" Such works include the American "International Critical Tables," the German "Landolt-Börnstein Zahlenwerke . . . ," the French "Tables annuelles . . . ," and various works limited to some specific kind of data, such as spectra, particularly volumes in the Standard Reference Data Series.

1 *Given:* The name (formula) of a substance (at a stated temperature and pressure)

   Examples: COS at 40mm; $CHClF_2$ at 10 mm

   Find: The vapor pressure; investigator; original reference

   Hint: Consult large works on physical constants.

2 *Given:* The formula (name) of an element or compound

   Examples: Ho (E); CH_4N_2S (A)

   Find: The emission spectrum (E) for the element, or the absorption spectrum (A) for the compound; the original reference

   Hint: Consult the Index of ICT; collections of spectral data.

3 *Given:* The kind of general information in the German tables

   Examples: Raman spectra; microwave spectra

   Find: The data; original references

   Hint: Consult bands I and II and supplements of the L-B tables.

4 *Given:* A kind of data for a specific system (electrolytic dissociation constant = DC; electric moment = EM; thermodynamic data = TD; dielectric properties = DP)

   Examples: NH_4OD (DC); SiH_4 (EM); C_6H_5F (DP); Co-As (TD)

   Find: The data; investigator; original reference

   Hint: Consult German tables.

5 *Given:* The name of an alkaloid

   Examples: Novacine; Jacoline; Vergiline

   Find: Various data for the compound; original reference

   Hint: Consult French tables.

6 *Given:* The formula (name) of a substance (solvent, temperature)
    Examples: COS (water, 15°); CO (C_2H_5OH, 20°)
  Find: Solubility data; original reference
  Hint: Consult solubility tables.
7 *Given:* The name (symbol) of a chemical element
    Examples: ^{60}Co; ^{13}C; ^{8}Be
  Find: Nuclear data; original reference
  Hint: Consult German tables (vol. I, new series).
8 *Given:* The name of a chemical compound
    Examples: Disulfur dichloride; ethyldibromoarsine
  Find: Constituent elements; finding numbers; chemical group numbers;
original reference
  Hint: Consult NAS-NRC Publn. 976, "Selected Property Values."

Library Problem 7

Organic Chemistry I   This problem deals with a variety of reference sources on organic chemistry. They range from single- to multivolume compilations.

1 *Given:* The name of an organic compound
    Examples: Caffeine; phenolphthalein; protein
  Find: The correct pronunciation
  Hint: Consult Weast, "Handbook of Chemistry and Physics."
2 *Given:* A named organic reaction (process, synthesis, test)
    Examples: Michael synthesis; Zincke reaction
  Find: The indicated transformation
  Hint: Consult chemical dictionaries. See this text, Chapter 9.
3 *Given:* The empirical formula of a new or unusual compound
    Examples: $C_{16}H_{13}ClF_3NO_4$; $C_{20}H_{56}N_{12}P_4$
  Find: A reference to it, abstracted during a specified time
  Hint: Consult formula indexes of *Chemical Abstracts*.
4 *Given:* The name of an organic compound
    Examples: Phenothiaarsine; 1,4-iminoisoquinoline
  Find: The ring structure and numbering of the atoms therein
  Hint: Consult "Index of Ring Systems" of *Chemical Abstracts*.
5 *Given:* The name of an organic ring complex
    Examples: Chroman (2 rings); ellipticine (4 rings)
  Find: The number of component rings; the number of elements in each ring; heterocyclic rings; carbocyclic rings; parent compounds having this configuration
  Hint: Consult "Index of Ring Systems" of *Chemical Abstracts*.
6 *Given:* The name (formula) of an organic group or radical
    Examples: Carbyl; CHN_4
  Find: The formula (name)

Hint: Consult "Index Guide" of *Chemical Abstracts*; "Nomenclature of Organic Chemistry" (IUPAC).

7 *Given:* The name of an organic compound to be synthesized
    Examples: Ferrocene; tetrolic acid
    Find: A recommended process for the synthesis
        Hint: Consult cumulative index for "Organic Syntheses"; Theilheimer's "Synthetic Methods of Organic Chemistry."

8 *Given:* A specified organic reaction
    Examples: Desulfurization; Schmidt reaction
    Find: Directions for using the reaction
        Hint: Consult "Organic Reactions."

9 *Given:* A specified laboratory operation
    Examples: Dialysis; salting out
    Find: Directions for performing the operation
        Hint: Consult treatises by E. Müller and by A. Weissberger.

10 *Given:* The author and title of an article abstracted in *Current Abstracts of Chemistry*
    Examples: N. Kuramoto and T. Kitac, Photochemical Oxidation of a Quinophthalone Dye; K. Friedrich and M. Zamkanei, Diels-Alder Reactions of Cyanothioformides
    Find: The structural diagrams of the important reaction(s); the new compounds involved; the analytical techniques used to identify the compound(s)
        Hint: See this text, Chapter 6B.

11 *Given:* The designation of an organic reaction
    Examples: Vinylation; nitrosamine (^{14}C) synthesis
    Find: Diagrams of the new reaction(s) involved in the process
        Hint: Consult *Current Chemical Reactions.*

Library Problem 8

Organic Chemistry II This problem involves the comprehensive general reference works, the ultimate in coverage being the Beilstein treatise.

1 *Given:* The name of an organic compound
    Examples: Uric acid; glycogen; thiamin
    Find: Information on this compound in the following works, with any original references cited:
    a Coffey, "Rodd's Chemistry of Organic Compounds"
    b Barton and Ollis, "Comprehensive Organic Chemistry"
    c Grignard, "Traité de Chimie Organique"
        Hint: Consult the indexes of the different works.

2 *Given:* The name of a compound (as found in the Beilstein treatise)
    Examples: 4-Chlorthiocarbanilid; Bis(2,4,6-trinitrophenyl)sulfid
    Find: Any other accepted name; the empirical formula (both the Hill and

Richter systems); structural formula; references in *Chemical Abstracts, Current Abstracts of Chemistry,* and *Chemisches Zentralblatt* (if searching prior to 1969)

> *Hint:* Consult the Beilstein treatise and abstracting journals; "Nomenclature of Organic Chemistry" (IUPAC).

3a *Given:* The name of the compound designated in part 2, above

Find: The location of the compound in the Beilstein classification system:

Division_____Subdivision_____
       (Skeleton name)   (Kind and number of heteroatoms)
Class_____
       (Number)   (Name and formula of functioning group)
Subclass_____
       (Letter)   (Name)
Rubric_____
       (Number)   (Name)   (General formula)
Series_____System number_____
   (Number of carbon atoms)
Index compound_____

> *Hint:* See this text, Chapter 10.

3b *Given:* The name of the compound designated in part 2, above

Find: A discussion of the chemistry of the compound in "Beilstein's Handbuch der Organischen Chemie," with the indicated original references. Check both the Hauptwerk and the Ergänzungswerk, I, II, III, and IV.

> *Hint:* See this text, Chapter 10.

4 *Given:* A specific subject

Examples: Carbanions; constitution of benzene

Find: A general discussion of the subject

> *Hint:* Consult advanced organic texts; monographs. See this text, Chapter 10.

5 *Given:* A CA Registry Number (name) of an organic compound

Examples: 16009-13-5; methyl red

Find: The corresponding name (Registry Number) for the compound; any other significant information included in the entry

> *Hint:* For the name, see "Chemical Substance Index" of *Chemical Abstracts.* For numbers: see "Registry Handbook—Number Section" of *Chemical Abstracts.*

Library Problem 9

Inorganic Chemistry

1 *Given:* The name (formula) of an inorganic compound

Examples: Tungsten carbide; $(NH_4)_2S_2O_8$; hydrazine

Find: Information on this compound in the following treatises, with original references:

    a "Gmelin's Handbuch der anorganischen Chemie"
    b Mellor, "Treatise on Inorganic and Theoretical Chemistry"
    c Pascal, "Traité de Chimie Minérale"
    d Trotman-Dickenson, "Comprehensive Inorganic Chemistry"
    Hint: See text for arrangement in each treatise.

2 *Given:* The same compound as in part 1, above
    Find: References in *Chemical Abstracts* (if any) for 1979
    Hint: Consult "Chemical Substance Index" of *Chemical Abstracts.*

3 *Given:* Chemical elements in an inorganic compound
    Examples: Zr, O, K, H, F; N, H, S, Be, O
    Find: The known compounds containing these elements (in any proportions)
    Hint: Consult "Formula Indexes" of *Chemical Abstracts.*

4 *Given:* The formula (name) of an inorganic radical
    Examples: CNSe; plutonyl; OS
    Find: The name (formula) of the radical
    Hint: Consult "Index Guide" for *Chemical Abstracts*; Introduction to the "Chemical Substances Index" of *Chemical Abstracts.*

5 *Given:* The formula (name) of an inorganic anion
    Examples: $UH_2O_7V^-$; nitroxylate$^-$; $AuC_4N_4S_4^-$
    Find: The name (formula) of the anion
    Hint: Same as part 4, above.

6 *Given:* The name (formula) of an inorganic substance
    Examples: BrF_3; potassium azide; NaSeCN
    Find: A laboratory method for synthesizing the substance
    Hint: Consult "Inorganic Syntheses."

7 *Given:* The 14 chemical elements having single-letter symbols are H, B, C, N, O, F, P, S, K, V, I, Y, W, U.[3] Automobile license plates using letters as part of the "number" may display one, two, or three letters. Thus, an element or a compound may be represented. Depending upon whether one, two, or three symbols are used, there seem to be the following possibilities:
    a One symbol only
        a′ An element; any of the 14 symbols, e.g., 79B338; 49F217
        b′ A molecule
            a″ Two atoms: O_2, as in OO7854; or N_2, as in NN4321
            b″ Three atoms; O_3, as in OOO357
    b Two elements
        a′ One of each; as in KI (or IK)
        b′ One of one symbol plus two of the other; as in HSH, SHH, HHS
    c Three symbols; as in KCN (KNC, CNK, CKN, NCK, NKC)
    Examples: BCH; VOF; UPS (Assignments might be, for example, W with each of two other specified elements; or C with any other two elements.)

---

[3]Two other symbols might be included, D (deuterium) and T (tritium). The symbol O for oxygen may not be seen because of possible confusion with the numerical symbol for zero. Likewise, the symbol I for iodine might be confused on plates with the numerical symbol for one.

Find: For a specified two or three elements, the known compounds represented on license plates containing these symbols.

    Hint: See formula indexes of *Chemical Abstracts.*

8a *Given:* The following symbols for elements, with the indicated naturally occurring isotopes of each: H—1, 2; C—12, 13; O—16, 17, 18; S—32, 33, 34, 36

    Examples: Possible compounds: $^2H_2{}^{18}O$; $^{13}C^{18}O$; $^2H_2{}^{32}S$

    Find: The known compounds containing H and O; C and O; H and S

    Hint: Consult formula indexes of *Chemical Abstracts.*

8b *Given:* The symbols H, C, B (the initials of Herbert C. Brown, a 1979 Nobel Prize winner in chemistry)

    Find: Known compounds containing all three symbols

    Hint: Consult formula indexes of *Chemical Abstracts.*

Library Problem 10

Analytical Chemistry

1 *Given:* The name of a substance to be detected and identified

    Examples: Hg (inorganic compounds); Nb; $=CH_2$ (functional group)

    Find: Procedures for the qualitative test

    a Characteristic reaction(s)

    b Method(s) of detection

    Hint: Consult works on qualitative chemical analysis; for elements, see the Gmelin treatise (under Identification).

2 *Given:* A specific system (or area, such as drugs or pollutants)

    Examples: $CuBr_2$ in ethanol (visible); pyridine (uv); 5-nitroisatin (ir)

    Find: The relevant spectral data for

    a Absorption (atomic, molecular) in ultraviolet, visible, infrared

    b Emission (arc, spark, flame excitation)

    c Mass

    d X-ray fluorescence

    e Others

    Hint: Consult works listed in Chapter 9 of this text.

3 *Given:* The name (symbol) of a chemical element

    Examples: Ti (determination of its impurities); Ti (in steel)

    Find:

    a Method(s) for its analysis (determination of its impurities)

    b Method(s) for determining it in a specified material classified to indicate the method(s) of separation and of measurement

    Hint: Consult general reference works on analytical chemistry.

4 *Given:* A specific kind of industrial product

    Examples: Drugs; foods; water; fertilizers; nonferrous alloys

    Find: Directions for analysis

    Hint: Consult works on applied analysis, including official methods, which deal with the kind of products indicated.

5 *Given:* A kind of general analytical information
   Examples: Analysis of gases; ion exchange separations; surface analysis
   Find: A general discussion of such information
      Hint: See Meites, "Handbook of Analytical Chemistry"; "General Subject Index" of *Chemical Abstracts.*

Library Problem 11

Biological Chemistry
1 *Given:* A topic in biochemistry
   Examples: Salt solutions; composition of keratins
   Find: General data available
      Hint: Consult Long, "Biochemists' Handbook."
2 *Given:* A biochemical preparation
   Examples: Insulin; glutathione
   Find: Laboratory directions for the process
      Hint: Consult "Biochemical Preparations."
3 *Given:* The name.of a substance to be analyzed (determined)
   Examples: Vitamin E; urine; blood sugar
   Find: Applicable analytical method(s)
      Hint: Consult Glick, "Methods of Biochemical Analysis."
4 *Given:* The trade name of a chelant
   Examples: Versene 100; kojic acid
   Find: The chemical name; manufacturer (supplier); ions chelated; stability
      Hint: Consult "Index Guide" of *Chemical Abstracts*; tabular compilations.
5 *Given:* A topic for general discussion
   Examples: Lipid metabolism; mechanism of hormone action
   Find: A presentation in a treatise
      Hint: Consult Florkin and Stolz, "Comprehensive Biochemistry."
6 *Given:* A topic involving enzymes
   Examples: Peptide bonds; RNA synthesis
   Find: A discussion of the subject
      Hint: Consult Boyer, "The Enzymes."
7 *Given:* A specific (general) assignment in biochemistry
   Examples: Natural ketoses; cyanogen bromide cleavage of peptides and proteins
   Find: Relevant data, with references
      Hint: Consult Fasman, "Handbook of Biochemistry and Molecular Biology."

Library Problem 12

Physical Chemistry and Physics
1 *Given:* A specific topic involving kinetics
   Examples: Decomposition of organic compounds; degradation of polymers

Find: A discussion of the process in a treatise
  Hint: Consult Bamford and Tipper, "Comprehensive Chemical Kinetics."
2 *Given:* A general topic
  Examples: Valency; liquid state; liquid crystals
  Find: A discussion of the topic
  Hint: See Eyring, Henderson, and Jost, "Physical Chemistry."
3 *Given:* A subject in physics
  Examples: Photoionization; atomic fission
  Find: A discussion of the topic
  Hint: See Flügge, "Handbuch der Physik."
4 *Given:* A topic on laboratory technique
  Examples: Hydrogen electrodes; standard cells
  Find: A description of the technique
  Hint: See Ewing, "Laboratory Instruments and Techniques"; a series of articles in *J. Chem. Educ.*
5 *Given:* A topic in physics
  Examples: Dielectrics; superfluids
  Find: General information on the topic
  Hint: See Condon and Odishaw, "Handbook of Physics."
6 *Given:* A subject recently reviewed
  Examples: Solid state chemistry; structure of liquid crystals
  Find: A review of the subject for a given year
  Hint: Consult "General Subject Index" of *Chemical Abstracts.*

Library Problem 10

Industrial Chemistry and Chemical Engineering
1 *Given:* A subject in industrial chemistry
  Examples: Bromine; heat transfer
  Find: Two books dealing with the subject
  Hint: Consult "General Subject Index" of *Chemical Abstracts; Technical Book Review Index*; publishers' catalogues; subject index of library card catalogue.
2 *Given:* The brand (trade) name of a chemical product
  Examples: Raysite; Zeotone
  Find: The name and location of a manufacturer (supplier)
  Hint: See "Index Guide" of *Chemical Abstracts*; a handbook of trade names.
3 *Given:* The name of a kind of industrial equipment
  Examples: Nitrators; rotary evaporators
  Find: The name and location of a manufacturer (supplier)
  Hint: Consult buyer's guides; current advertising in industrial journals.
4 *Given:* A named item (effect, process, law, test)
  Examples: Schiff test; Moseley law; Castner process

Find: A statement (explanation) of the item
    Hint: Consult a dictionary of named effects.

5 *Given:* A kind of laboratory equipment
    Examples: Gas chromatographs; atomic absorption spectrometers
    Find: The cost and availability of such an item
    Hint: Consult catalogues of supply houses; manufacturers' technical bulletins.

6 *Given:* The name of a chemical product
    Examples: Cyclopropane; vitamin C
    Find: Two sources of supply
    Hint: Consult dealers' catalogues; buyers' guides.

7 *Given:* The name and location of a chemical manufacturer
    Examples: Hercules; Ethyl Corporation
    Find: The nature and number of its products
    Hint: See works listed in this text, Chapter 12 (trade catalogues).

8 *Given:* The name (formula) of a chemical product
    Examples: Butadiene; BHI_3
    Find: Commercial method(s) of preparation
    Hint: Consult works on industrial chemistry; chemical encyclopedias.

9 *Given:* The name of a resin
    Examples: Permutit; Dowex-3
    Find: Type of resin; ions exchanged; capacity; stability
    Hint: Chemical encyclopedias; tabular compilations

10 *Given:* An industrial topic
    Examples: Acetylation of aniline; colorants for ceramics
    Find: A description (if any) in the following works:
    a "Ullmann's Encyklopädie der Technischen Chemie"
    b "Kirk-Othmer's Encyclopedia of Chemical Technology"
    c McKetta and Cunningham, "Encyclopedia of Chemical Processing and Design"

11 *Given:* A topic on pollution or environmental control
    Examples: Solid wastes; air pollution control
    Find: A discussion of the topic in
    a Young and Cherimisinoff, "Pollution Engineering and Technology"
    b Bond and Straub, "Handbook of Environmental Control"

12 *Given:* A subject on materials
    Examples: Illium B; American woods—properties and uses
    Find: A discussion of the subject
    Hint: See Lynch, "Handbook of Materials Science."

13 *Given:* A subject involving chemical engineering
    Examples: Effluent treatments; plastics as construction materials
    Find: A discussion of the subject
    Hint: See Cremer, "Chemical Engineering Practice."

14 *Given:* A topic recently reviewed

Examples: Heat-insulating glass; polypropylene
Find: A review of specific developments
  Hint: Consult "General Subject Index" of *Chemical Abstracts*. See this text, Chapter 6C.

Library Problem 14

Metallurgy and Metallography

1 *Given:* The name (symbol) of a chemical element
  Examples: Li; uranium; W
  Find:
  a Two common ores
  b the most important commercial source(s)
    Hint: Consult chemical encyclopedias; works on production metallurgy
2 *Given:* The name of an ore
  Examples: Monazite; bauxite
  Find: Its composition; two methods of treating it
    Hint: Consult works on production metallurgy.
3 *Given:* A named alloy
  Examples: Bain steel; Wood's metal
  Find: Its composition; properties; uses; manufacturer
    Hint: Consult Woldman, "Engineering Alloys"; "Index Guide" of *Chemical Abstracts*.
4 *Given:* A metallurgical subject
  Examples: Metal fatigue; rare earth metals
  Find: Two books on this subject
    Hint: Consult "General Subject Index" of *Chemical Abstracts*; publishers' catalogues.
5 *Given:* An alloy containing specified metals
  Examples: **Ni**,[4] Fe, Mo; **Cu**, Si, Sn
  Find: Its name; specific gravity; coefficient of thermal expansion; melting point
    Hint: Consult tabular compilations.
6 *Given:* An alloy containing specified elements
  Examples: Cu-Pd; Co-W
  Find: A phase equilibrium diagram
    Hint: Consult "Metals Handbook"; indexes of *Chemical Abstracts*.
7 *Given:* The name (symbol) of a rare metal
  Examples: Yt; gallium; Nd
  Find: General information on the element
    Hint: See Hampel, "Rare Metals Handbook."

[4]Boldface symbol indicates major component.

8 *Given:* The name of an element or alloy
    Examples: White gold (air); Rh (acids)
  Find: Its resistance to corrosion
    Hint: Consult "Metals Handbook."
9 *Given:* The name (symbol) of a chemical element
    Examples: Hg; lead; W
  Find: The current price; supplier(s)
    Hint: Consult current market reports in journals.
10 *Given:* A specified metal or alloy
    Examples: Stainless steel; Dirilite
  Find: The nature of its microstructure
    Hint: Consult books on metallography.

Library Problem 15

Medicinal and Pharmaceutical Chemistry

1 *Given:* A name of a drug
    Examples: Eutron; Valium
  Find: Two sources of supply; essential composition; most important physiological action(s); side effects; commercial method(s) of treatment or preparation; pharmaceutical use
    Hint: Consult chemical encyclopedias; "Merck Index"; "Dispensatory of the United States."
2 *Given:* The name of a drug or medicinal
    Examples: Diethyl ether; ascorbic acid
  Find: Specifications for, and method of analysis of, the material
    Hint: Consult "Merck Index"; "Pharmacopeia of the United States."
3 *Given:* The name of a drug product
    Examples: Menthol; Bacitracin
  Find: Two manufacturers; current prices
    Hint: Consult buyers' guides; *Chemical Marketing Reporter.*
4 *Given:* The name of a drug preparation
    Examples: Myrcia spirit; Senna powder
  Find: Its chemical composition
    Hint: Consult "National Formulary"; "Pharmacopeia of the United States."
5 *Given:* The name of a pharmaceutical product
    Examples: Sulfanilamide; oxytetracycline
  Find: A method of analysis (testing)
    Hint: Same as part 4, above.
6 *Given:* A medicinal product available to physicians
    Examples: Lasix; Aldomet
  Find: Recommendations for use. Note side effects
    Hint: See Huff, "Physicians' Desk Reference."

Library Problem 16

Miscellaneous Information   This problem illustrates the assignment of miscellaneous items. Other kinds of data may be selected for particular situations. Thus, a problem could be based on any of the 80 sections of *Chemical Abstracts* not covered in the preceding problems.

1 *Given:* The name of a specific substance
   Examples: Metal borides; Bi telluride; Sc
   Find: The kinds of thermophysical data available; references
      Hint: Consult works on thermodynamic and thermophysical properties.
2 *Given:* A specific isotope of a chemical element
   Examples: Pu^{239}; I^{127}
   Find: The thermal neutron cross section; the reaction; the scattering
      Hint: Consult tabular compilations.
3 *Given:* The name of a mineral
   Examples: Topaz; celestite
   Find: Index of refraction; crystal form; color; optical axial angle
      Hint: Consult tabular compilations.
4 *Given:* The formula (name) of a chemical compound
   Examples: UF_6; silver hydride
   Find: The thermodynamic properties
      Hint: Consult tabular compilations.
5 *Given:* The formula for an organometallic compound.
   Examples: BCH_3S; $TiC_4H_{10}O_3$
   Find: The usual formula; characteristics; physical data; references
      Hint: Consult "Handbook of Organometallic Compounds."
6 *Given:* The trade name of an industrial organic compound
   Examples: Furan; Arlax
   Find: The chemical name; formula; purity; specific gravity; shipping container; melting point
      Hint: Consult tabular compilations.
7 *Given:* A kind of industrial construction material
   Examples: Abrasives; graphite
   Find: Any data of interest
      Hint: Consult chemical encyclopedias; buyers' guides.
8 *Given:* The formula (name) of a substance
   Examples: Fused $K_2S_2O_7$; 35% HCl
   Find: Chemically resistant materials for handling this substance
      Hint: Consult works on chemical properties of materials; specific works on corrosion.
9 *Given:* The name of a hazardous chemical waste substance
   Examples: Isophorone; aldrin; dioxin
   Find: The nature of the hazard; possible protective treatment
      Hint: Consult handbooks on hazardous chemicals.

Library Problem 17

Economic Survey of a Chemical Commodity[5,6]

Data for the Commodity

Raw materials
    Sources of supply
    Specifications for intended quality
    Tonnage requirements for economic handling
Manufacture
    Methods (with flow sheets for a commercial plant)
    Control tests
    Production costs (itemized statement)
    By-products (including utilization or disposal)
Finished product
    Usual purity
    Grades
    Shipping containers
    Shipping regulations
    Specifications required
    Physical properties
    Chemical properties
Items of commercial importance
    Domestic production
    Domestic consumption
    Imports (amount and sources)
        Tariff
    Exports (amount and destinations)
        Tariff
    Names and locations of five manufacturers
    Graphical comparison of monthly selling price with weighted index of prices of all commodities

Financial Data for One Manufacturer

Firm name
    Subsidiary companies

---

[5]In industrial chemistry, instead of confining the problem to collecting individual technical facts, as in Library Problems 11 to 15, it is often desirable to make an economic survey in the form of a more general report. To compile the information will require the use of some of the sources recommended for the preceding specialized problems, but certain others will be necessary for financial facts. If this problem is to be a finished term report rather than merely a compilation of facts, it might well be combined with Library Problem 19. The outline followed is based on a paper by Kobe, *J. Chem. Educ.*, **10**:738 (1933).

[6]Rather than numbered questions here, only an outline is suggested to follow in presenting the data.

Rating
  Dun and Bradstreet[7]
  Moody[8]
Stocks and bonds outstanding
Stocks listed on exchange at. . . .
Graph of earnings and dividends per share (5-year period)
Graph of stock activity (10-year period)
Important items in last financial report (date)
Research activity
  Size of laboratory (including number of personnel)
  Location
  Program
Conclusions (evaluation of the company)

Library Problem 18

Searching by Computer Chapter 15 deals with the general problem of conducting a literature search by means of computer-readable data bases. An individual instructor, wishing to demonstrate this procedure, may simply secure a printout product of a batch search from a commercial vendor, or show an on-line process, with its accompanying printout. What is desirable and practical depends upon the local situation.

In any case, successful searching by this means depends upon the searcher's knowledge and experience. That is, one should "know the business." What is the specific question, or just what is wanted? What searching profile or searching terms are to be used? Which data bases are most likely to yield relevant data? These and other questions are very important.

Taking as an example the problem discussed in Chapter 15, an instructor might assign suitable subjects for a class. Then each member would prepare a searching profile. Using the entries in Table XI helps one decide which data bases listed in Table X are probably worth searching. One or more of these projects could be used to obtain a printout of a batch search, or to demonstrate on-line searching.

Success in computer searching probably will require considerable advice and guidance by one experienced in such searching.

Library Problem 19

Preparation of a Bibliography The previous problems are designed to familiarize the student, to some extent at least, with the general nature of the

---

[7]"The Mercantile Agency Reference Book," Dun and Bradstreet, New York, latest edition.

[8]"Moody's Manual of Investments," volume on industrial securities, Moody's Investors Service, New York, latest edition.

different kinds of chemical publications and with their use in finding information relating to the various kinds of questions which a chemist takes to the library. An individual should be able not only to locate special facts and data but also to make a general survey of the recorded information on a given subject and to prepare a report on the important points revealed by the survey.

In general, the first step in making such a survey is the compilation of a partial bibliography. As mentioned in Chapter 8, good lists of references are already available on many subjects as a starting point; but for this problem it is assumed that such is not the case. For the preparation of a bibliography as a library exercise, several points should be considered.

Selection of the Subject The student should select her or his own subject, as it is then likely to have a personal interest. Several examples of such personal interests may be noted. Thus, one student, captain of the university golf team, selected "The Chemistry of a Golf Ball." Another chose "Embalming Liquids," as his father was an undertaker. A third thought it wise to find information on a small chemical business, as his girl friend's father was president of the company.

Graduate students have a different kind of motivation if they can compile their bibliography in connection with their M.S. or Ph.D. theses. This can develop into a kind of ongoing project as the work on the thesis progresses.

For those who have in mind no subjects of their own, a list of suitable subjects should be available from which to make selections. In many cases suitable subjects yield three to five references per year in the volume indexes of *Chemical Abstracts*. Then it is usually necessary to go back to at least one cumulative index to secure a workable number of references. To avoid having students choose subjects which will yield too few or too many references, the instructor may require the subjects to be submitted for approval.

Determination of the Scope of the Subject Many subjects are either too limited or too broad in their range; that is, too many references will be found for the student to handle in the time available, or too few can be located to give any opportunity for working out a classified bibliography. Although a complete bibliography, even on a comparatively little-used element, such as selenium, is rather extensive, the list is much less imposing when the subject is confined to the organic compounds of selenium.

Determination of the Scope of the Bibliography If a bibliography is not to be complete, one must decide which portion of the whole field of the literature is to be covered. When the compilation of a bibliography is merely a part of a course on chemical literature, a large amount of time is not available. The location of approximately 100 references will give sufficient experience in this type of work.

Determination of the Bibliographical Details to Include Ordinarily the following items should be recorded for each entry: author (or patentee); original reference, including, if a periodical, its official abbreviation, series, volume, page, and year; or, of some other kind of publication, such information as will indicate unmistakably the source of the material; abstract reference, the abbreviation, volume, and page being sufficient; title of article (or other publication); and annotations if the title does not give a sufficient clue to the nature of the contents.

Method of Searching General directions for searching are included in Chapter 14. It is valuable experience for a student to be turned loose at this point, with an occasional guiding suggestion, in order to determine what he or she can find on a given topic. The report should contain a list of the index headings examined during the search for references.

Arrangement of Material The usual forms of arrangement used in bibliographies are discussed in Chapter 8. Here again, students should probably be left to their own devices not only in the arrangement of the details on the page for each entry but also in the classification or listing of the references as a whole. If the various bibliographies mentioned in Chapter 8 are available for examination, the merits of the different schemes can be ascertained.

The report submitted should be a readily usable bibliography. If it is arranged alphabetically by authors, there should be a subject index. If it is classified, the divisions and subdivisions should be clearly indicated. In card form this means the use of appropriately marked index or separator cards.

Many students find it helpful to see a few examples of bibliographies prepared by former students.

Library Problem 20

Preparation of a Critical Report The previous assignments have dealt primarily with the problem of locating required information in the various sources available. Very often one needs not only to find specific facts but also to make effective use of them in a report.

Let us assume that an ordinary problem of chemical research is undertaken. The general scope of the problem is first determined. Then one usually wants to ascertain what is already known concerning it, in order to begin somewhere near the point where others quit. This involves searching chemical publications and the collection of material. On the basis of this knowledge, experiments are planned, if necessary, and executed, resulting in the accumulation of more data. A final report of the work will frequently include the following points: a statement of the problem; a review of others' work bearing upon it; new experimental work performed, including the data secured; a general discussion of the results, including their connection with previous work; and the general conclusions.

In preparing the review of others' work, the compilation of a bibliography, as given in Library Problem 19, is only a part of the whole task. True, it may require much time and considerable ingenuity in searching; but something more is required if one is to avoid being submerged in the waves of facts and is to get one's bearings by gaining a perspective of a considerable portion of the sea around. The references must be examined; and the material in them must be sifted, classified, correlated, and evaluated. With respect to the facts involved, we must, in the words of Glenn Frank, find them, filter them, focus them, face them, and follow them.

The preparation of a concise review of the information available on some simple problem is suggested as the logical assignment to follow the preparation of a bibliography. The field of the bibliography submitted for Library Problem 19, if not too extensive, or some definite portion of it, may well be taken.

The problem and the technique of preparing the manuscript for such reports have been considered in detail in several other publications. For general information of this kind the student is referred to the sources listed below. They deal not only with the type of report just discussed but also with the preparation of articles or books, including the final process of printing.

Assignments for Library Problem 20   A report covering general information available may be prepared upon almost any chemical subject. However, not all subjects are equally suitable for student use, since beginners lack experience and perspective. Also the material on different subjects varies widely. In general, an assignment that is satisfactory for a bibliography in Problem 19 should be workable for a report. In any case a bibliography forms the basis of a report. For industrial chemistry, assignments such as those in Problem 13 may be desirable.

When one broadens report writing to include the preparation of manuscripts in a form suitable for publication as articles or books, so many details are involved that a weekly problem is insufficient for their consideration. Consequently, the author expanded Problem 20 into a 1-hour semester's course on technical writing.[9]

References on Technical Writing   One could easily assemble a "5-foot" shelf of books dealing with the various aspects of technical writing. The overall problem is learning to use five means of communication: words, formulas and equations, tables, graphs, and pictures. In using these means there are five general requirements or goals: clarity, accuracy, brevity, consistency, and as far as possible, interest.

In the fourth edition of this book (pages 271–272) nearly 50 books were listed. The essentials of technical writing are not subject to changes of fall models, and these sources remain relevant.That list is reprinted here.

General Technical Writing

Allen, E. M.: "Author's Handbook," International Textbook Co., Scranton, Pa., 1938
Baker, C.: "Technical Publications: Their Purpose, Preparation, and Production," John Wiley & Sons, Inc., New York, 1955.

[9]See *J. Chem. Doc.*, **4**:1 (1964).

Emberger, M. R., and M. R. Hall: "Scientific Writing," Harcourt, Brace, & World, Inc., New York, 1955.

Fieser, L. F., and Mary Fieser: "Style Guide for Chemists," Reinhold Publishing Corporation, New York, 1960.

Fishbein, M.: "Medical Writing: The Technique and the Art," McGraw-Hill Book Company, New York, 1948.

Flesch, R.: "The Art of Readable Writing," Harper & Row, Publishers, Incorporated, New York, 1949.

Flesch, R., and A. H. Lass: "The Way to Write," McGraw-Hill Book Company, New York, 1955.

Gensler, W. J., and K. D. Gensler: "Writing Guide for Chemists," McGraw-Hill Book Company, New York, 1961.

Gowers, E.: "Plain Words: Their ABC," Alfred A. Knopf, Inc., New York, 1954.

Gunning, R.: "The Technique of Clear Writing," McGraw-Hill Book Company, New York, 1952.

Harwell, G. C.: "Technical Communication," Reinhold Publishing Corporation, New York; 1960.

Howell, A. C.: "Handbook of English in Engineering Usage," John Wiley & Sons, Inc., New York, 1940.

Kapp, R. O.: "The Presentation of Technical Information," The Macmillan Company, New York, 1958.

Kirkpatrick, T. W., and M. H., Breese: "Better English for Technical Authors," The Macmillan Company, New York, 1961.

Lewis, N.: "The New Roget's Thesaurus of the English Language," Garden City Books, New York, 1961.

Mellon, M. G.: *J. Chem. Doc.,* **4**:1 (1964).

Mills, G. H., and J. A. Walter: "Technical Writing," Holt, Rinehart and Winston, Inc., New York, 1954.

Norgaard, Margaret: "A Technical Writer's Handbook," Harper & Row, Publishers, Incorporated, New York, 1959.

Rathbone, R. R., and J. B. Stone: "Writer's Guide for Engineers and Scientists," Prentice-Hall, Inc., Englewood Cliffs, NJ, 1962.

Reisman, S. S.: "A Style Manual for Technical Writers and Editors," The Macmillan Company, New York, 1962.

Sherman, T. A.: "Modern Technical Writing," Prentice-Hall, Inc., Englewood Cliffs, NJ, 1955.

Shidle, N. G.: "Clear Writing for Easy Reading," McGraw-Hill Book Company, New York, 1951.

Skillen, M. E., R. M. Gay et al.: "Words into Type," 3d ed., Prentice-Hall, Inc., Englewood Cliffs, NJ, 1974.

Smith, R. W.: "Technical Writing," Barnes & Noble, Inc., New York, 1963.

Strunk, W., Jr., and E. B. White: "The Elements of Style," 3d ed., The Macmillan Company, New York, 1979.

Welborn, G. P., L. B. Green, and K. A. Nall: "Technical Writing," Houghton Mifflin Company, Boston, 1961.

"Author's Guide," John Wiley & Sons, Inc., New York, 1950.

Graphs

Arkin, H., and R. R. Colton: "Graphs," Harper & Row, Publishers, Incorporated, New York, 1940.

ASTM, "Manual on Presentation of Data," American Society for Testing and Materials, Philadelphia, 1940.

French, T. E., and C. J. Vierck: "Fundamentals of Engineering Drawing," McGraw-Hill Book Company, New York, 1960.

Schmidt, C. F.: "Handbook of Graphic Presentation," The Ronald Press Company, New York, 1954.

Worthing, A. G., and J. Geffner: "Treatment of Experimental Data," John Wiley & Sons, Inc., New York, 1943.

Reports, Research Papers, Theses

Davis, D. S.: "Elements of Engineering Reports," Chemical Publishing Company, Inc., New York, 1963.

Kerekes, F., and R. Winfrey: "Report Preparation," Iowa State University Press, Ames, IA, 1951.

Kobe, K. A.: "Chemical Engineering Reports," Interscience Publishers, Inc., New York, 1957.

Linton, C. D.: "How to Write Reports," Harper & Row, Publishers, Incorporated, New York, 1954.

Nelson, J. R.: "Writing the Technical Report," McGraw-Hill Book Company, New York, 1952.

Trelease, S. F.: "The Scientific Paper," The Williams & Wilkins Company, Baltimore, MD, 1947.

Tuttle, R. E., and C. A. Brown: "Writing Useful Reports," Appleton-Century-Crofts, Inc., New York, 1956.

Ulman, J. N.: "Technical Reporting," Holt, Rinehart & Winston, Inc., New York, 1952.

Waldo, W. H.: "Better Report Writing," Reinhold Publishing Corporation, New York, 1957.

Weil, B. H.: "The Technical Report," Reinhold Publishing Corporation, New York, 1954.

Williams, C. B., and J. Ball: "Report Writing," The Ronald Press Company, New York, 1955.

As a supplement to this list the following later books may be added:

Barrass, R.: "Scientists Must Write," John Wiley & Sons, Inc., New York, 1978.

Blicq, R. S.: "Technically–Write," Prentice-Hall, Inc., Englewood Cliffs, NJ, 1972.

Brogan, J.: "Clear Technical Writing," McGraw-Hill Book Company, New York, 1973.

Day, R. A.: "How to Write and Publish a Scientific Paper," ISI Press, Philadelphia, PA, 1979.

Erlich, E., and D. Murphy: "Basic Grammar for Writing," McGraw-Hill Book Company, New York, 1971.

Gould, J. R.: "Practical Technical Writing," American Chemical Society, Washington, DC, 1973.

Jordan, L.: "The New York Times Manual of Style," New York Times Book Company, New York, 1976.

Mitchell, J. H.: "Writing for Professional and Technical Journals," John Wiley and Sons, Inc., New York, 1968.

Nickles, H. G.: "The Dictionary of Do's and Don'ts," McGraw-Hill Book Company, New York, 1974.

O'Connor, H.: "The Scientist as Editor," John Wiley & Sons, Inc., New York, 1979.

Tracy, R. C., and H. L. Jennings: "Writing for Industry," American Technical Society, Chicago, 1973.

Vardaman, G. T., and P. B. Vardaman: "Successful Writing," Halsted Press, New York, 1977.

Weisman, H. M.: "Basic Technical Writing," C. S. Merrill Publishing Company, Columbus, OH, 1974.

"Bibliographic Guide for Editors and Authors," American Chemical Society, Washington, DC, 1974.

"A Manual of Style," The University of Chicago Press, Chicago, 1969.

"Government Printing Office Style Manual," Government Printing Office, Washington, DC, 1973.

"Handbook for Authors of Papers in American Chemical Society Publications," American Chemical Society, Washington, DC, 1978.

"The McGraw-Hill Author's Book," McGraw-Hill Book Company, New York, 1968.

"Council of Biology Editors Style Manual," American Institute of Biological Sciences, Arlington, VA, 1978.

INDEXES

Of many large volumes, the index is the best portion and the most useful.

M. Wilmotte

AUTHOR INDEX

The following list contains the names of the authors, editors, and compilers of the various works described, or cited as references, in this text. Among these works are bibliographies, dictionaries, encyclopedias, formularies, handbooks, monographs, textbooks, and multivolume treatises. In many cases such publications are known better by their authors' surnames than by the titles. Many of the books have two or more authors.

Not included are the names of the authors of the incidental quotations and of the many articles cited as references in periodicals.

SUBJECT INDEX

In this combined subject, author, and title index, subject entries are by far the most numerous. Only a few authors' names are included for a few publications well known by their authors' names. Specific titles are included for publications best known by their titles. Page numbers followed by *n.* indicate footnotes.